Chapman & Hall/CRC Mathematical and Computational Biology Series

AN INTRODUCTION TO SYSTEMS BIOLOGY

DESIGN PRINCIPLES OF BIOLOGICAL CIRCUITS

CHAPMAN & HALL/CRC
Mathematical and Computational Biology Series

Aims and scope:

This series aims to capture new developments and summarize what is known over the whole spectrum of mathematical and computational biology and medicine. It seeks to encourage the integration of mathematical, statistical and computational methods into biology by publishing a broad range of textbooks, reference works and handbooks. The titles included in the series are meant to appeal to students, researchers and professionals in the mathematical, statistical and computational sciences, fundamental biology and bioengineering, as well as interdisciplinary researchers involved in the field. The inclusion of concrete examples and applications, and programming techniques and examples, is highly encouraged.

Series Editors

Alison M. Etheridge
Department of Statistics
University of Oxford

Louis J. Gross
Department of Ecology and Evolutionary Biology
University of Tennessee

Suzanne Lenhart
Department of Mathematics
University of Tennessee

Philip K. Maini
Mathematical Institute
University of Oxford

Shoba Ranganathan
Research Institute of Biotechnology
Macquarie University

Hershel M. Safer
Weizmann Institute of Science
Bioinformatics & Bio Computing

Eberhard O. Voit
The Wallace H. Couter Department of Biomedical Engineering
Georgia Tech and Emory University

Proposals for the series should be submitted to one of the series editors above or directly to:
CRC Press, Taylor & Francis Group
24-25 Blades Court
Deodar Road
London SW15 2NU
UK

Published Titles

Cancer Modeling and Simulation
Luigi Preziosi

Computational Biology: A Statistical Mechanics Perspective
Ralf Blossey

Computational Neuroscience: A Comprehensive Approach
Jianfeng Feng

Data Analysis Tools for DNA Microarrays
Sorin Draghici

Differential Equations and Mathematical Biology
D.S. Jones and B.D. Sleeman

Exactly Solvable Models of Biological Invasion
Sergei V. Petrovskii and Lian-Bai Li

An Introduction to Systems Biology: Design Principles of Biological Circuits
Uri Alon

Knowledge Discovery in Proteomics
Igor Jurisica and Dennis Wigle

Modeling and Simulation of Capsules and Biological Cells
C. Pozrikidis

Normal Mode Analysis: Theory and Applications to Biological and Chemical Systems
Qiang Cui and Ivet Bahar

Stochastic Modelling for Systems Biology
Darren J. Wilkinson

The Ten Most Wanted Solutions in Protein Bioinformatics
Anna Tramontano

Chapman & Hall/CRC Mathematical and Computational Biology Series

An Introduction to Systems Biology
Design Principles of Biological Circuits

Uri Alon

Chapman & Hall/CRC
Taylor & Francis Group
Boca Raton   London   New York

Chapman & Hall/CRC is an imprint of the
Taylor & Francis Group, an informa business

Chapman & Hall/CRC
Taylor & Francis Group
6000 Broken Sound Parkway NW, Suite 300
Boca Raton, FL 33487-2742

International Standard Book Number-10: 1-58488-642-0 (Softcover)
International Standard Book Number-13: 978-1-58488-642-6 (Softcover)

Library of Congress Cataloging-in-Publication Data

Alon, Uri.
    Introduction to systems biology: design principles of biological circuits / by Uri Alon.
        p. cm. -- (Chapman and Hall/CRC mathematical & computational biology series ; 10)
        Includes bibliographical references (p.) and index.
    ISBN 1-58488-642-0
    1. Computational biology. 2. Biological systems--Mathematical models. I. Title. II. Series.

QH324.2.A46 2006
570.285--dc22   2005056902

Visit the Taylor & Francis Web site at
http://www.taylorandfrancis.com

and the CRC Press Web site at
http://www.crcpress.com

For Pnina and Hanan

Acknowledgments

It is a pleasure to thank my teachers. First my mother, Pnina, who gave much loving care to teaching me, among many things, math and physics throughout my childhood, and my father, Hanan, for humor and humanism. To my Ph.D. adviser Dov Shvarts, with his impeccable intuition, love of depth, and pedagogy, who offered, when I was confused about what subject to pursue after graduation, the unexpected suggestion of biology. To my second Ph.D. adviser, David Mukamel, for teaching love of toy models and for the freedom to try to make a mess in the labs of Tsiki Kam and Yossi Yarden in the biology building. To my postdoctoral adviser Stan Leibler, who introduced me to the study of design principles in biology with caring, generosity, and many inspiring ideas. To Mike Surette and Arnie Levine for teaching love of experimental biology and for answers to almost every question. And to my other first teachers of biology, Michael Elowitz, Eldad Tzahor, and Tal Raveh, who provided unforgettable first experiences of such things as centrifuge and pipette.

And not less have I learned from my wonderful students, much of whose research is described in this book: Ron Milo, Shai Shen-Orr, Shalev Itzkovitz, Nadav Kashtan, Shmoolik Mangan, Erez Dekel, Guy Shinar, Shiraz Kalir, Alon Zaslaver, Alex Sigal, Nitzan Rosenfeld, Michal Ronen, Naama Geva, Galit Lahav, Adi Natan, Reuven Levitt, and others. Thanks also to many of the students in the course "Introduction to Systems Biology," upon which this book is based, at the Weizmann Institute from 2000 to 2006, for questions and suggestions. And special thanks to Naama Barkai for friendship, inspiration, and for developing and teaching the lectures that make up Chapter 8 and part of Chapter 7.

To my friends for much laughter mixed with wisdom, Michael Elowitz, Tsvi Tlusty, Yuvalal Liron, Sharon Bar-Ziv, Tal Raveh, and Arik and Uri Moran. To Edna and Ori, Dani and Heptzibah, Nili and Gidi with love. To Galia Moran with love.

For reading and commenting on all or parts of the manuscript, thanks to Dani Alon, Tsvi Tlusty, Michael Elowitz, Ron Milo, Shalev Itzkovitz, Hannah Margalit, and Ariel Cohen. To Shalev Itzkovitz for devoted help with the lectures and book, and to Adi Natan for helping with the cover design.

To the Weizmann Institute, and especially to Benny Geiger, Varda Rotter, and Haim Harari, and many others, for keeping our institute a place to play.

Contents

CHAPTER 1

Introduction

When I first read a biology textbook, it was like reading a thriller. Every page brought a new shock. As a physicist, I was used to studying matter that obeys precise mathematical laws. But cells are matter that dances. Structures spontaneously assemble, perform elaborate biochemical functions, and vanish effortlessly when their work is done. Molecules encode and process information virtually without errors, despite the fact that they are under strong thermal noise and embedded in a dense molecular soup. How could this be? Are there special laws of nature that apply to biological systems that can help us to understand why they are so different from nonliving matter?

We yearn for laws of nature and simplifying principles, but biology is astoundingly complex. Every biochemical interaction is exquisitely crafted, and cells contain networks of thousands of such interactions. These networks are the result of evolution, which works by making random changes and selecting the organisms that survive. Therefore, the structures found by evolution are, to some degree, dependent on historical chance and are laden with biochemical detail that requires special description in every case.

Despite this complexity, scientists have attempted to discern generalizable principles throughout the history of biology. The search for these principles is ongoing and far from complete. It is made possible by advances in experimental technology that provide detailed and comprehensive information about networks of biological interactions.

Such studies led to the discovery that one can, in fact, formulate general laws that apply to biological networks. Because it has evolved to perform functions, biological circuitry is far from random or haphazard. It has a defined style, the style of systems that must function. Although evolution works by random tinkering, it converges again and again onto a defined set of circuit elements that obey general design principles.

The goal of this book is to highlight some of the design principles of biological systems, and to provide a mathematical framework in which these principles can be used to understand biological networks. The main message is that biological systems contain an inherent simplicity. Although cells evolved to function and did not evolve to be comprehensible, simplifying principles make biological design understandable to us.

This book is written for students who have had a basic course in mathematics. Specialist terms and gene names are avoided, although detailed descriptions of several well-studied biological systems are presented in order to demonstrate key principles. This book presents one path into systems biology based on mathematical principles, with less emphasis on experimental technology. The examples are those most familiar to the author. Other directions can be found in the sources listed at the end of this chapter, and in the extended bibliography at the end of this book.

The aim of the mathematical models in the book is not to precisely reproduce experimental data, but rather to allow intuitive understanding of general principles. This is the art of "toy models" in physics: the belief that a few simple equations can capture some essence of a natural phenomenon. The mathematical descriptions in the book are therefore simplified, so that each can be solved on the blackboard or on a small piece of paper. We will see that it can be very useful to ask, "*Why* is the system designed in such a way?" and to try to answer with simplified models.

We conclude this introduction with an overview of the chapters. The first part of the book deals with transcription regulation networks. Elements of networks and their dynamics are described. We will see that these networks are made of repeating occurrences of simple patterns called network motifs. Each network motif performs a defined information processing function within the network. These building block circuits were rediscovered by evolution again and again in different systems. Network motifs in other biological networks, including signal transduction and neuronal networks, are also discussed. The main point is that biological systems show an inherent simplicity, by employing and combining a rather small set of basic building-block circuits, each for specific computational tasks.

The second part of the book focuses on the principle of robustness: biological circuits are designed so that their essential function is insensitive to the naturally occurring fluctuations in the components of the circuit. Whereas many circuit designs can perform a given function on paper, we will see that very few can work robustly in the cell. These few robust circuit designs are nongeneric and particular, and are often aesthetically pleasing. We will use the robustness principle to understand the detailed design of well-studied systems, including bacterial chemotaxis and patterning in fruit fly development.

The final chapters describe how constrained evolutionary optimization can be used to understand optimal circuit design, and how kinetic proofreading can minimize errors made in biological information processing.

These features of biological systems, reuse of a small set of network motifs, robustness to component tolerances, and constrained optimal design, are also found in a completely different context: systems designed by human engineers. Biological systems have additional features in common with engineered systems, such as modularity and hierarchical design. These similarities hint at a deeper theory that can unify our understanding of evolved and designed systems.

This is it for the introduction. A glossary of terms is provided at the end of the book, and some of the solved exercises after each chapter provide more detail on topics not discussed in the main text. I wish you enjoyable reading.

FURTHER READING

Fall, C., Marland E., Wagner J., and Tyson J. (2005). *Computational Cell Biology*, Springer.

Fell, D., (1996). *Understanding the Control of Metabolism*. Portland Press.

Heinrich, R. and Schuster, S. (1996). *The Regulation of Cellular Systems*. Kluwer Academic Publishers.

Klipp, E., Herwig, R., Kowald, A., Wierling, C., and Lehrach, H. (2005). *Systems Biology in Practice: Concepts, Implementation and Application*. Wiley.

Kriete, A. and Eils, R. (2005). *Computational Systems Biology*. Academic Press.

Palsson, B.O. (2006). *Systems Biology: Properties of Reconstructed Networks*. Cambridge University Press.

Savageau, M.A. (1976). *Biochemical Systems Analysis: A Study of Function and Design in Molecular Biology*. Addison Wesley.

Transcription Networks: Basic Concepts

2.1   INTRODUCTION

The cell is an integrated device made of several thousand types of interacting proteins. Each protein is a nanometer-size molecular machine that carries out a specific task with exquisite precision. For example, the micron-long bacterium *Escherichia coli* is a cell that contains a few million proteins, of about 4000 different types (typical numbers, lengths, and timescales can be found in Table 2.1).

Cells encounter different situations that require different proteins. For example, when sugar is sensed, the cell begins to produce proteins that can transport the sugar into the cell and utilize it. When damaged, the cell produces repair proteins. The cell therefore continuously monitors its environment and calculates the amount at which each type of protein is needed. This information-processing function, which determines the rate of production of each protein, is largely carried out by **transcription networks**.

The first few chapters in this book will discuss transcription networks. The present chapter defines the elements of transcription networks and examines their dynamics.

2.2   THE COGNITIVE PROBLEM OF THE CELL

Cells live in a complex environment and can sense many different signals, including physical parameters such as temperature and osmotic pressure, biological signaling molecules from other cells, beneficial nutrients, and harmful chemicals. Information about the internal state of the cell, such as the level of key metabolites and internal damage (e.g., damage to DNA, membrane, or proteins), is also important. Cells respond to these signals by producing appropriate proteins that act upon the internal or external environment.

TABLE 2.1 Typical Parameter Values for the Bacterial *E. coli* Cell, the Single-Celled Eukaryote
Saccharomyces cerevisae (Yeast), and a Mammalian Cell (Human Fibroblast)

Property	E. coli	Yeast (S. cerevisae)	Mammalian (Human Fibroblast)
Cell volume	~1 µm^3	~1000 µm^3	~10,000 µm^3
Proteins/cell	~4 10^6	~4 10^9	~4 10^{10}
Mean size of protein	5 nm		
Size of genome	4.6 10^6 bp	1.3 10^7 bp	3 10^9 bp
	4500 genes	6600 genes	~30,000 genes
Size of: Regulator binding site	~10 bp	~10 bp	~10 bp
Promoter	~100 bp	~1000 bp	~10^4 to 10^5 bp
Gene	~1000 bp	~1000 bp	~10^4 to 10^6 bp (with introns)
Concentration of one protein/cell	~1 nM	~1 pM	~0.1 pM
Diffusion time of protein across cell	~0.1 sec $D = 10$ µm^2/sec	~10 sec	~100 sec
Diffusion time of small molecule across cell	~1 msec, $D = 1000$ µm^2/sec	~10 msec	~0.1 sec
Time to transcribe a gene	~1 min 80 bp/sec	~1 min	~30 min (including mRNA processing)
Time to translate a protein	~2 min 40 aa/sec	~2 min	~30 min (including mRNA nuclear export)
Typical mRNA lifetime	2–5 min	~10 min to over 1 h	~10 min to over 10 h
Cell generation time	~30 min (rich medium) to several hours	~2 h (rich medium) to several hours	20 h — nondividing
Ribosomes/cell	~10^4	~10^7	~10^8
Transitions between protein states (active/inactive)	1–100 µsec	1–100 µsec	1–100 µsec
Timescale for equilibrium binding of small molecule to protein (diffusion limited)	~1 msec (1 µM affinity)	~1 sec (1 nM affinity)	~1 sec (1 nM affinity)
Timescale of transcription factor binding to DNA site	~1 sec		
Mutation rate	~10^{-9} /bp/generation	~10^{-10} /bp/generation	~10^{-8}/bp/year

bp: base-pair (DNA letter).

To represent these environmental states, the cell uses special proteins called **transcription factors** as symbols. Transcription factors are usually designed to transit rapidly between active and inactive molecular states, at a rate that is modulated by a specific environmental signal (input). Each active transcription factor can bind the DNA to regulate the rate at which specific target genes are read (Figure 2.1). The genes are read (transcribed) into mRNA, which is then translated into protein, which can act on the environment. The activities of the transcription factors in a cell therefore can be considered an internal representation of the environment. For example, the bacterium *E. coli* has an internal representation with about 300 degrees of freedom (transcription factors). These regulate the rates of production of *E. coli*'s 4000 proteins.

The internal representation by a set of transcription factors is a very compact description of the myriad factors in the environment. It seems that evolution selected internal representations that symbolize states that are most important for cell survival and growth. Many different situations are summarized by a particular transcription factor activity that signifies "I am starving." Many other situations are summarized by a different transcription factor activity that signifies "My DNA is damaged." These transcription factors regulate their target genes to mobilize the appropriate protein responses in each case.

2.3   ELEMENTS OF TRANSCRIPTION NETWORKS

The interaction between transcription factors and genes is described by transcription networks. Let us begin by briefly describing the elements of the network: genes and transcription factors. Each gene is a stretch of DNA whose sequence encodes the information

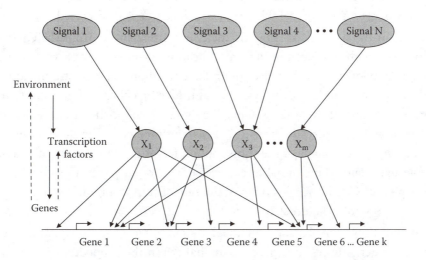

FIGURE 2.1 The mapping between environmental signals, transcription factors inside the cell, and the genes that they regulate. The environmental signals activate specific transcription factor proteins. The transcription factors, when active, bind DNA to change the transcription rate of specific target genes, the rate at which mRNA is produced. The mRNA is then translated into protein. Hence, transcription factors regulate the rate at which the proteins encoded by the genes are produced. These proteins affect the environment (internal and external). Some proteins are themselves transcription factors that can activate or repress other genes.

needed for production of a protein. Transcription of a gene is the process by which **RNA polymerase** (RNAp) produces mRNA that corresponds to that gene's coding sequence. The mRNA is then translated into a protein, also called the **gene product** (Figure 2.2a).

The rate at which the gene is transcribed, the number of mRNA produced per unit time, is controlled by the **promoter**, a regulatory region of DNA that precedes the gene (Figure 2.2a). RNAp binds a defined site (a specific DNA sequence) at the promoter (Figure 2.2a). The quality of this site specifies the transcription rate of the gene.[1]

Whereas RNAp acts on virtually all of the genes, changes in the expression of specific genes are due to transcription factors. Each transcription factor modulates the transcription rate of a set of target genes. Transcription factors affect the transcription rate by binding specific sites in the promoters of the regulated genes (Figure 2.2b and c). When bound, they change the probability per unit time that RNAp binds the promoter and produces an mRNA molecule.[2] The transcription factors thus affect the rate at which RNAp initiates transcription of the gene. Transcription factors can act as **activators** that increase the transcription rate of a gene, or as **repressors** that reduce the transcription rate (Figure 2.2b and c).

Transcription factor proteins are themselves encoded by genes, which are regulated by other transcription factors, which in turn may be regulated by yet other transcription factors, and so on. This set of interactions forms a **transcription network** (Figure 2.3). The transcription network describes all of the regulatory transcription interactions in a cell (or at least those that are known). In the network, the **nodes** are genes and **edges** represent transcriptional regulation of one gene by the protein product of another gene. A directed edge X → Y means that the product of gene X is a transcription factor protein that binds the promoter of gene Y to control the rate at which gene Y is transcribed.

The inputs to the network are **signals** that carry information from the environment. Each signal is a small molecule, protein modification, or molecular partner that directly affects the activity of one of the transcription factors. Often, external stimuli activate biochemical signal-transduction pathways that culminate in a chemical modification of specific transcription factors. In other systems, the signal can be as simple as a sugar molecule that enters the cells and directly binds the transcription factor. The signals usually cause a physical change in the shape of the transcription factor protein, causing it to assume an active molecular state. Thus, signal S_x can cause X to rapidly shift to its active state X^*, bind the promoter of gene Y, and increase the rate of transcription, leading to increased production of protein Y (Figure 2.2b).

The network thus represents a dynamical system: after an input signal arrives, transcription factor activities change, leading to changes in the production rate of proteins. Some of the proteins are transcription factors that activate additional genes, and so on.

---

[1] The sequence of the site determines the chemical affinity of RNAp to the site.

[2] When RNAp binds the promoter, it can transit into an open conformation. Once RNAp is in an open conformation, it initiates transcription: RNAp races down the DNA and transcribes one mRNA at a rate of tens of DNA letters (base-pairs) per second (Table 2.1). Transcription factors affect the probability per unit time of transcription initiation from the promoter.

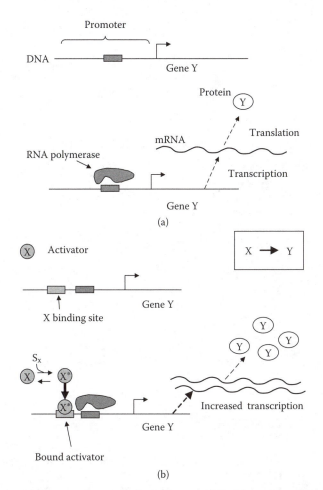

FIGURE 2.2 Gene transcription regulation, the basic picture. (a) Each gene is usually preceded by a regulatory DNA region called the promoter. The promoter contains a specific site (DNA sequence) that can bind RNA polymerase (RNAp), a complex of several proteins that forms an enzyme that can synthesize mRNA that corresponds to the gene coding sequence. The process of forming the mRNA is called transcription. The mRNA is then translated into protein. (b) An activator, X, is a transcription factor protein that increases the rate of mRNA transcription when it binds the promoter. The activator typically transits rapidly between active and inactive forms. In its active form, it has a high affinity to a specific site (or sites) on the promoter. The signal, S_x, increases the probability that X is in its active form, X^*. X^* binds a specific site in the promoter of gene Y to increase transcription and production of protein Y. (c) A repressor, X, is a transcription factor protein that decreases the rate of mRNA transcription when it binds the promoter. The signal, S_x, increases the probability that X is in its active form, X^*. X^* binds a specific site in the promoter of gene Y to decrease transcription and production of protein Y.

The rest of the proteins are not transcription factors, but rather carry out the diverse functions of the living cells, such as building structures and catalyzing reactions.

2.3.1 Separation of Timescales

Transcription networks are designed with a strong **separation of timescales**: the input signals usually change transcription factor activities on a sub-second timescale. Binding

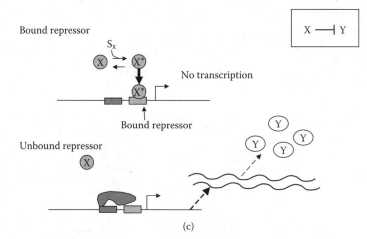

Bound repressor

S_x

No transcription

Bound repressor

Unbound repressor

X ⊣ Y

(c)

FIGURE 2.2 (continued)

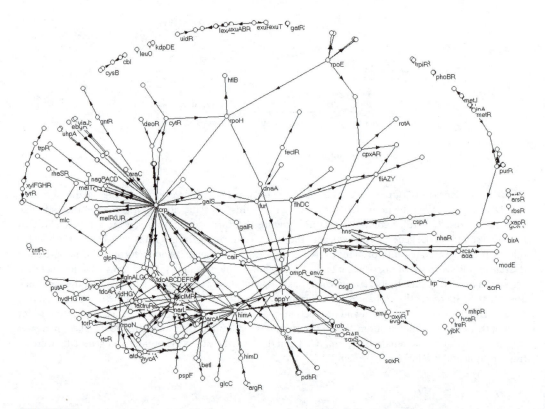

FIGURE 2.3 A transcription network that represents about 20% of the transcription interactions in the bacterium *E. coli*. Nodes are genes (or groups of genes coded on the same mRNA called operons). An edge directed from node X to node Y indicates that the transcription factor encoded in X regulates operon Y. This network describes direct transcriptional interactions based on experiments in many labs, compiled in databases such as regulonDB and Ecocyc. (From Shen-Orr et al., 2002.)

of the active transcription factor to its DNA sites often reaches equilibrium in seconds. Transcription and translation of the target gene takes minutes, and the accumulation of the protein product can take many minutes to hours (Table 2.1). Thus, the different steps between the signal and the accumulation of the protein products have very different time-scales. Table 2.2 gives typical approximate timescales for *E. coli*.

Thus, the transcription factor activity levels can be considered to be at steady state within the equations that describe network dynamics on the slow timescale of changes in protein levels.

In addition to transcription networks, the cell contains several other networks of inter-actions, such as signal-transduction networks made of interacting proteins, which will be discussed in later chapters. These networks typically operate much faster than transcription networks, and thus they can be considered to be approximately at steady state on the slow timescales of transcription networks.

There is a rich variety of mechanisms by which transcription factors regulate genes. Here, biology shows its full complexity. Transcription factors display ingenious ways to bind DNA at strategically placed sites. When bound, they block or recruit each other and RNAp (and, in higher organisms, many other accessory proteins) to control the rate at which mRNA is produced. However, on the level of transcription network dynamics, and on the slow timescales in which they operate, we will see that one can usually treat all of these mechanisms within a unifying and rather simple mathematical description.

One additional remarkable property of transcription networks is the modularity of their components. One can take the DNA of a gene from one organism and express it in a different organism. For example, one can take the DNA coding region for green fluores-cent protein (GFP) from the genome of a jellyfish and introduce this gene into bacteria. As a result, the bacteria produce GFP, causing the bacteria to turn green. Regulation can also be added by adding a promoter region. For example, control of the GFP gene in the bacterium can be achieved by pasting in front of the gene a DNA fragment from the pro-moter of a different bacterial gene, say, one that is controlled by a sugar-inducible tran-scription factor. This causes *E. coli* to express GFP and turn green only in the presence of the sugar. Promoters and genes are generally interchangeable. This fact underlies the use of GFP as an experimental tool, employed in the coming chapter to illustrate the dynam-ics of gene expression.

Modular components make transcription networks very plastic during evolution and able to readily incorporate new genes and new regulation. In fact, transcription networks can evolve rapidly: the edges in transcription networks appear to evolve on a faster

TABLE 2.2 Timescales for the Reactions in the Transcription Network of the Bacterium *E. coli* (Order of Magnitude)

Binding of a small molecule (a signal) to a transcription factor, causing a change in transcription factor activity	~1 msec
Binding of active transcription factor to its DNA site	~1 sec
Transcription + translation of the gene	~5 min
Timescale for 50% change in concentration of the translated protein (stable proteins)	~1 h (one cell generation)

timescale than the coding regions of the genes. For example, related animals, such as mice and humans, have very similar genes, but the transcription regulation of these genes, which governs when and how much of each protein is made, is evidently quite different. In other words, many of the differences between animal species appear to lie in the differences in the edges of the transcription networks, rather than in the differences in their genes.

2.3.2   The Signs on the Edges: Activators and Repressors

As we just saw, each edge in a transcription network corresponds to an interaction in which a transcription factor directly controls the transcription rate of a gene. These interactions can be of two types. **Activation**, or **positive control**, occurs when the transcription factor increases the rate of transcription when it binds the promoter (Figure 2.2b). **Repression**, or **negative control**, occurs when the transcription factor reduces the rate of transcription when it binds the promoter (Figure 2.2c). Thus, each edge in the network has a sign: + for activation, – for repression.[1] Transcription networks often show comparable numbers of plus and minus edges, with more positive (activation) interactions than negative interactions (e.g., 60 to 80% activation interactions in organisms such as *E. coli* and yeast). In Chapter 11, we will discuss principles that can explain the choice of mode of control for each gene.

Can a transcription factor be an activator for some genes and a repressor for others? Typically, transcription factors act primarily as either activators or repressors. In other words, the signs on the interaction edges that go out from a given node, and thus represent genes regulated by that node, are highly correlated. Some nodes send out edges with mostly minus signs. These nodes represent repressors. Other nodes, that represent activators, send out mostly plus-signed edges. However, most activators that regulate many genes act as repressors for some of their target genes. The same idea applies to many repressors, which can positively regulate a fraction of their target genes.[2]

Thus, transcription factors tend to employ one mode of regulation for most of their target genes. In contrast, the signs on the edges that go into a node, which represent the transcription interactions that regulate the gene, are less correlated. Many genes controlled by multiple transcription factors show activation inputs from some transcription factors and repression inputs from other transcription factors. In short, the signs on outgoing edges (edges that point out from a given node) are rather correlated, but the signs on incoming edges (edges that point into a given node) are not.[3]

---

[1] Some transcription factors, called dual transcription factors, can act on a given gene as activators under some conditions and repressors under other conditions.

[2] For example, a bacterial activator can readily be changed to a repressor by shifting its binding site so that it overlaps with the RNAp binding site. In this position, the binding of the activator protein physically blocks RNAp, and it therefore acts as a repressor.

[3] A similar feature is found in neuronal networks, where X → Y describes synaptic connections between neuron X and neuron Y (Chapter 6). In many cases, the signs (activation or inhibition) are more highly correlated on the outgoing synapses than the signs of incoming synapses. This feature, known as Dale's rule, stems from the fact that many neurons primarily use one type of neurotransmitter, which can be either excitatory or inhibitory for most outgoing synaptic connections.

2.3.3 The Numbers on the Edges: The Input Function

The edges not only have signs, but also can be thought to carry numbers that correspond to the strength of the interaction. The strength of the effect of a transcription factor on the transcription rate of its target gene is described by an **input function**. Let us consider first the production rate of protein Y controlled by a single transcription factor X. When X regulates Y, represented in the network by X → Y, the number of molecules of protein Y produced per unit time is a function of the concentration of X in its active form, X^*:

$$\text{rate of production of Y} = f(X^*) \tag{2.3.1}$$

Typically, the input function $f(X^*)$ is a monotonic, S-shaped function. It is an increasing function when X is an activator and a decreasing function when X is a repressor (Figure 2.4). A useful function that describes many real gene input functions is called the **Hill function**. The Hill function can be derived from considering the equilibrium binding of the transcription factor to its site on the promoter (see Appendix A for further details).

The Hill input function for an activator is a curve that rises from zero and approaches a maximal saturated level (Figure 2.4a):

$$f(X^*) = \frac{\beta X^{*n}}{K^n + X^{*n}} \qquad \textit{Hill function for activator} \tag{2.3.2}$$

The Hill function has three parameters, K, β, and n. The first parameter, K, is termed the **activation coefficient,** and has units of concentration. It defines the concentration of active X needed to significantly activate expression. From the equation it is easy to see that half-maximal expression is reached when $X^* = K$ (Figure 2.4a). The value of K is related to the chemical affinity between X and its site on the promoter, as well as additional factors.

The second parameter in the input function is the *maximal expression level* of the promoter, β. Maximal expression is reached at high activator concentrations, $X^* \gg K$, because at high concentrations, X^* binds the promoter with high probability and stimulates RNAp to produce many mRNAs per unit time. Finally, the **Hill coefficient** n governs the steepness of the input function. The larger is n, the more step-like the input function (Figure 2.4a). Typically, input functions are moderately steep, with n = 1 – 4.

As do many functions in biology, the Hill function approaches a limiting value at high levels of X^*, rather than increasing indefinitely. This saturation of the Hill function at high X^* concentration is fundamentally due to the fact that the probability that the activator binds the promoter cannot exceed 1, no matter how high the concentration of X^*. The Hill equation often describes empirical data with good precision.

For a repressor, the Hill input function is a decreasing S-shaped curve, whose shape depends on three similar parameters:

$$f(X^*) = \frac{\beta}{1 + \left(\dfrac{X^*}{K}\right)^n} \qquad \textit{Hill input function for repressor} \tag{2.3.3}$$

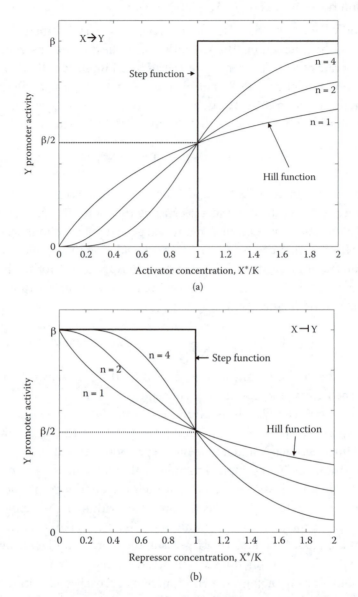

FIGURE 2.4 (a) Input functions for activator X described by Hill functions with Hill coefficient n = 1, 2, and 4. Promoter activity is plotted as a function of the concentration of X in its active form (X*). Also shown is a step function, also called a logic input function. The maximal promoter activity is β, and K is the threshold for activation of a target gene (the concentration of X* needed for 50% maximal activation). (b) Input functions for repressor X described by Hill functions with Hill coefficient n = 1, 2, and 4. Also shown is the corresponding logic input function (step function). The maximal unrepressed promoter activity is β, and K is the threshold for repression of a target gene (the concentration of X* needed for 50% maximal repression).

Since a repressor allows strong transcription of a gene only when it is not bound to the promoter, this function can be derived by considering the probability that the promoter is unbound by X^* (see Appendix A). The maximal production rate β is obtained when the repressor does not bind the promoter at all (Figure 2.2c), that is, when $X^* = 0$. Half-maximal repression is reached when the repressor activity is equal to K, the gene's **repression coefficient**. The Hill coefficient n determines the steepness of the input function (Figure 2.4b).

Hence, each edge in the network can be thought to carry at least three numbers, β, K, and n. These numbers can readily be tuned during evolution. For example, K can be changed by mutations that alter the DNA sequence of the binding site of X in the promoter of gene Y. Even a change of a single DNA letter in the binding site can strengthen or weaken the chemical bonds between X and the DNA and change K. The parameter K can also be varied if the position of the binding site is changed, as well as by changes in sequence outside of the binding site (the latter effects are currently not fully understood). Similarly, the maximal activity β can be tuned by mutations in the RNAp binding site or many other factors. Laboratory evolution experiments show that when placed in a new environment, bacteria can accurately tune these numbers within several hundred generations to reach optimal expression levels (Chapter 10). In other words, these numbers are under selection pressure and can heritably change over many generations if environments change.

The input functions we have described range from a transcription rate of zero to a maximal transcription rate β. Many genes have a nonzero minimal expression level. This is called the genes' **basal expression level**. A basal level can be described by adding to the input function a term β_0.

2.3.4 Logic Input Functions: A Simple Framework for Understanding Network Dynamics

Hill input functions are useful for detailed models. For mathematical clarity, however, it is often useful to use even simpler functions that capture the essential behavior of these input functions. The essence of input functions is transition between low and high values, with a characteristic threshold K. In the coming chapters, we will often approximate input functions in transcription networks using the **logic approximation** (Figure 2.4) (Glass and Kauffman, 1973; Thieffry and Thomas, 1998). In this approximation, the gene is either OFF, $f(X^*) = 0$, or maximally ON, $f(X^*) = \beta$. The threshold for activation is K. Hence, logic input functions are step-like approximations for the smoother Hill functions. For activators, the logic input function can be described using a **step-function** θ that makes a step when X^* exceeds the threshold K:

$$f(X^*) = \beta\, \theta(X^* > K) \qquad \textit{logic approximation for activator} \qquad (2.3.4)$$

where θ is equal to 0 or 1 according to the logic statement in the parentheses. The logic approximation is equivalent to a very steep Hill function with Hill coefficient $n \to \infty$ (Figure 2.4a).

Similarly, for repressors, a decreasing step function is appropriate:

$$f(X^*) = \beta \, \theta(X^* < K) \qquad \textit{logic approximation for repressor} \qquad (2.3.5)$$

We will see in the next chapters that by using a logic input function, dynamic equations become easy to solve graphically.

2.3.5   Multi-Dimensional Input Functions Govern Genes with Several Inputs

We just saw how Hill functions and logic functions can describe input from a single transcription factor. Many genes, however, are regulated by multiple transcription factors. In other words, many nodes in the network have two or more incoming edges. Their promoter activity is thus a multi-dimensional input function of the different input transcription factors (Yuh et al., 1998; Pilpel et al., 2001; Buchler et al., 2003; Setty et al., 2003). Appendix B describes how input functions can be modeled by equilibrium binding of multiple transcription factors to the promoter.

Often, multi-dimensional input functions can be usefully approximated by logic functions, just as in the case of single-input functions. For example, consider genes regulated by two activators. Many genes require binding of *both* activator proteins to the promoter in order to show significant expression. This is similar to an AND gate:

$$f(X^*, Y^*) = \beta \, \theta \, (X^* > K_x) \, \theta \, (Y^* > K_y) \sim X^* \text{ AND } Y^* \qquad (2.3.6)$$

For other genes, binding of either activator is sufficient. This resembles an OR gate:

$$f(X^*, Y^*) = \beta \, \theta \, (X^* > K_x \text{ OR } Y^* > K_y) \sim X^* \text{ OR } Y^* \qquad (2.3.7)$$

Not all genes have Boolean-like input functions. For example, some genes display a SUM input function, in which the inputs are additive (Kalir and Alon, 2004):

$$f(X^*, Y^*) = \beta_x X^* + \beta_y Y^* \qquad (2.3.8)$$

Other functions are also possible. For example, a function with several plateaus and thresholds was found in the *lac* system of *E. coli* (Figure 2.5) (See color insert following page 112). Genes in multi-cellular organisms often display input functions that can calculate elaborate functions of a dozen or more inputs (Yuh et al., 1998; Davidson et al., 2002; Beer and Tavazoie, 2004).

The functional form of input functions can be readily changed by means of mutations in the promoter of the regulated gene. For example, the *lac* input function of Figure 2.5 can be changed to resemble pure AND or OR gates with a few mutations in the *lac* promoter (Mayo et al., 2006). It appears that the precise form of the input function of each gene is under selection pressure during evolution.

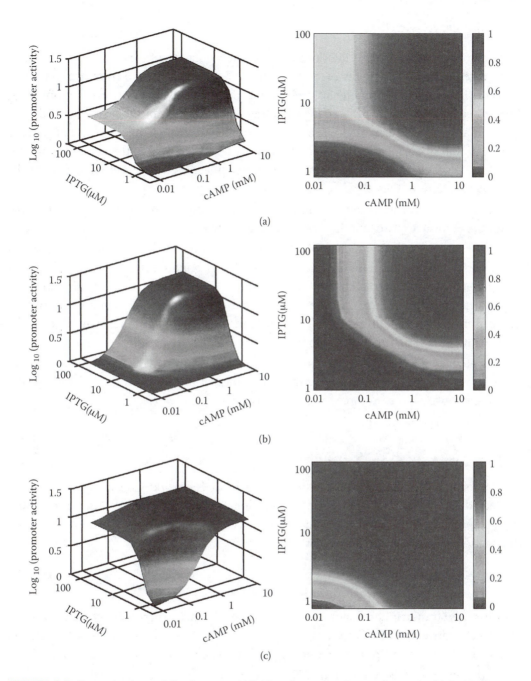

FIGURE 2.5 (See color insert following page 112) Two-dimensional input functions. (a) Input function measured in the *lac* promoter of *E. coli,* as a function of two input signals, the inducers cAMP and IPTG. (b) An AND-like input function, which shows high promoter activity only if both inputs are present. (c) An OR-like input function that shows high promoter activity if either input is present. (From Setty et al., 2003.)

2.3.6 Interim Summary

Transcription networks describe the transcription regulation of genes. Each node represents a gene.[1] Edges denoted X → Y mean that gene X encodes for a transcription factor protein that binds the promoter of gene Y and modulates its rate of transcription. Thus, the protein encoded by gene X changes the rate of production of the protein encoded by gene Y. Protein Y, in turn, might be a transcription factor that changes the rate of production of Z, and so on, forming an interaction network. Most nodes in the network stand for genes that encode proteins that are not transcription factors. These proteins carry out the various functions of the cell.

The inputs to the network are signals that carry information from the environment and change the activity of specific transcription factors.

The active transcription factors bind specific DNA sites in the promoters of their target genes to control the rate of transcription. This is quantitatively described by input functions: the rate of production of gene product Y is a function of the concentration of active transcription factor X^*. Genes regulated by multiple transcription factors have multi-dimensional input functions. The input functions are often rather sharp and can be approximated by Hill functions or logic gates.

Every edge and input function is under selection pressure. A nonuseful edge would rapidly be lost by mutations. It only takes a change of one or a few DNA letters in the binding site of X in the promoter of Y to abolish the edge X → Y.

Now, we turn to the dynamics of the network.

2.4   DYNAMICS AND RESPONSE TIME OF SIMPLE GENE REGULATION

Let us focus on the dynamics of a single edge in the network. Consider a gene that is regulated by a single regulator, with no additional inputs (or with all other inputs and post-transcriptional modes of regulation held constant over time[2]). This transcription interaction is described in the network by

$$X \rightarrow Y$$

which reads "transcription factor X regulates gene Y." Once X becomes activated by a signal, Y concentration begins to change. Let us calculate the **dynamics** of the concentration of the gene product, the protein Y, and its **response time**.

In the absence of its input signal, X is inactive and Y is not produced (Figure 2.2b). When the signal S_x appears, X rapidly transits to its active form X^* and binds the promoter of gene Y. Gene Y begins to be transcribed, and the mRNA is translated, resulting

---

[1] In bacteria, each node represents an operon: a set of one or more genes that are transcribed on the same mRNA. An edge X → Y means that one of the genes in operon X encodes a transcription factor that regulates operon Y.

[2] Proteins are potentially regulated in every step of their synthesis process, including the following post-transcriptional regulation interactions: (1) rate of degradation of the mRNA, (2) rate of translation, controlled primarily by sequences in the mRNA that bind the ribosomes and by mRNA-binding regulatory proteins and regulatory RNA molecules and (3) rate of active and specific protein degradation. In eukaryotes, regulation also occurs on the level of mRNA splicing and transport in the cell. Many other modes of regulation are possible.

in accumulation of protein Y. The cell produces protein Y at a constant rate, which we will denote β (units of concentration per unit time).

The production of Y is balanced by two processes, protein **degradation** (its specific destruction by specialized proteins in the cell) and **dilution** (the reduction in concentration due to the increase of cell volume during growth). The degradation rate is α_{deg}, and the dilution rate is α_{dil}, giving a total degradation/dilution rate (in units of 1/time) of

$$\alpha = \alpha_{dil} + \alpha_{deg} \tag{2.4.1}$$

The change in the concentration of Y is due to the difference between its production and degradation/dilution, as described by a dynamic equation[1]:

$$dY/dt = \beta - \alpha\,Y \tag{2.4.2}$$

At steady state, Y reaches a constant concentration Y_{st}. The steady-state concentration can be found by solving for $dY/dt = 0$. This shows that the steady-state concentration is the ratio of the production and degradation/dilution rates:

$$Y_{st} = \beta/\alpha \tag{2.4.3}$$

This makes sense: the higher the production rate β, the higher the protein concentration reached, Y_{st}. The higher the degradation/dilution rate α, the lower is Y_{st}.

What happens if we now take away the input signal, so that production of Y stops ($\beta = 0$)? The solution of Equation 2.4.2 with $\beta = 0$ is an exponential decay of Y concentration (Figure 2.6a):

$$Y(t) = Y_{st}\,e^{-\alpha t} \tag{2.4.4}$$

How fast does Y decay? An important measure for the speed at which Y levels change is the **response time**. The response time, $T_{1/2}$, is generally defined as the time to reach halfway between the initial and final levels in a dynamic process. For the decay process of Equation 2.4.4, the response time is the time to reach halfway down from the initial level Y_{st} to the final level, $Y = 0$. The response time, therefore, is given by solving for the time when $Y(t) = Y_{st}/2$, which, using Equation 2.4.4, shows an inverse dependence on the degradation/dilution rate:

$$T_{1/2} = \log(2)/\alpha \tag{2.4.5}$$

---

[1] This dynamic equation has been used since the early days of molecular biology (for example, Monod et al., 1952). It gives excellent agreement with high-resolution dynamics experiments done under conditions of protein activation during exponential growth of bacteria (Rosenfeld et al., 2002; Rosenfeld and Alon, 2003). Note that in the present treatment we assume that the concentration of the regulator, active X, is constant throughout, so that $\beta = f(X^*)$ is constant. Furthermore, the time for transcription and translation of the protein is neglected because it is small compared to the response time of the protein-level dynamics (Table 2.2).

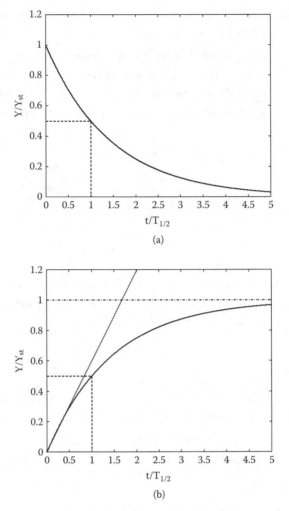

(a)

(b)

FIGURE 2.6 (a) Decay of protein concentration following a sudden drop in production rate. The response time, the time it takes the concentration to reach half of its variation, is $T_{1/2} = \log(2)/\alpha$. The response time can be found graphically by the time when the curve crosses the horizontal dashed line placed halfway between the initial point and the steady-state point of the dynamics. (b) Rise in protein concentration following a sudden increase in production rate. The response time, the time it takes the dynamics to reach half of its variation, is $T_{1/2} = \log(2)/\alpha$. At early times, the protein accumulation is approximately linear with time, $Y = \beta t$ (dotted line).

Note that the degradation/dilution rate α directly determines the response time: fast degradation/dilution allows rapid changes in concentration. The production rate β affects the steady-state level but not the response time.

Some proteins show rapid degradation rates (large α). At steady-state, this leads to a seemingly futile cycle of production and destruction. To maintain a given steady-state, $Y_{st} = \beta/\alpha$, requires high production β to balance the high degradation rate α. The benefit of such futile cycles is fast response times once a change is needed.

We have seen that loss of input signal leads to an exponential decay of Y. Let us now consider the opposite case, in which an unstimulated cell with Y = 0 is provided with a signal, so that protein Y begins to accumulate. If an unstimulated gene becomes suddenly stimulated by a strong signal S_x, the dynamic equation, Equation 2.4.2, results in an approach to steady state (Figure 2.6b):

$$Y(t) = Y_{st} (1 - e^{-\alpha t}) \qquad (2.4.6)$$

The concentration of Y rises from zero and gradually converges on the steady-state Y_{st} = β/α. Note that at early times, when $\alpha t \ll 1$, we can use a Taylor expansion[1] to find a linear accumulation of Y:

$$Y \sim \beta t \qquad \qquad \textit{early times, } \alpha t \ll 1 \qquad (2.4.7)$$

This makes sense: the concentration of protein Y accumulates at early times with a slope equal to its production rate. Later, as Y levels increase, the degradation term $-\alpha Y$ begins to be important and Y converges to its steady-state level.

The response time, the time to reach $Y_{st}/2$, can be found by solving for the time when $Y(t) = Y_{st}/2$. Using Equation 2.4.6, we find the same response time as in the case of decay:

$$T_{1/2} = \log(2)/\alpha \qquad (2.4.8)$$

The response time for both increase and decrease in protein levels is the same and is governed only by the degradation/dilution rate. The larger the degradation/dilution rate α, the more rapid the changes in concentration.

2.4.1 The Response Time of Stable Proteins Is One Cell Generation

Many proteins are not actively degraded in growing cells ($\alpha_{deg} = 0$). These are termed stable proteins. The production of stable proteins is balanced by dilution due to the increasing volume of the growing cell, $\alpha = \alpha_{dil}$. For such stable proteins, the *response time is equal to one cell generation time*. To see this, imagine that a cell produces a protein, and then suddenly production stops ($\beta = 0$). The cell grows and, when it doubles its volume, splits into two cells. Thus, after one **cell generation time** τ, the protein concentration has decreased by 50%, and therefore:

$$T_{1/2} = \log(2)/\alpha_{dil} = \tau \qquad \textit{response time is one cell generation} \quad (2.4.9)$$

This is an interesting result. Bacterial cell generation times are on the order of 30 min to a few hours, and eukaryotic generation times are even longer. One would expect that transcription networks that are made to react to signals such as nutrients and stresses should respond at least as rapidly as the cell generation time. But for stable proteins, the response time, as we saw, is one cell generation time. Thus, *response time can be a limiting factor that poses a constraint for designing efficient gene circuits.*

---

[1] Using $e^{-\alpha t} \sim 1 - \alpha t$, and $Y_{st} = \beta/\alpha$.

In summary, we have seen that the response time of simple gene regulation is determined by the degradation and dilution rates of the protein product. In the next chapter, we will discuss simple transcriptional circuits that can help speed the response time.

FURTHER READING

Molecular Mechanisms of Transcriptional Regulation

Ptashne, M. (1986). *A Genetic Switch*. Cell Press and Scientific Publications.
Ptashne, M. and Gann, A. (2002). *Genes and Signals*. Cold Spring Harbor Laboratory Press.

Overview of Transcription Networks

Alon, U. (2003). Biological networks: the tinkerer as an engineer. *Science*, 301: 1866–1867.
Levine, M. and Davidson, E.H. (2005). Gene regulatory networks for development. *Proc. Natl. Acad. Sci. U.S.A.* 102: 4936–4942.
Thieffry, D., Huerta, A.M., Perez-Rueda, E., and Collado-Vides, J. (1998). From specific gene regulation to genomic networks: a global analysis of transcriptional regulation in *Escherichia coli*. *Bioessays*, 20: 433–440.

Ecocyc Database

www.ecocyc.org

Dynamics of Gene Networks

Monod, J., Pappenheimer, A.M., Jr., and Cohen-Bazire, G. (1952). The kinetics of the biosynthesis of beta-galactosidase in *Escherichia coli* as a function of growth. *Biochem. Biophys. Acta*, 9: 648–660.
Rosenfeld, N. and Alon, U. (2003). Response delays and the structure of transcription networks. *J. Mol. Biol.*, 329: 645–654.

EXERCISES

2.1. *A change in production rate.* A gene Y with simple regulation is produced at a constant rate β_1. The production rate suddenly shifts to a different rate β_2.

    a. Calculate and plot the gene product concentration Y(t).

    b. What is the response time (time to reach halfway between the steady states)?

Solution (for part a):

    a. Let us mark the time when the shift occurs as t = 0. Before the shift, Y reaches steady state at a level $Y(t = 0) = Y_{st} = \beta_1/\alpha$. After the shift,

$$dY/dt = \beta_2 - \alpha Y \qquad (P2.1)$$

The solution of such an equation is generally $Y = C_1 + C_2 e^{-\alpha t}$, where the constants C_1 and C_2 need to be determined so that $Y(t = 0) = \beta_1/\alpha$, and Y at long times reaches its new steady state, β_2/α. This yields the following sum of an exponential and a constant:

$$Y(t) = \beta_2/\alpha + (\beta_1/\alpha - \beta_2/\alpha)\, e^{-\alpha t} \qquad (P2.2)$$

Take the derivative with respect to time, dY/dt, and verify that Equation P2.1 is fulfilled.

2.2. *mRNA dynamics.* In the main text, we considered the activation of transcription of a gene (mRNA production) and used a dynamical equation to describe the changes in the concentration of the gene product, the protein Y. In this equation, $dY/dt = \beta - \alpha Y$, the parameter β describes the rate of protein production. In reality, mRNA needs to be translated to form the protein, and mRNA itself is also degraded by specific enzymes.

a. Derive dynamical equations for the rate of change of mRNA and the rate of change of the protein product, assuming that mRNA is produced at rate β_m and degraded at rate α_m, and that each mRNA produces on average p protein molecules over its lifetime. The protein is degraded/diluted at rate α.

b. Note that mRNA is often degraded at a much faster rate than the protein product $\alpha_m \gg \alpha$. Can this be used to form a quasi-steady-state assumption that mRNA levels are at steady state with respect to slower processes? What is the effective protein production rate β in terms of β_m, α_m, and p? What would be the response time if the mRNA lifetime were much longer than the protein lifetime?

Solution:

a. The dynamic equation for the concentration of mRNA of gene Y, Y_m, is:

$$dY_m/dt = \beta_m - \alpha_m Y_m \qquad (P2.3)$$

The dynamical equation for the protein product is due to production of p copies per mRNA and degradation/dilution at rate α:

$$dY/dt = p\, Y_m - \alpha Y \qquad (P2.4)$$

b. In the typical case that mRNA degradation is faster than the degradation/dilution of the protein product, we can assume that Y_m reaches steady state quickly in comparison to the protein levels. The reason is that the typical time for the mRNA to reach steady state is the response time $\log(2)/\alpha_m$, which is much shorter than the protein response time $\log(2)/\alpha$ because $\alpha_m \gg \alpha$. The steady-state mRNA level is found by setting $dY_m/dt = 0$ in Equation P2.3, yielding

$$Y_{m,st} = \beta_m/\alpha_m \qquad (P2.5)$$

Using this for Y_m in Equation P2.4 yields the following equation for the protein production rate:

$$dY/dt = p\, \beta_m/\alpha_m - \alpha Y \qquad (P2.6)$$

In other words, the effective protein production rate, which is the first term on the right-hand side of the equation, is equal to the steady-state mRNA level times the number of proteins translated from each mRNA:

$$\beta = p \, \beta_m / \alpha_m \qquad \text{(P2.7)}$$

2.3. *Time-dependent production and decay.* A gene Y with simple regulation has a time-dependent production rate $\beta(t)$ and a time-dependent degradation rate $\alpha(t)$. Solve for its concentration as a function of time.

Solution:

Verify by taking the time derivative that the following is correct:

$$Y(t) = \exp(-\textstyle\int \alpha(t')\,dt')\,[Y(0) + \textstyle\int \beta(t')\,\exp(\textstyle\int \alpha(t'')\,dt'')\,dt'] \qquad \text{(P2.8)}$$

2.4. *Cascades.* Consider a cascade of three activators, $X \to Y \to Z$. Protein X is initially present in the cell in its inactive from. The input signal of X, S_x, appears at time $t = 0$. As a result, X rapidly becomes active and binds the promoter of gene Y, so that protein Y starts to be produced at rate β. When Y levels exceed a threshold K_y, gene Z begins to be transcribed. All proteins have the same degradation/dilution rate α. What is the concentration of protein Z as a function of time? What is its response time with respect to the time of addition of S_x? What about a cascade of three repressors? Compare your solution to the experiments shown in Figure 2.7.

2.5. *Fan-out.* Transcription factor X regulates two genes, Y_1 and Y_2. Draw the resulting network, termed a fan-out with two target genes. The activation thresholds for these genes are K_1 and K_2. The activator X begins to be produced at time $t = 0$ at rate β. Its signal is degraded/diluted at rate α, and its signal S_x is present throughout. What are the times at which the gene products, the stable proteins Y_1 and Y_2, reach halfway to their maximal expression?

2.6. Pulse of activation: Consider the cascade of exercise 2.4. The input signal S_x appears at time t=0 for a pulse of duration D, and then vanishes.

(a) What is the concentration Y(t)?

(b) What is the minimal pulse duration needed for the activation of gene Z?

(c) Plot the maximal level reached by the protein Z as a function of pulse duration D.

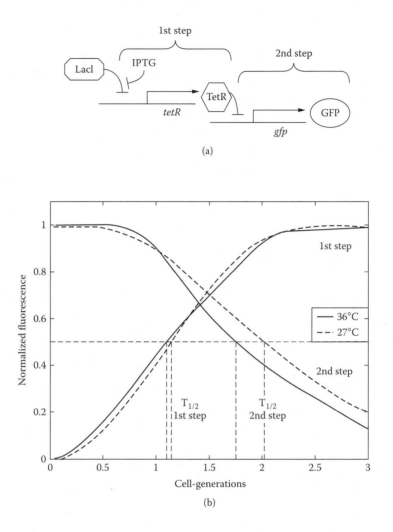

FIGURE 2.7 (a) A transcriptional cascade made of repressors in *E. coli*. The transcription cascade (X ⊣ Y ⊣ Z) is made of two well-studied repressors, LacI and TetR, both of which have negligible degradation rates in *E. coli*. The cascade was built using genetic engineering, by combining the appropriate promoter DNA fragments to the appropriate genes. TetR was made to also repress the green fluorescent protein gene, acting as a reporter for the second cascade step. The bacteria thus turn green in proportion to the promoter activity regulated by TetR. In a separate *E. coli* strain prepared for this experiment, LacI represses a green fluorescent protein gene, acting as a reporter for the first cascade step. (b) Response time is about one cell generation per cascade step. The first step in the cascade rises in response to the inducer IPTG that inactivates the repressor LacI. This inactivation leads to the production of the repressor TetR that, when present at sufficient amounts, causes a decrease in the activity of the second-step promoter. The experiment was carried out at two temperatures, which show different cell generation times (the generation time is about two-fold longer at 27°C compared to 36°C). The x-axis shows time in units of the cell generation time in each condition. (From Rosenfeld and Alon, 2003.)

CHAPTER 3

Autoregulation: A Network Motif

3.1   INTRODUCTION

In the previous chapter we learned the basic dynamics of a single interaction in a transcription network. Now, let us take a look at a real, live transcription network made of many interaction edges (Figure 3.1). As an example, we will use a network of transcription interactions of *Escherichia coli* that includes about 20% of the organism's genes (Shen-Orr et al., 2002).

This network looks very complex. Our goal will be to define understandable patterns of connections that serve as building blocks of the network. Ideally, we would like to understand the dynamics of the entire network based on the dynamics of the individual building blocks. In this chapter, we will:

1. Define a way to detect building-block patterns in complex networks, called **network motifs.**

2. Examine the simplest network motif in transcription networks, **negative autoregulation.**

3. Show that this motif has useful functions: speeding up the response time of transcription interactions and stabilizing them.

3.2   PATTERNS, RANDOMIZED NETWORKS, AND NETWORK MOTIFS

The transcription network of *E. coli* contains numerous patterns of nodes and edges. Our approach will be to look for meaningful patterns on the basis of statistical significance. To define statistical significance, we compare the network to an ensemble of **randomized networks**. The randomized networks are networks with the same characteristics as the real network, (e.g., the same number of nodes and edges as the real network), but where the connections between nodes and edges are made at random. Patterns that occur in the real network significantly more often than in randomized networks are called **network motifs** (Milo et al., 2002; Shen-Orr et al., 2002).

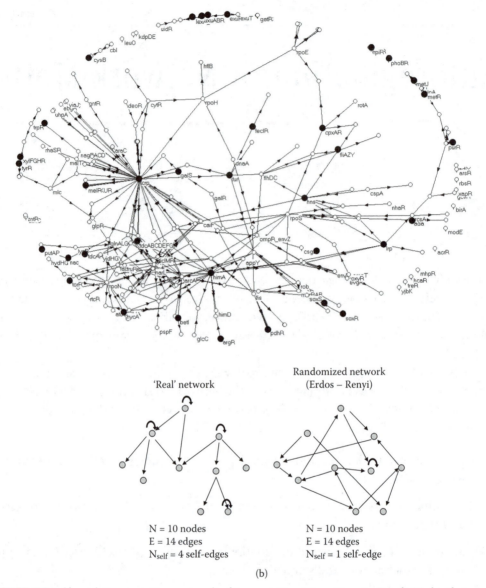

FIGURE 3.1 Self-regulating genes in a network of transcription interactions in *E. coli*. Nodes that correspond to genes that encode transcription factor proteins that regulate their own promoters (self-regulating genes, represented by self-edges) are shown in black. This network, which we will use as an example in the coming chapters, has about N = 420 nodes, E = 520 edges, and N_self = 40 self-edges. (b) Example of a small network and its randomized Erdos–Renyi version, with the same number of nodes and edges.

The basic idea is that patterns that occur in the real network much more often than in randomized networks must have been preserved over evolutionary timescales against mutations that randomly change edges. To appreciate this, note that edges are easily lost in a transcription network. As previously mentioned, a mutation that changes a single DNA letter in a promoter can abolish binding of a transcription factor and cause the loss of an edge in the network.

Such mutations can occur at a comparatively high rate, as can be appreciated by the following example. A single bacterium placed in 10 ml of liquid nutrient, grows and divides to reach a saturating population of about 10^{10} cells within less than a day. This population therefore underwent 10^{10} DNA replications. Since the mutation rate is about 10^{-9} per letter per replication, the population will include, for each letter in the genome, 10 different bacteria with a mutation in that letter. Thus, a change of any DNA letter can be reached many times over very rapidly in bacterial populations. A similar rate of mutations per generation per genome occurs in multi-cellular organisms (for genome sizes and mutation rates, see Table 2.1).

Similarly, new edges can be added to the network by mutations that generate a binding site for transcription factor X in the promoter region of gene Y. Such sites can be generated, for example, by mutations or by events that duplicate or reposition pieces of a genome, or that insert into the genome pieces of DNA from other cells (Shapiro, 1999). Hence, *edges in network motifs must be constantly selected* in order to survive randomization forces.

This suggests that if a network motif appears in a network much more often than in randomized networks, it must have been selected based on some advantage it gives to the organism. If the motif did not offer a selective advantage, it would be washed out and occur about as often as in randomized networks.

3.2.1   Detecting Network Motifs by Comparison to Randomized Networks

To detect network motifs, we need to compare the real network to an ensemble of randomized networks. We will first consider the simplest ensemble of randomized networks, introduced by Erdos and Renyi (Erdos and Renyi, 1959; Bollobas, 1985). This makes calculations easy and gives the same qualitative answers as more elaborate random network models (this will be discussed in Chapter 4).

For a meaningful comparison, the randomized networks should share the basic features of the real network. The real transcription network has N nodes and E edges. To compare it to the Erdos–Renyi (ER) model, one builds a random network with the same number of nodes and edges. In the random network, defined by the ER model, directed edges are assigned *at random* between pairs of nodes.

Since there are N nodes, there are $N(N - 1)/2$ possible pairs of nodes that can be connected by an edge. Each edge can point in one of two directions, for a total of $N(N - 1)$ possible places to put a directed edge between two nodes. In addition, an edge can begin and end at the same node, forming a self-edge (total of N possible self-edges). The total number of possible edges is therefore:

$$N (N - 1) + N = N^2 \qquad (3.2.1)$$

In the ER model, the E edges are placed at random in the N^2 possible positions, and therefore each possible edge position is occupied with probability $p = E/N^2$. Figure 3.1b compares a small network to a corresponding random ER network with the same number of nodes and edges.

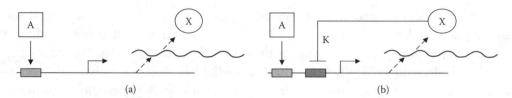

(a)   (b)

FIGURE 3.2 Simple regulation and negative autoregulation. (a) Gene X is simply regulated by A. (b) A gene X that is negatively autoregulated; that is, it is repressed by its own gene product, the repressor X. The gene is also simply regulated by A. Repressor X binds a site in its own promoter and thus acts to repress its own transcription. The symbol ⊣ stands for repression. The repression threshold is K, defined as the concentration of X needed to repress the promoter activity by 50%.

3.3   AUTOREGULATION: A NETWORK MOTIF

Now we can begin to compare features of the real *E. coli* transcription network with the randomized networks. Let us start with **self-edges**, edges that originate and end at the same node. The *E. coli* network that we use as an example has 40 self-edges (Figure 3.1a). These self-edges correspond to transcription factors that regulate the transcription of their own genes.

Regulation of a gene by its own gene product is known as autogenous control, or **autoregulation**. Thirty-four of the autoregulatory proteins in the network are repressors that repress their own transcription: **negative autoregulation** (Figure 3.2).

Is autoregulation significantly more frequent in the real network than at random? To decide, we need to calculate the probability of having k self-edges in an ER random network. To form a self-edge, an edge needs to choose its node of origin as its destination, out of the N possible target nodes. This probability is thus:

$$p_{self} = 1/N \tag{3.3.1}$$

Since E edges are placed at random to form the random network, the probability of having k self-edges is approximately binomial (throwing a coin E times and getting k heads):

$$P(k) = \binom{E}{k} p_{self}^{k} (1 - p_{self})^{E-k} \tag{3.3.2}$$

The average number of self-edges is equal to the number of edges E times the probability that an edge is a self-edge (just as the expected number of heads is the number of times the coin is thrown multiplied by the probability of heads):

$$\langle N_{self} \rangle_{rand} \sim E\, p_{self} \sim E/N \tag{3.3.3}$$

with a standard deviation that is approximately the square root of the mean (again, similar to a coin-tossing experiment with a small probability p_{self} for heads, which approximates a Poisson process):

$$\sigma_{rand} \sim \sqrt{E/N}. \tag{3.3.4}$$

In the *E. coli* transcription network of Figure 3.1, the numbers of nodes and edges are N = 424 and E = 519. Thus, according to Equations 3.3.3 and 3.3.4, a corresponding ER network with the same N and E would be expected to have only about one self-edge, plus minus one:

$$\langle N_{self}\rangle_{rand} \sim E/N \sim 1.2, \qquad \sigma_{rand} \sim \sqrt{1.2} \sim 1.1 \qquad (3.3.5)$$

In contrast, the real network has 40 self-edges, which exceeds the random networks by many standard deviations. This significant difference in the number of self-edges can be described by the number of standard deviations by which the real network exceeds the random ensemble:

$$Z = \frac{\langle N_{self}\rangle_{real} - \langle N_{self}\rangle_{rand}}{\sigma_{rand}} \qquad (3.3.6)$$

Self-edges show $Z \sim 32$, which means they occur far more often than at random. Note that 32 standard deviations mark a very high statistical significance.

Thus, self-edges, and in particular negatively autoregulated genes, are a network motif. A **network motif** is a recurring pattern in the network that occurs far more often than at random.

The next question is: Why is negative autoregulation a network motif? Does it have a useful function?

To answer this, we will compare a negatively autoregulated gene to a simply (non-auto) regulated gene (Figure 3.2). Our criterion for comparison will be the **response time** of the system.

As we saw in the previous chapter, the response time of a simply regulated gene is governed by its degradation/dilution rate α:

$$T_{1/2} = \log(2)/\alpha \qquad (3.3.7)$$

For stable proteins that are not appreciably degraded in the cell, the response time is equal to the cell generation time. We will now see how the negative autoregulation network motif can help speed up transcription responses.

3.4 NEGATIVE AUTOREGULATION SPEEDS THE RESPONSE TIME OF GENE CIRCUITS[1]

Negative autoregulation occurs when a transcription factor X represses its own transcription (Figure 3.2b). This self-repression occurs when X binds its own promoter to inhibit production of mRNA. As a result, the higher the concentration of X, the lower its production rate.

As we saw in the previous chapter, the dynamics of X are described by its production rate f(X) and degradation/dilution rate:[2]

---

[1] See Savageau, 1974a; Rosenfeld et al., 2002.

[2] To understand the dynamics of a negatively autoregulated system, recall the separation of timescales in transcription networks. The production rate of X is governed by the probability that X binds the promoter of its own gene. The binding and unbinding of X to the promoter rapidly reaches equilibrium (usually on the order of seconds or less). The concentration of protein X, on the other hand, changes much more slowly, on the timescale of tens of minutes. Therefore, it makes sense to describe the production rate by an input function, f(X), equal to the mean promoter activity at a given level of X, averaged over many repressor binding–unbinding events (in other words, this is a quasi-steady-state description of promoter activity).

$$dX/dt = f(X) - \alpha X \qquad (3.4.1)$$

where f(X) is a decreasing function of X. As mentioned in Chapter 2, a good approximation for many promoters is a decreasing Hill function (Figure 2.4):

$$f(X) = \frac{\beta}{1+(X/K)^n} \qquad (3.4.2)$$

In this input function, when X is much smaller than the repression coefficient K, the promoter is free and the production rate reaches its maximal value, β. On the other hand, when repressor X is at high concentration, no transcription occurs, $f(X) \sim 0$. Recall that the repression coefficient K has units of concentration and defines the concentration at which X represses the promoter activity by 50%.

To solve the dynamics in the most intuitive way, let us use the logic approximation, where production is zero if X > K, and production is maximal, namely, $f(X) = \beta$, when X is smaller than K. This was described in Chapter 2.3.4 using the step function θ:

$$f(X) = \beta \, \theta \, (X < K) \qquad (3.4.3)$$

In exercise 3.1 we will also solve the dynamics with a Hill function, to find that the logic approximation is reasonable.

To study the response time, consider the case where X is initially absent, and its production starts at t = 0. At early times, while X concentration is low, the promoter is unrepressed and production is full steam at rate β, as described by the production–degradation equation:

$$dX/dt = \beta - \alpha X \qquad \textit{while } X < K \quad (3.4.4)$$

This results in an approach to a high steady-state value, as described in Section 2.4 of the previous chapter. At early times, in fact, we can neglect degradation ($\alpha X \ll \beta$) to find linear accumulation of X with time (Equation 2.4.7):

$$X(t) \sim \beta \, t \qquad \textit{while } X < K \textit{ and } X \ll \beta/\alpha \quad (3.4.5)$$

However, production stops when X levels reach the self-repression threshold, X = K (recall that production is zero when X exceeds K) (Figure 3.3). Small oscillations will occur around X = K if there are any delays in the system. Delays may cause X to overshoot beyond K slightly, but then production stops and X levels decline until they decrease below K, upon which production starts again, etc. These oscillations are generally damped if f(X) is not strictly a logic function, but rather a smoother function like a Hill function. Thus, X effectively locks into a *steady-state level equal to the repression coefficient of its own promoter*:

$$X_{st} = K \qquad (3.4.6)$$

The resulting dynamics shows a rapid rise and a sudden saturation, as shown in Figure 3.3.

The response time, $T_{1/2}$, can be found by asking when X reaches half steady state so that $X(T_{1/2}) = X_{st}/2$. For simplicity, let us calculate the response time using linear

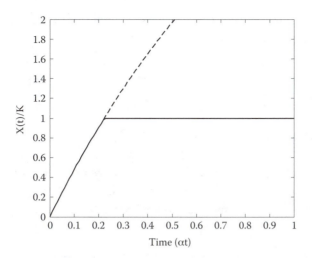

FIGURE 3.3 Dynamics of a negatively autoregulated gene product. Production starts at t = 0. Full line: negatively autoregulated gene with maximal production rate β = 5, autorepression threshold K = 1, and degradation/dilution rate α = 1. Dashed line: Dynamics of the same gene if autoregulation is removed, resulting in simple regulation that approaches a higher, unrepressed steady state, $X_{st} = β/α = 5$.

accumulation[1] of X (Equation 3.4.5), in which X = β t. The response time $T_{1/2}^{(n.a.r.)}$, where n.a.r stands for negative autoregulation, is given by finding the time when X reaches half of the steady-state level, $β T_{1/2}^{(n.a.r.)} = X_{st}/2 = K/2$, so that:

$$T_{1/2}^{(n.a.r.)} = \frac{K}{2β} \qquad \textit{response time for negative autoregulation} \qquad (3.4.7)$$

The stronger the maximal unrepressed promoter activity β, the shorter the response time. Negative autoregulation can therefore use a strong promoter to give an initial fast production, and then use autorepression to stop production at the desired steady state.

Note that evolutionary selection can easily tune the parameters β and K independently. The repression threshold K can be modified, for example, by mutations in the binding site of X in the promoter, whereas β can be tuned by mutations in the binding site of RNAp in the promoter. Thus, the steady state ($X_{st} = K$) and the response time can in principle be separately determined.

Let us compare this design with a simply regulated gene (a gene without negative autoregulation, as described in Section 2.4), which is produced at rate $β_{simple}$ and degraded at rate $α_{simple}$. To make a meaningful comparison, we must compare the two designs with the same steady-state levels. This is because the steady-state level of the protein is usually important for its optimal function. In addition, the designs should have as many of the same biochemical parameters as possible. In the present case, the two designs will have the same protein degradation/dilution rate $α = α_{simple}$. Such a comparison that is carried out with equivalence of as many internal and external parameters as possible between the alternative designs is termed a **mathematically controlled comparison** (Savageau, 1976).

---

[1] This is a good approximation for strong autorepression, in which X_{st} is much smaller than what it would be without autorepression, $X_{st} = K \ll β/α$.

In simple gene regulation, the steady state is a balance of production and degradation, $X_{st} = \beta_{simple}/\alpha_{simple}$ (Equation 2.4.3). In contrast, as we saw above, in the negative autoregulation case, the steady state is equal to the repression threshold, $X_{st} = K$ (Equation 3.4.6). We can tune K so that both designs reach the same steady-state expression level:

$$K = \beta_{simple}/\alpha_{\,simple} \qquad \textit{mathematically controlled comparison} \qquad (3.4.8)$$

What is the response time of the two designs? The response time of simple regulation is governed by the degradation/dilution rate as described in Chapter 2, so that $T_{1/2}^{simple} = \log(2)/\alpha_{simple}$. A much faster response can be achieved by the corresponding negative autoregulated circuit by making β large, because the response time, $T_{1/2}^{(n.a.r.)} = K/2\beta$, is inversely proportional to β. Using Equation 3.4.8, we find that the ratio of the response times in the two designs can be made very small by making β large:

$$T_{1/2}^{(n.a.r.)}/T_{1/2}^{(simple)} = (\beta_{simple}/\beta)/2\log(2) \qquad (3.4.9)$$

An example is shown in Figure 3.4, in which the response time of the negative autoregulation design is about sevenfold faster than simple regulation.[1]

Qualitatively, the same type of speed-up is found when using the Hill input function (solved exercise 3.1, Figure 3.5). The accelerated response of a negative autoregulatory circuit compared to simple regulation was experimentally demonstrated using high-resolution dynamic gene expression measurements (Figure 3.6).

In summary, negative autoregulation gets the best of both worlds: a strong promoter can give rapid production, and a suitable repression coefficient provides the desired steady state. The same strong promoter on a simple-regulation circuit would reach a much higher steady state, leading to undesirable overexpression of the gene product (Figure 3.3).

3.5   NEGATIVE AUTOREGULATION PROMOTES ROBUSTNESS TO FLUCTUATIONS IN PRODUCTION RATE

In addition to speeding the response time, negative autoregulation confers a second important benefit. This benefit is increased robustness of the steady-state expression level with respect to fluctuations in the production rate β. This property was experimentally demonstrated using measurements of protein levels in individual cells (Becskei and Serrano, 2000).

The production rate of a given gene, β, fluctuates over time due to overall fluctuations in the metabolic capacity of the cell and its regulatory systems, and to stochastic effects in the production of the protein (see Appendix D). Hence, twin cells usually show differences in the production rates β of most proteins. These cell–cell differences are typically on the order of a few percent to tens of percents. The differences can last over the entire generation time of the cells. Thus, a snapshot of several genetically identical cells grown under identical conditions will generally show cell–cell differences in the expression of every

---

[1] There are limits to the smallest response time achievable. For example, response time cannot be shorter than the time for transcription and translation of the gene product, delays that are not taken into account in the present equations. Also, β cannot be arbitrarily large — it is limited, for example, by the maximal ribosomal capacity to produce proteins (see Chapter 10).

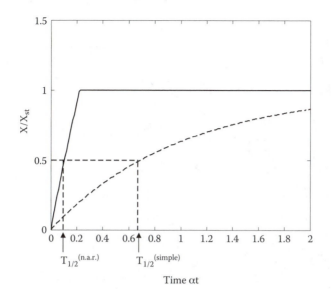

FIGURE 3.4 Dynamics of negatively autoregulated gene product (full line) and simply regulated gene product (dashed line), which reach the same steady-state level and have equal degradation/dilution rates, α. The response time is the time that the protein level reaches 50% of the steady state, denoted $T_{1/2}$ (n.a.r) and $T_{1/2}$ (simple) for the negatively autoregulated (n.a.r) and simply regulated gene products, respectively. The parameters $\beta = 5$, $\alpha = 1$, and $\beta_{simple} = 1$ were used.

protein. On the other hand, parameters such as the repression threshold K vary much less from cell to cell, because they are specified by the strength of the chemical bonds between X and its DNA binding site and the position and number of the X binding sites in the promoter.

Simple gene regulation is affected quite strongly by fluctuations in production rate β. The steady-state level is linearly dependent on the production rate:

$$X_{st} = \beta/\alpha \tag{3.5.1}$$

and therefore a change in β leads to a proportional change in X_{st}.

In contrast, negative autoregulation can buffer fluctuations in the production rate. In the case of sharp autorepression that we have just discussed, the steady-state level depends only on the repression threshold of X for its own promoter:

$$X_{st} = K \tag{3.5.2}$$

As mentioned above, the repression threshold K is determined by hardwired factors such as the chemical bonds between X and its DNA site. Such parameters vary much less from cell to cell than production rates. Therefore, negative autoregulation increases the robustness of steady-state protein levels with respect to the most likely fluctuations, namely, fluctuations in production rate (see exercise 3.2 for more details).

Robustness of key properties of a biological system is a general design principle. Much more will be said about robustness in Chapters 7 and 8.

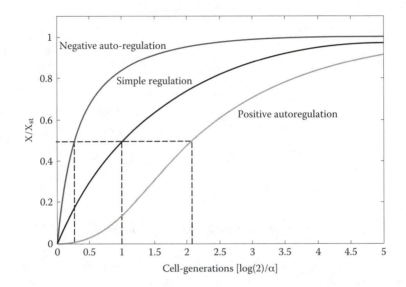

FIGURE 3.5 Dynamics of a negatively autoregulated gene, a simply regulated gene and a positively autoregulated gene. The negatively and positively autoregulated genes have a Hill input function with Hill coefficient n = 1. Shown is protein concentration normalized by its steady-state value, X/X_{st}, following an increase in production rate. Time is in cell generations, or for actively degraded proteins, $\log(2)/\alpha$, where α is the protein degradation/dilution rate. The response time is found by the intersect of the dynamics with a horizontal line at $X/X_{st} = 0.5$.

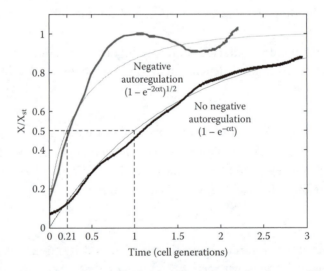

FIGURE 3.6 Experiment on negatively autoregulated and simply regulated genes. The experiment used green fluorescent protein fused to the TetR repressor as a reporter and automated fluorescence measurements on growing *E. coli* cells. Protein concentration was normalized to its steady-state level. Shown also are the analytical solutions for a simply autoregulated gene and for a negatively autoregulated gene with a Hill input function with n = 1 in the limit of strong autorepression (thin lines). (From Rosenfeld et al., 2002.)

3.5.1   Positive Autoregulation Slows Responses and Can Lead to Bi-Stability

Some transcription factors show positive autoregulation, in which they activate their own transcription. This occurs, for example, in about 10% of the known transcription factors in *E. coli*. Exercise 3.4 shows that positive autoregulation slows the response time relative to simple regulation (Figure 3.5). The dynamics are initially slow, but as the levels of X build up, its production rate increases due to the positive autoregulation loop. This results in a concave curve that reaches 50% of its steady-state value at a delay relative to simple regulation.

Thus, positive autoregulation has an effect that is opposite to that of negative autoregulation. The former slows response times, whereas the latter speeds response times. The slow dynamics provided by positive autoregulation can be useful in processes that take a relatively long time, such as developmental processes (see Chapter 6). Such slow processes can benefit from prolonged delays between the production of proteins responsible for different stages of the process.

In addition, when the rate of positive autoregulation is strong compared to the degradation/dilution rate, the system can become bi-stable (exercise 3.4). Once the gene is activated, it is locked into a state of high expression and keeps itself ON, even after the original activation input has vanished (Carrier and Keasling, 1999; Demongeot et al., 2000; Becskei et al., 2001; Ferrell, 2002; and Isaacs et al., 2003). This type of memory circuit is used in developmental transcription networks to make irreversible decisions that lock a cell into a particular fate (e.g., to determine the type of tissue the cell will become in a multi-cellular organism; see Chapter 6).

3.6   SUMMARY

Negative autoregulation is a network motif, a pattern that recurs throughout the network at numbers much higher than expected in random networks.

To understand why negative autoregulation is a network motif, we analyzed its dynamic behavior. The dynamic analysis can be phrased as an engineering story. Think of evolution as an engineer working to design a gene circuit that would reach a desired steady-state concentration X_{st}. One possible design, design A, is simple regulation with a production rate set to reach X_{st}. Design B is negative autoregulation, with a stronger initial production rate, which, as X builds up, is suppressed to result in the desired steady-state.

The second design has the advantage that the goal, X_{st}, is reached faster. Furthermore, the fluctuations around X_{st} due to variations in production rate are reduced in the second, autoregulated design. In an imaginary competition between two species, identical except that one uses circuit A and the second uses circuit B, the latter would have a selective advantage.

Over evolutionary times, structures that have engineering advantages would tend to be selected and appear as network motifs.

FURTHER READING

Becskei, A. and Serrano, L. (2000). Engineering stability in gene networks by autoregulation. *Nature*, 405: 590–593.

Rosenfeld, N., Elowitz, M.B., and Alon, U. (2002). Negative auto-regulation speeds the response time of transcription networks. *J. Mol. Biol.*, 323: 785–793.

Savageau, M.A. (1976). *Biochemical Systems Analysis: A study of Function and Design in Molecular Biology.* Addison-Wesley. Chap. 16.

Savageau, M.A. (1974). Comparison of classical and auto-genous systems of regulation in inducible operons. *Nature*, 252: 546–549.

EXERCISES

3.1. *Autorepression with Hill input function.* What is the response time for a repressor that cooperatively represses its own promoter (described by a Hill function with Hill coefficient n)?

$$dX/dt = \beta/(1 + (X/K)^n) - \alpha X \tag{P3.1}$$

How much faster is the response than in non-autoregulated circuits? Use the approximation of strong autorepression, that is, $(X/K)^n \gg 1$.

Solution:

In the limit of strong autorepression, we can neglect the 1 in the denominator of the input function as soon as $(X/K)^n \gg 1$, and we have[1]:

$$dX/dt = \beta\, K^n/X^n - \alpha X \tag{P3.2}$$

To solve this equation, multiply both sides by X^n and switch to the new variable, $u = X^{n+1}$. Note that $du/dt = (n + 1)\, X^n\, dX/dt$. The equation now reads:

$$du/dt = (n + 1)\,\beta\, K^n - (n + 1)\,\alpha\, u \tag{P3.3}$$

The solution of this linear equation is simple exponential convergence to steady state, the same as in Chapter 2:

$$u = u_{st}\, (1 - e^{-(n + 1)\,\alpha\, t}) \tag{P3.4}$$

Switching back to the original variable X, we have:

$$X = X_{st}\, (1 - e^{-(n + 1)\,\alpha\, t})^{1/(n+1)} \tag{P3.5}$$

The response time is found by $X(T_{1/2}) = X_{st}/2$. This yields:

---

[1] When is this approximation valid? Note that the steady state is, according to Equation P3.2, $X_{st} = K\,(\beta/\alpha\, K)^{1/(n+1)}$. Thus, when the unrepressed steady state is much larger than the repression coefficient, that is, when $\beta/\alpha \gg K$, we have $X_{st} \gg K$. This means that to describe the dynamics that occur when X exceeds K and begins to approach its steady state, we can neglect the 1 in the denominator of the input function.

$$T_{1/2} = [(n + 1)\, \alpha]^{-1} \log(2^{n+1}/[2^{n+1} - 1]) \tag{P3.6}$$

The response time decreases with n. For n = 1, 2, 3, the ratio of $T_{1/2}$ to the response time of simply regulated genes ($T_{1/2}^{(\text{simple})} = \log(2)/\alpha$) is about $T_{1/2}^{(\text{n.a.r})}/T_{1/2}^{(\text{simple})} = 0.2$, 0.06, and 0.02, respectively. See Figure 3.5 for the dynamics of a strongly autoregulated gene with n = 1. The sharper the negative autoregulation (higher n), the more the system approaches the sharp logic function limit discussed in this chapter, and the faster it responds.

3.2. *Parameter sensitivity.* Analyze the robustness of the steady-state level of X with respect to cell–cell variations in the production rate β for the system of problem 3.1. Calculate the parameter sensitivity coefficient (Savageau, 1976; Goldbeter and Koshland, 1981; Heinrich and Schuster, 1996) of the steady-state concentration with respect to β. The **parameter sensitivity coefficient** of property A with respect to parameter B, denoted S(A, B), is defined as the relative change in A for a given small relative change in B, that is, S:

$$S(A, B) = (\Delta A/A)/(\Delta B/B) = (B/A)\, dA/dB \tag{P3.7}$$

Solution:

The steady-state level is found from Equation P3.2 using dX/dt = 0, yielding:

$$X_{st} = K\, (\beta/\alpha\, K)^{1/(n+1)} \tag{P3.8}$$

The parameter sensitivity, which describes relative changes in steady state due to changes in production rate, is:

$$S(X_{st}, \beta) = (\beta/X_{st})\, dX_{st}/d\beta = 1/(n + 1) \tag{P3.9}$$

Thus, sensitivity decreases with Hill coefficient n. The higher n is, the weaker the dependence of the steady state on β. In other words, robustness to variations in production rates increases with the Hill coefficient.

For Hill coefficient n = 4, for example, $S(X_{st}, \beta) = 1/5$, which means that a 10% change in β yields only a 2% change in X_{st}. In the limit of very high n, the steady-state does not depend at all on production or degradation rates, $X_{st} = K$. This is the steady-state solution found in the main text for the logic input function. Simple regulation is equivalent to n = 0, so that $S(X^{st}, \beta) = 1$. This means that a small change of x% in production leads to the same change of x% in steady-state.

3.3. *Autoregulated cascade.* Gene X encodes a repressor that represses gene Y, which also encodes a repressor. Both X and Y negatively regulate their own promoters.

   a. At time t = 0, X begins to be produced at rate β, starting from an initial concentration of X = 0. What are the dynamics of X and Y? What are the response times of X and Y? Assume logic input functions, with repression thresholds K_{xx}, K_{xy} for the action of X on its own promoter and on the Y promoter, and K_{yy} for the action of Y on its own promoter.

b. At time t = 0, production of X stops after a long period of production, and X concentration decays from its initial steady-state level. What are the dynamics of X and Y? What are the response times of X and Y?

3.4. *Positive feedback.* What is the effect of *positive* autoregulation on the response time? Use as a model the following linear equation:

$$dX/dt = \beta + \beta_1 X - \alpha X$$

with $\beta_1 < \alpha$. Explain each term and solve for the response time. When might such a design be biologically useful? What happens when $\beta_1 > \alpha$?

3.5. *Turning off auto-regulation.* What is the dynamics of a negatively auto-regulation gene once its maximal promoter activity is suddenly reduced from β_1 to $\beta_2 = 0$? What is the response time, and how does it compare to simple regulation?

3.6. *Two-node positive feedback for decision making.* During development from an egg to an embryo, cells need to make irreversible decisions to express the genes appropriate to their designated tissue types and repress other genes. One common mechanism is positive transcriptional feedback between two genes. There are two types of positive feedback made of two transcription factors. The first type is of two positive interactions X → Y and Y → X. The second type has two negative interactions X ⊣ Y and Y ⊣ X. What are the stable steady states in each type of feedback? Which type of feedback would be useful in situations where genes regulated by both X and Y belong to the same tissue? Which would be useful when genes regulated by X belong to different tissues than the genes regulated by Y?

The Feed-Forward Loop Network Motif

4.1 INTRODUCTION

In this chapter, we will continue to discover network motifs in transcription networks and discuss their function. The main point is that out of the many possible patterns that could appear in the network, only a few are found significantly and are network motifs. The network motifs have defined information processing functions. The benefit of these functions may explain why the same network motifs are rediscovered by evolution again and again in diverse systems.

To find significant patterns, we will first calculate the number of appearances of different patterns in real and random networks. We will focus in this chapter on patterns with three nodes (such as triangles). There are 13 possible three-node patterns (Figure 4.1). Patterns with two nodes and patterns with more than three nodes will be discussed in the next chapters. We will see that of the 13 possible three-node patterns, only one, the feed-forward loop (FFL), is a **network motif**.

To understand the possible functions of the feed-forward loop, we need to understand the regulation described by each of its three edges. Each of these edges can be an activation or a repression interaction. There are therefore eight possible FFL types. We will see that of the eight possible types of FFLs, only two appear in large numbers in transcription networks. We will analyze the dynamical functions of these circuits. We will see that the common types of FFLs can carry out interesting functions such as the filtering of noisy input signals, pulse generation, and response acceleration.

After discussing the common FFL types, we will ask why the other six types of FFLs occur much more rarely. Asking why will lead us to consider functional differences in the common and rare FFL types. Finally, we will discuss the evolution of the FFLs.

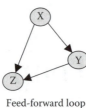
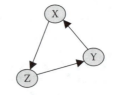

Feed-forward loop   3-node feedback loop (cycle)

(a)

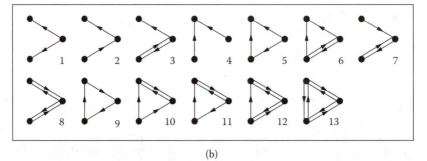

(b)

FIGURE 4.1 (a) The feed-forward loop and the feedback loop, two examples of subgraphs with three nodes. (b) The 13 connected three-node directed subgraphs. The feed-forward loop is subgraph 5, and the feedback loop is subgraph 9.

4.2   THE NUMBER OF APPEARANCES OF A SUBGRAPH IN RANDOM NETWORKS

In the previous chapter we discussed the simplest network motif, self-regulation, a pattern that had one node. Let us now consider larger patterns of nodes and edges. Such patterns are also called **subgraphs**. Two examples of three-node subgraphs are shown in Figure 4.1a: the three-node feedback loop and the three-node **feed-forward loop**. In total there are 13 possible ways to connect three nodes with directed edges, shown in Figure 4.1b. There are 199 possible directed four-node subgraphs (Figure 5.5), 9364 five-node subgraphs, etc.

To find which of these subgraphs are significant, we need to compare the subgraphs in the real network to those in randomized networks. The rest of this section is for readers interested in mathematical analysis of random networks. Other readers can safely skip to Section 4.3.

We begin by calculating the number of times that a given subgraph G appears in a random Erdos–Renyi (ER) network (ER networks were defined in Section 3.2). The subgraph G that we are interested in has n nodes and g edges. The feed-forward loop, for example, has n = 3 nodes and g = 3 edges (Figure 4.1a). Other three-node patterns have between two and six edges (Figure 4.1b). Recall that in the ER random network model, E edges are placed randomly between N nodes (Section 3.2). Since there are N^2 possible places to put a directed edge (Equation 3.2.1), the probability of an edge in a given direction between a given pair of nodes is:

$$p = E/N^2 \tag{4.2.1}$$

It is important to note that most biological networks are **sparse**, which is to say that only a tiny fraction of the possible edges actually occur. Sparse networks are defined by $p \ll 1$. For example, in the *Escherichia coli* network we use as an example, there are about 400 nodes and 500 edges, so that $p \sim 0.002$. One reason that biological networks are sparse is that each interaction in the network is selected by evolution against mutations that would rapidly abolish the interaction. Thus, only useful interactions are maintained.

We want to calculate the mean number of times that subgraph G occurs in the random network. To generate an instance of subgraph G in the random network, we need to choose n nodes and place g edges in the proper places. Thus, the average number of occurrences of subgraph G in the network, denoted $<N_G>$, is approximately equal to the number of ways of choosing a set of n nodes out of N: about N^n for large networks (because there are N ways of choosing the first node, times $N - 1 \approx N$ ways of choosing the second node, etc.), multiplied by the probability to get the g edges in the appropriate places (each with probability p):

$$<N_G> \approx a^{-1} N^n p^g \tag{4.2.2}$$

where a is a number that includes combinatorial factors related to the structure and symmetry of each subgraph,[1] equal, for example, to $a = 1$ for the FFL and $a = 3$ for the three-node feedback loop.

Let us now recast this equation in terms of the **mean connectivity** of the network, defined as the average number of edges per node:

$$\lambda = E/N \qquad \textit{mean connectivity} \tag{4.2.3}$$

In terms of the mean connectivity, we find, using $p = \lambda/N$ (from Equation 4.2.1), a simple equation in which the higher the mean connectivity of the network λ, the higher the mean number of appearances of subgraph G:

$$<N_G> \approx a^{-1} \lambda^g N^{n-g} \tag{4.2.4}$$

Hence, densely connected networks with high λ have, in general, more subgraphs than sparse ones.

We will now see that many patterns G are very rare in random networks, in the sense that they occur in vanishingly small numbers in large random networks. If any of these patterns occur in the real biological network, they are likely to be network motifs. To see this, let us ask how $<N_G>$, the number of times that subgraph G appears in the network,

---

[1] The factor a is the number of permutations of the n nodes in the subgraph G that gives an isomorphic subgraph. Thus, dividing by a avoids overcounting of subgraphs. For example, in the three-node feedback loop, any of the three possible cyclic permutations of the nodes leaves the subgraph intact, and thus $a = 3$. In the feed-forward loop, $a = 1$, since there are no permutations of the nodes that make an isomorphic pattern. For more details, see Itzkovitz et al., 2003.

scales with the network size, N. Imagine a series of larger and larger random ER networks, all with the same mean connectivity λ. The dependence of $\langle N_G \rangle$ on network size N is described by a **scaling relation**. This scaling relation describes the way that the number of subgraphs in Equation 4.2.4 depends on the size of the network (ignoring prefactors):

$$\langle N_G \rangle \sim N^{n-g} \tag{4.2.5}$$

The scaling relation tells us that the scaling of subgraph numbers in ER networks depends *only on the difference between the number of nodes and edges* in the subgraph, n – g.

For example, V-shaped patterns, such as patterns 1 and 2 in Figure 4.1b, have n = 3 nodes and g = 2 edges. Their number, therefore, grows linearly with network size:

$$N_{\text{V-shaped}} \sim N^{n-g} = N \tag{4.2.6}$$

If we double the number of nodes and edges in the random network, the number of V-shaped subgraphs will also double. These patterns are very common in random networks. In contrast, the fully connected clique (the last pattern in Figure 4.1b) has six edges, g = 6, but only three nodes, n = 3. This subgraph scales as $N^{n-g} = N^{-3}$, and therefore occurs very rarely in large random networks.

Let us now consider the case of our two triangle-shaped patterns, the three-node feedback loop and feed-forward loop (Figure 4.1a). Both have three nodes, n = 3, and three edges, g = 3, and so, using Equation 4.2.4 and the appropriate symmetry factors (a = 1 for feed-forward and a = 3 for feedback loops), we find

$$\langle N_{\text{FFL}} \rangle \sim \lambda^3 N^0 \tag{4.2.7}$$

$$\langle N_{\text{3loop}} \rangle \sim 1/3 \, \lambda^3 N^0 \tag{4.2.8}$$

This result is remarkable. The scaling of the number of these triangles with the network size goes as $N^{n-g} = N^0$. In other words, *the numbers of these triangle patterns are constant in ER networks and do not increase with network size.*

The reason for the fact that triangle numbers do not depend on the size of the random network is that the number of V-shaped pairs of edges in the network scales linearly with network size N (Equation 4.2.6), but the probability that a V-shaped pattern will close to form a triangle scales as 1/N (because an edge that emerges from a node at one arm of the V and closes it into a triangle by pointing to the node at the other arm needs to choose the one target node out of N possibilities). This yields a total of $N \cdot 1/N = N^0$ triangles. This

TABLE 4.1 Number of Feed-Forward Loops and Feedback Loops with Three Nodes in the Transcription Network of *E. coli* Used as an Example in this Book, and in Randomized Networks

Feed-Forward Loop (FFL)	Feedback Loop with 3 Nodes

E. coli	42		0
ER random nets	1.7 ± 1.3	(Z = 31)	0.6 ± 0.8
Degree-preserving random nets	7 ± 5	(Z = 7)	0.2 ± 0.6

The parameter Z is the number of standard deviations that the real network exceeds the randomized networks. An algorithm called Mfinder, which generates randomized networks, counts subgraphs, and detects network motifs, can be found at www.weizmann.ac.il/MCB/UriAlon.

means that triangles and more complex patterns occur rarely in random networks. We now turn back to the real transcription networks.

4.3 THE FEED-FORWARD LOOP IS A NETWORK MOTIF

How do the numbers of patterns in transcription networks compare to the numbers expected in random networks?

In the *E. coli* transcription network that we use as an example in this book, there are 42 feed-forward loops and no feedback loops made of a cycle of three nodes (Table 4.1). In contrast, in the corresponding randomized ER networks with the same mean connectivity $\lambda = 500/400 \sim 1.2$, the mean number of feed-forward loops is only about 2 (Equation 4.2.7),

$$\langle N_{FFL} \rangle_{rand} = 1.2^3 \sim 1.7$$

and the mean number of feedback loops is less than 1 (Equation 4.2.8),

$$\langle N_{feedback} \rangle_{rand} = 1.2^3/3 \sim 0.6$$

The standard deviations of these numbers are generally the square roots of the means, $\sqrt{N_G}$, because in many cases the number of subgraphs follows a Poisson distribution in ER random networks. The comparison between real and random networks is shown in Table 4.1.

We see that *the feed-forward loop (FFL) is a strong network motif.* It occurs much more often than at random. Its frequency is greater than its frequency in the ensemble of randomized networks by more than 30 standard deviations. In contrast, the three-node feedback loop is not a network motif (it is actually an *anti-motif* in many biological networks).

In fact, in sensory transcription networks such as those of *E. coli* and yeast (Lee et al., 2002; Milo et al., 2002), as well as higher organisms, *the feed-forward loop is the only significant network motif of the 13 possible three-node patterns.* In this sense, these networks are much simpler than they could have been. The same conclusions apply also when comparing transcription networks to more stringent ensembles of randomized networks that more closely preserve the properties of the real network.[1]

---

[1] One important property is the **degree sequence** of the network: the number of incoming and outgoing edges for each node in the network (see Appendix C). Real transcription networks have several nodes with many more

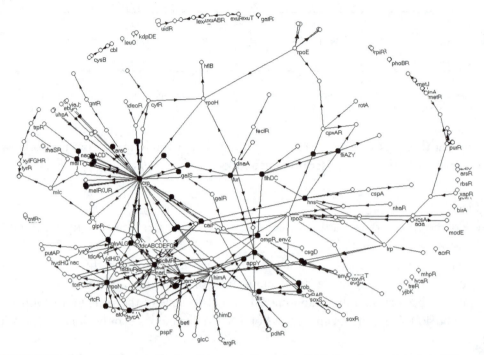

FIGURE 4.2 Feed-forward loops in the *E. coli* transcription network. Black nodes participate in FFLs.

The massive overabundance of feed-forward loops raises the question: Why are they selected against randomizing forces? Do they perform a function that confers an advantage to the organism?

To address this question, let us now analyze the structure and function of the feed-forward loop network motif.

4.4   THE STRUCTURE OF THE FEED-FORWARD LOOP GENE CIRCUIT

The feed-forward loop is composed of transcription factor X that regulates a second transcription factor, Y, and both X and Y regulate gene Z (Figure 4.1a). Thus, the feed-forward loop has two parallel regulation paths, a direct path from X to Z and an indirect path that goes through Y. The direct path consists of a single edge, and the indirect path is a cascade of two edges.

---

outgoing edges than the average node: these are global regulators that regulate many genes in response to key environmental stimuli. To include this property in the random network model, one can compare the real network to random networks that not only preserve the total number of nodes N and edges E, but also preserve the number of incoming and outgoing edges for *each* node in the network. Despite the fact that the degree sequence is the same, the identity of which transcription factor regulates which gene is randomized. These **degree-preserving random networks** can be generated on the computer by randomly switching pairs of edge, repeating the switching operation many times until the network is randomized. For a given real network, many thousands of different randomized degree-preserving networks can be readily generated. These randomized networks serve as a more stringent random model for comparison to the real networks. Degree-preserving random networks have more FFLs than ER random networks (Itzkovitz et al., 2003), but vastly fewer FFLs than the real networks (Table 4.1). The FFL is the only significant three-node pattern.

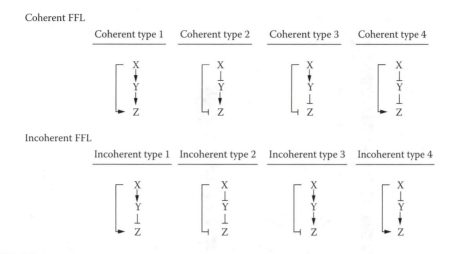

FIGURE 4.3 The eight sign combinations (types) of feed-forward loops. Arrows denote activation and ⊣ symbols denote repression.

Each of the three edges in the FFL can correspond to activation (plus sign) or repression (minus sign). There are therefore $2^3 = 8$ possible types of FFLs (Figure 4.3).

The eight FFL types can be classified into two groups: coherent and incoherent. This grouping is based on comparing the sign of the direct path from X to Z to the sign of the indirect path that goes through Y. In **coherent** FFLs, the indirect path has the same overall sign as the direct path. The overall sign of a path is given by the multiplication of the sign of each arrow on the path (so that two minus signs give an overall plus sign). For example, in type-1 coherent FFLs, X activates Z, and also activates an activator of Z, so that both paths are positive.

In **incoherent** FFLs, the sign of the indirect path is *opposite* to that of the direct path. For example, in the type-1 incoherent FFL, the direct path is positive and the indirect path is negative. The two paths have antagonistic effects. Note that incoherent FFLs have an odd number of minus edges (one or three).

Not all the FFL types appear with equal frequency in transcription networks (Figure 4.4). The most abundant FFL is the type-1 coherent FFL (C1-FFL), in which all three regulations are positive (Mangan and Alon, 2003). The C1-FFL will be studied in detail in this chapter. The second most abundant type of FFL across biological networks is the incoherent type-1 FFL (I1-FFL) (Ma et al., 2004; Mangan et al., 2006), which we will also study in detail. The six other FFL types seem to appear much less frequently than the C1-FFL and the I1-FFL. Toward the end of the chapter, we will try to understand why the frequencies of the FFL types are so different.

In addition to the signs on the edges, to understand the dynamics of the FFL we must also know how the inputs from the two regulators X and Y are integrated at the promoter of gene Z. That is, we need to know the input function of gene Z. We will consider two biologically reasonable logic functions: AND logic, in which *both* X and Y activities need to be high in order to turn on Z expression, and OR logic, in which *either* X or Y is suffi-

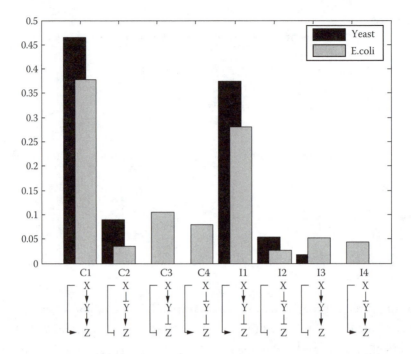

FIGURE 4.4 Relative abundance of the eight FFL types in the transcription networks of yeast and *E. coli*. FFL types are marked C and I for coherent and incoherent. The *E. coli* network is based on the Ecocyc and RegulonDB databases and has about twice as many edges as in the network of Figure 2.3. (From Mangan et al., 2006.)

cient. Thus, there are eight types of FFL sign combinations, each of which can appear with at least two types of input functions (AND, OR).

After noting the signs and input functions, we need to consider the input signals to this circuit. The transcription factors X and Y in the FFL usually respond to external stimuli. These input stimuli are represented by the input signals S_x and S_y (Figure 4.5). In some systems the signals are molecules that directly bind the transcription factors, and in other systems the signals are modifications of the transcription factor caused by signal transduction pathways activated by the external stimuli. The effect of the signals, which carry information from the external world, usually operates on a much faster timescale than the transcriptional interactions in the FFL. When S_x appears, transcription factor X rapidly becomes active, X*, binds to specific DNA sites in the promoters of genes Y and Z in a manner of seconds, and changes the transcription rate so that the concentration of the protein Z changes on the timescale of minutes to hours.

We will next discuss the dynamics of the proteins that make up the FFL as a function of time following a change in an external signal. We will begin with the most common FFL type in which all three interactions are positive (Figure 4.5). As for the input function of the Z promoter, we will first consider AND logic. This is the case in which both activators X and Y need to bind the promoter of Z in order to initiate the production of protein Z.

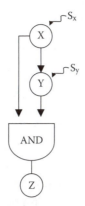

FIGURE 4.5 The coherent type-1 FFL with an AND input function: transcription factor X activates the gene encoding transcription factor Y, and both X and Y jointly activate gene Z. The two input signals are S_x and S_y. An input function integrates the effects of X and Y at the Z promoter (an AND gate in this figure).

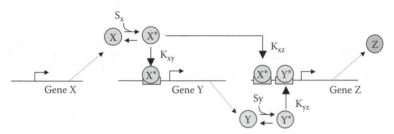

FIGURE 4.6 The molecular interactions in the coherent FFL of Figure 4.5. The transcription factor protein X is activated by signal S_x, which causes it to assume the active conformation X*. It then binds its sites in the promoters of genes Y and Z. As a result, protein Y accumulates and, in the presence of its signal S_y, is active, Y*. When Y* concentration crosses the activation threshold, K_{yz}, Y* binds the promoter of gene Z. Protein Z is produced when both X* and Y* bind the promoter of gene Z (in the case of an AND input function).

4.5 DYNAMICS OF THE COHERENT TYPE-1 FFL WITH AND LOGIC

Suppose that the cell expresses numerous copies of protein X, the top transcription factor in the FFL. The input to X is the signal S_x (Figure 4.6). Without the signal, X is in its inactive form. Now, at time t = 0, a strong signal S_x triggers the activation of X. This is known as a **step-like stimulation** of X. As a result, the transcription factor X rapidly transits to its active form X*. The active protein X* binds the promoter of gene Y, initiating production of protein Y, the second transcription factor in the FFL. In parallel, X* also binds the promoter of gene Z. However, since the input function at the Z promoter is AND logic, X* alone cannot activate Z production.

Production of Z requires binding of both X* and Y*. This means that the concentration of Y must build up to sufficient levels to cross the **activation threshold** for gene Z. This activation threshold is denoted K_{yz}. In addition, Z activation requires that the second input signal, S_y, is present, so that Y is in its active form, Y* (Figure 4.6). Thus, once the signal S_x appears, Y needs accumulate in order to activate Z. This results in a **delay** in Z production.

We will now mathematically describe the FFL dynamics, in order to see how a simple mathematical model can be used to gain an intuitive understanding of the function of a gene circuit. To describe the FFL, let us use logic input functions. Production of Y occurs at rate β_y when X^* exceeds the activation threshold K_{xy}, as described by the step function θ:

$$\text{production rate of Y} = \beta_y \, \theta \, (X^* > K_{xy}) \tag{4.5.1}$$

When the signal S_x appears, X rapidly shifts to its active conformation X^*. If the signal is strong enough, X^* exceeds the activation threshold K_{xy} and rapidly binds the Y promoter to activate transcription. Thus, Y production begins shortly after S_x. The accumulation of Y is described by our now familiar dynamic equation with a term for production and another term for degradation/dilution:

$$dY/dt = \beta_y \, \theta \, (X^* > K_{xy}) - \alpha_y \, Y \tag{4.5.2}$$

The promoter of Z in our example is governed by an AND gate input function. Thus, the production of Z can be described by a product of two step functions, each indicating whether the appropriate regulator crossed the activation threshold:

$$\text{production of Z} = \beta_z \, \theta \, (X^* > K_{xz}) \, \theta \, (Y^* > K_{yz}) \tag{4.5.3}$$

Thus, the C1-FFL gene circuit has three activation thresholds (numbers on the arrows in Figure 4.6). In the case of strong step-like stimulation, X^* rapidly crosses the two thresholds[1] K_{xy} and K_{xz}. The delay in the production of Z is due to the time it takes Y^* to accumulate and cross its threshold K_{yz}. Only after Y^* crosses the threshold can Z production proceed at rate β_z. The dynamics of Z are governed by a degradation/dilution term and a production term with an AND input function:

$$dZ/dt = \beta_z \, \theta \, (X^* > K_{xz}) \, \theta \, (Y^* > K_{yz}) - \alpha_z \, Z \tag{4.5.4}$$

We now have the equations needed to analyze the dynamics of the C1-FFL. We next analyze its dynamics as a sign-sensitive delay element.

4.6   THE C1-FFL IS A SIGN-SENSITIVE DELAY ELEMENT

To describe the dynamics of the C1-FFL, we will consider the response to steps of S_x, in which the signal S_x is first absent and then saturating S_x suddenly appears (**ON steps**). We will also consider **OFF steps**, in which S_x is at first present and is then suddenly removed. For simplicity, we will assume throughout that the signal S_y is present, so that the transcription factor Y is in its active form:

---

[1] The values of the thresholds K_{xy} and K_{xz} can affect the function of the FFL for subsaturating S_x signals (exercise 4.7).

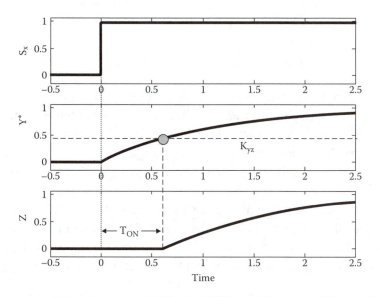

FIGURE 4.7 Dynamics of the coherent type-1 FFL with AND logic following an ON step of S_x at time t = 0 in the presence of S_y. The activation threshold of Z by Y is K_{yz} (horizontal dashed line). The production and degradation rates are $\alpha_y = \alpha_z = 1$, $\beta_y = \beta_z = 1$. The delay in Z production is T_{ON}.

$$Y^* = Y \tag{4.6.1}$$

4.6.1 Delay Following an ON Step of S_x

Following an ON step of S_x, Y^* begins to be produced at rate β_y. Hence, as we saw in Chapter 2, the concentration of Y begins to exponentially converge to its steady-state level (Figure 4.7):

$$Y^*(t) = Y_{st} (1 - e^{-\alpha_y\, t}) \tag{4.6.2}$$

Recall that the steady-state concentration of Y is equal to the ratio of its production and degradation/dilution rates:

$$Y_{st} = \beta_y/\alpha_y \tag{4.6.3}$$

What about Z? Production of Z is governed by an AND input function, in which one input, X^*, crosses its threshold as soon as S_x is added. But one input is not enough to activate an AND gate. The second input, Y^*, takes some time to accumulate and to cross the activation threshold, K_{yz}. Therefore, Z begins to be expressed only after a delay (Figure 4.7). The delay, T_{ON}, is the time needed for Y^* to reach its threshold and can be seen graphically as the time when the Y concentration crosses the horizontal line at height K_{yz}. The delay, T_{ON}, can be found using Equation 4.6.2:

$$Y^*(T_{ON}) = Y_{st} (1 - e^{-\alpha_y\, T_{ON}}) = K_{yz} \tag{4.6.4}$$

This equation can be solved for T_{ON}, yielding:

$$T_{ON} = 1/\alpha_y \log [1/(1 - K_{yz}/Y_{st})] \qquad (4.6.5)$$

This equation describes how the duration of the delay depends on the biochemical parameters of the protein Y (Figure 4.8a). These parameters are the lifetime of the protein, α_y, and the ratio between Y_{st} and the activation threshold K_{yz}. The delay can therefore be tuned over evolutionary timescales by mutations that change these biochemical parameters.

Note that the delay T_{ON} diverges when the activation threshold K_{yz} exceeds the steady-state level of Y, because protein Y can never reach its threshold to activate Z. Recall that Y_{st} is prone to cell–cell fluctuations due to variations in protein production rates. Hence, a robust design will have a threshold K_{yz} that is significantly lower than Y_{st}, to avoid these fluctuations. In bacteria, K_{yz} is typically at least 3 to 10 times lower than Y_{st}, and typical parameters give delays T_{ON} that range from a few minutes to a few hours.

4.6.2   No Delay Following an OFF Step of S_x

We just saw that Z shows a delay following ON steps of S_x. We now consider OFF steps of S_x, in which S_x is suddenly removed (Figure 4.8b). Following an OFF step, X rapidly becomes inactive and unbinds from the promoters of genes Y and Z. Recall that Z is governed by an AND gate that requires binding of both X* and Y*. It therefore only takes one input to go off for the AND gate to stop Z expression. Therefore, after an OFF step of S_x, Z production stops at once. There is no delay in Z dynamics after an OFF step.

4.6.3   The C1-FFL Is a Sign-Sensitive Delay Element

We saw that the C1-FFL with AND logic shows a delay following ON steps of S_x. It does not show a delay following OFF steps. This type of behavior is called sign-sensitive delay, where sign-sensitive means that the delay depends on the sign of the step, ON or OFF.

A sign-sensitive delay element can also be considered as a kind of asymmetric filter. For example, consider a pulse of S_x that appears only briefly (an ON pulse) (Figure 4.8c). An ON pulse that is shorter than the delay time, T_{ON}, does not lead to any Z expression in the C1-FFL. That is because Y does not have time to accumulate and cross its activation threshold during the pulse. Only persistent pulses (longer than T_{ON}) result in Z expression. Thus, this type of FFL is a **persistence detector** for ON pulses. On the other hand, it responds immediately to OFF pulses. In contrast to the FFL, simple regulation (with no FFL) does not filter out short input pulses, but rather shows production of Z that lasts as long as the input pulse is present.

4.6.4   Sign-Sensitive Delay Can Protect against Brief Input Fluctuations

Why might sign-sensitive delay be useful?

For clues, we can turn to the uses of sign-sensitive delays in engineering. In engineering, sign-sensitive delay is commonly used in situations where the cost of an error is not symmetric. A familiar example occurs in elevators: consider the beam of light used to sense obstructions in the elevator door. If you obstruct the light with your hand, the door opens. If you remove your hand for only a short pulse, nothing happens (that is, a

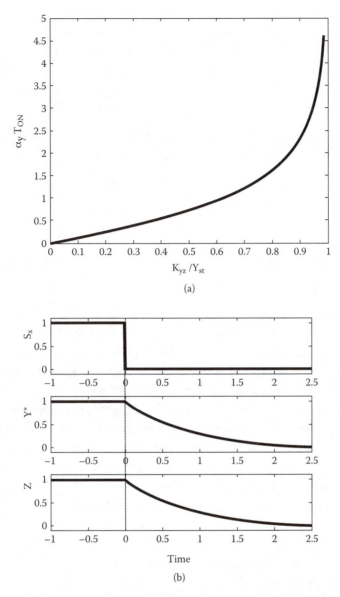

FIGURE 4.8 (a) Delay in Z production in the C1-FFL as a function of the biochemical parameters of the transcription factor Y. The delay T_{ON}, made dimensionless by multiplying with the degradation/dilution rate of Y, α_y, is shown as a function of the ratio of the activation threshold K_{yz} and the maximal (steady-state) level of Y, denoted Y_{st}. (b) Dynamics of the C1-FFL following an OFF step of S_x at time $t = 0$. All production and degradation rates are equal to 1. (c) The coherent type-1 FFL with AND logic as a persistence detector. A brief pulse of signal S_x does not give Y enough time to accumulate and cross its activation threshold for Z. Hence, Z is not expressed. A persistent pulse yields Z production at a delay. Z production stops with no delay when S_x is removed. (From Shen-Orr et al., 2002.)

short pulse of light is filtered out). Only if you remove your hand for a persistent length of time do the doors close (a persistent pulse of light leads to a response). Put your hand back in and the doors open immediately. Again, the cost of an error (doors closing or opening at the wrong time) is asymmetric: the design aims to respond quickly to a person

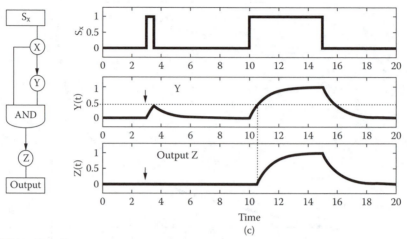

FIGURE 4.8 (continued)

in the beam and make sure that the person has moved away for a persistent period of time before closing the doors. The sign-sensitive delay serves a protective function.

In transcription networks, evolutionary selection may have placed the C1-FFL in diverse systems in the cell that require such a protection function. Indeed, the environment of cells is often highly fluctuating, and sometimes stimuli can be present for brief pulses that should not elicit a response. The C1-FFL can offer a filtering function that is advantageous in these types of fluctuating environments. The conditions for the natural selection of the FFL based on its filtering function are discussed in more detail in Chapter 10.

4.6.5   Sign-Sensitive Delay in the Arabinose System of *E. coli*

Our discussion of the function of the FFL has dealt with this gene circuit in isolation. In reality, this network motif is always embedded within a network of additional interactions. It is therefore crucial to perform experiments on the FFL within living cells, to see whether it actually performs the expected dynamical functions.

Experiments have demonstrated that sign-sensitive delays are carried out by the C1-FFL in living cells. For example, dynamic behavior of an FFL was experimentally studied in a well-characterized gene system in *E. coli*, the system that allows the cells to grow on the sugar arabinose. The arabinose system consists of proteins that transport the sugar arabinose into the cell and break it down for use as an energy and carbon source. Arabinose is only used by the cells when the sugar glucose is not present, because glucose is a superior energy source and is used in preference to most other sugars. Thus, the arabinose system needs to make a decision based on two inputs: the sugars arabinose and glucose. The proteins in this system are only made when the following condition is met by the sugars in the environment of the cell: arabinose AND NOT glucose.

The absence of glucose is symbolized within the cell by the production of a small molecule called cAMP. To make its decision, the arabinose system has two transcription activators, one called CRP that senses cAMP, and the other called araC that senses arabi-

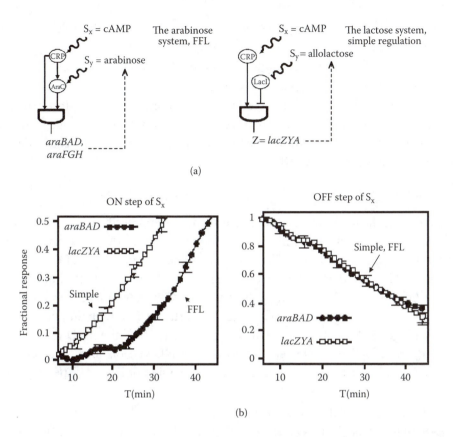

FIGURE 4.9 Experimental dynamics of the C1-FFL in the arabinose system of *E. coli*. The arabinose (*ara*) system encodes enzymes that utilize the sugar arabinose (*araBAD*) and transport it into the cell (*araFGH*, *araE*). The system is activated by the activator X = CRP (signal S_x = cAMP, a molecule produced within the cell upon glucose starvation) and in the presence of S_y = arabinose by the activator Y = AraC. The input function is an AND gate. As a control system with no FFL (simple regulation), the experiment used the *lac* operon, in which same activator X = CRP regulates the lactose operon, but X does not regulate Y_1 = LacI. Dashed arrows: Rapid, non-transcriptional feedback loops in which the output gene products affect the signal (e.g., by transporting the sugar S_y into the cell and degrading it). (b) Dynamics of the promoter activity of the *ara* and *lac* operons were monitored at high temporal resolution in growing cells by means of green flourescent protein (GFP) expressed from the relevant promoter, in the presence of S_y. The experiments followed the dynamics after ON and OFF steps of S_x. Shown is GFP per cell normalized to its maximal level. A delay occurs in the FFL after ON steps, but not after OFF steps. (Based on Mangan et al., 2003.)

nose. These regulators are connected in a C1-FFL with an AND input function (Figure 4.9a). The input signals in this system are S_x = cAMP and S_y = arabinose.

Experiments on this system used steps of S_x and monitored the dynamics of the promoter of the arabinose degradation genes that act as node Z in the FFL. A delay was found after ON steps of S_x, but not after OFF steps (Figure 4.9b). The delay in this FFL following ON steps of S_x is $T_{ON} \sim 20$ min under the conditions of the experiment.

The observed delay in the arabinose FFL is on the same order of magnitude as the duration of spurious pulses of the input signal S_x in the environment of *E. coli*. These spurious pulses of S_x occur when *E. coli* transits between different growth conditions.

Thus, the FFL in this system may have 'learned' the typical timescale of short fluctuations in the input signal, and can filter them out. It responds only to persistent stimuli, such as persistent periods of glucose starvation that require utilization of the sugar arabinose.

Note that the FFL in the arabinose system shows sign-sensitive delay despite the fact that it is embedded in additional interactions, such as protein-level feedback loops[1] (Figure 4.9a). Thus, although the theory we have discussed concerns a three-gene FFL circuit in isolation, the arabinose FFL shows the expected dynamics also when embedded within the interaction networks of the cell.

4.6.6   The OR Gate C1-FFL Is a Sign-Sensitive Delay for OFF Steps of S_x

What happens if the C1-FFL has an OR gate at the Z promoter instead of an AND gate? With an OR gate, Z is activated immediately upon an ON step of S_x, because it only takes one input to activate an OR gate. Thus, there is no delay following an ON step of S_x. In contrast, Z is deactivated at a delay following an OFF step, because both inputs need to go off for the OR gate to be inactivated: Y^* can activate Z even without X^* and it takes time for Y^* to decay away after an OFF step of S_x. Thus, the C1-FFL with an OR gate is also a sign-sensitive delay element, but with signs *opposite* to those of the AND version (exercise 4.2). It shows a delay after OFF steps, whereas the AND version shows a delay after ON steps. Hence, the OR gate C1-FFL can maintain expression of Z even if the input signal is momentarily lost.

Such dynamics were demonstrated experimentally in the flagella system of *E. coli* using high-resolution expression measurements (Figure 4.10). This FFL controls the production of proteins that self-assemble into a motor that rotates the flagella that allow *E. coli* to swim. We will discuss this system in more detail in Chapter 5. The delay observed in this FFL after removal of S_x is about one cell generation time — about 1 h under the conditions of the experiment. This delay is on the same order of magnitude as the time it takes to assemble a flagella motor. The OR gate FFL provides continued expression for about an hour after the input signal goes off, and can thus protect this gene system against transient loss of input signal.

4.6.7   Interim Summary

We have seen that of the 13 possible three-node patterns, only one is a significant network motif in sensory transcription networks that need to respond to external stimuli. This network motif is the feed-forward loop. The FFL has eight possible types, each corresponding to a specific combination of positive and negative regulations. Two of the FFL types are far more common than others in transcription networks. The most common form, called coherent type-1 FFL, is a sign-sensitive delay element that can protect against unwanted responses to fluctuating inputs. The magnitude of the delay in the FFL can be

---

[1] For example, some of the genes in the arabinose system encode for protein transporters (*araE* and *araFGH*) that pump the sugar arabinose into the cell. Thus, once arabinose is present, these proteins are expressed and more arabinose is pumped into the cell. Other proteins (*araBAD*) degrade arabinose and lower intracellular arabinose levels. These types of rapid interactions generally affect the input signals (in this case, S_y = intracellular arabinose) of the system on a timescale of seconds.

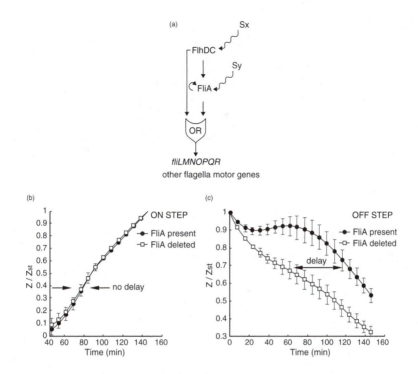

FIGURE 4.10 (a) the C1-FFL with OR logic in the flagella system of *E.coli*. The output genes, such as *fliLM-NOPQR*, make up the flagella motor. The input signals S_x are environmental factors such as glucose limitation, osmotic pressure, and temperature that affect the promoter of the activator FlhDC. The input signal S_y to the second activator, FliA, is a check point that is triggered when the first motors are completed (a protein inhibitor of FliA called FlgM is exported through the motors out of the cells). (b) Experiments on the promoter activity of the output genes, measured by means of a green-fluorescent protein expressed as a reporter from the fliL promoter, after an ON step of S_x. (c) Promoter dynamics after an OFF step of S_x, in the presence of S_y. The results are shown for the wild-type bacterium, and for a bacterium in which the gene for FliA was deleted from the genome. The FFL generates a delay after an OFF step of S_x. (From Kalir et al., 2005.)

tuned over evolutionary timescales by varying the biochemical parameters of regulator protein Y, such as its lifetime, maximal level, and activation threshold.

4.7   THE INCOHERENT TYPE-1 FFL

We will now turn from the coherent FFL to study the function of the incoherent feed-forward loop network motif. We will see that it can function as a pulse generator and sign-sensitive accelerator.

4.7.1   The Structure of the Incoherent FFL

Let us analyze the second most common FFL type, the incoherent type-1 FFL (I1-FFL). The I1-FFL motif makes up about a third of the FFLs in the transcription networks of *E. coli* and yeast (Figure 4.4).

The I1-FFL is made of two parallel but antagonistic regulation paths. In the I1-FFL, activator X activates Z, but it also activates Y — a repressor of Z (Figure 4.11a). Thus, the two arms of the I1-FFL act in opposition: the direct arm activates Z, and the indirect arm

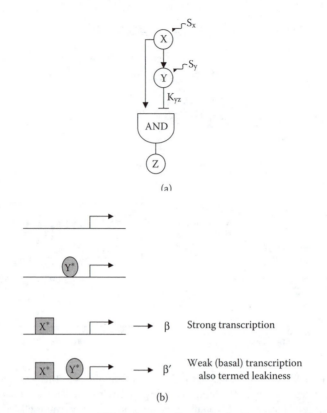

FIGURE 4.11 (a) The incoherent type-1 FFL with an AND gate at the Z promoter. The inputs are signals S_x and S_y. The repression threshold of gene Z by repressor Y is K_{yz}. (b) The four binding states of a simple model for the promoter region of Z, regulated by activator X and repressor Y. Transcription occurs when the activator X* is bound, and to a much lesser extent when both activator and repressor Y* are bound. The AND input function thus corresponds to X* AND NOT Y*.

represses Z. The gene Z shows high expression when the activator X* is bound, and much weaker expression when the repressor Y* binds (Figure 4.11b).

To analyze the dynamics of this motif, we will continue to use logic input functions. Hence, the dynamics will be composed of transitions between exponential approaches to steady states and exponential decays. As we saw above, these piecewise exponential dynamics make graphical analysis and analytical solutions rather easy.

4.7.2 Dynamics of the I1-FFL: A Pulse Generator

The I1-FFL responds to the input signals S_x and S_y (Figure 4.11a). Upon a step of S_x, protein X becomes activated, binds the promoter of gene Z, initiating transcription and causing protein Z to begin to be produced (Figure 4.12a). In parallel, X activates the production of Y. Therefore, after a delay, enough protein Y accumulates to repress Z production and Z levels decrease. Thus, the I1-FFL can generate a pulse of Z production.

Let us analyze this in more detail. Consider the response to a step addition of the signal S_x, in the presence of the second signal S_y. When the signal S_x appears, protein X rapidly transits to its active conformation, X*. The active transcription factor X* binds its DNA site in the Y promoter within seconds, and Y begins to be produced. Since S_y is present,

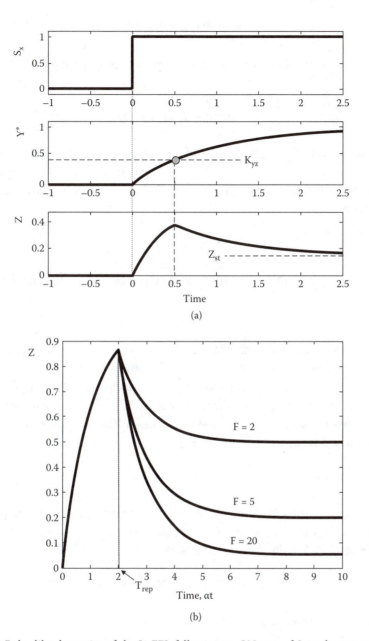

FIGURE 4.12 (a) Pulse-like dynamics of the I1-FFL following an ON step of S_x in the presence of S_y. The input step occurs at $t = 0$, and X rapidly transits to its active form, X^*. As a result, Z begins to be expressed. In addition, the repressor protein Y is produced, and eventually represses Z production when it crosses the repression threshold K_{yz}. In this figure, all production and decay rates are equal to 1. (b) Effect of repression strength on the pulse-like dynamics of the I1-FFL. Shown are the dynamics of Z in an incoherent type-1 FFL with repression coefficients F = 2, 5, and 20. The repression coefficient is the ratio of the steady-state expression in the absence of repressor to the steady-state expression with active repressor. T_{rep} is the time when repression begins.

the protein Y is in its active form Y^* and accumulates over time according to the production and degradation equation:

$$dY^*/dt = \beta_y - \alpha_y Y^* \qquad (4.7.1)$$

Hence, Y shows the familiar exponential convergence to its steady-state $Y_{st} = \beta_y/\alpha_y$, Figure 4.12a:

$$Y^*(t) = Y_{st} (1 - e^{-\alpha_y t}) \qquad (4.7.2)$$

In addition to activating Y, molecules of X^* also bind the Z promoter. As a result, protein Z^* begins to be produced at a rapid rate β_z, since its promoter is occupied by the activator X^* but there is not yet enough repressor Y^* in the cell to inhibit production (Figure 4.12a). In this phase,

$$dZ/dt = \beta_z - \alpha_z Z \qquad (4.7.3)$$

and Z accumulates, beginning an exponential convergence to a high level $Z_m = \beta_z/\alpha_z$:

$$Z(t) = Z_m (1 - e^{-\alpha_z t}) \qquad while\ Y^* < K_{yz} \quad (4.7.4)$$

This fast production of Z lasts until the repressor Y^* crosses its repression threshold for Z, K_{yz}. At this time, the production rate of Z (in our logic approximation) suddenly drops to a low value β'_z. In the extreme case of no leakiness, it drops to $\beta'_z = 0$. The onset of repression occurs at the moment that Y^* reaches K_{yz}. This repression time, T_{rep}, can be found from Equation 4.7.2 by finding the time when $Y^*(t) = K_{yz}$, showing that T_{rep} depends on the biochemical parameters of protein Y:

$$T_{rep} = 1/\alpha_y \log [1/(1 - K_{yz}/Y_{st})] \qquad (4.7.5)$$

At times after T_{rep}, the Z promoter is bound by the repressor Y^* and the production rate of Z is reduced. Figure 4.12a shows how Z concentration decays exponentially to a new lower steady-state $Z_{st} = \beta'_z/\alpha_z$ (see solved exercise 2.1):

$$Z(t) = Z_{st} + (Z_o - Z_{st}) e^{-\alpha_z (t - T_{rep})} \qquad (4.7.6)$$

where Z_o is the level reached at time T_{rep}, given by Equation 4.7.4 at $t = T_{rep}$:

$$Z_o = Z_m (1 - e^{-\alpha_z T_{rep}}) \qquad (4.7.7)$$

and Z_{st} is the final steady-state Z level, due the low expression level when both X^* and Y^* bind the Z promoter:

$$Z_{st} = \beta'_z/\alpha_z \qquad (4.7.8)$$

The shape of the dynamics generated by the I1-FFL depends on β'_z, the basal production rate of Z. This basal rate corresponds to the low rate of transcription from the repressed promoter. The effect of different basal levels on the dynamics is shown in Figure 4.12b for

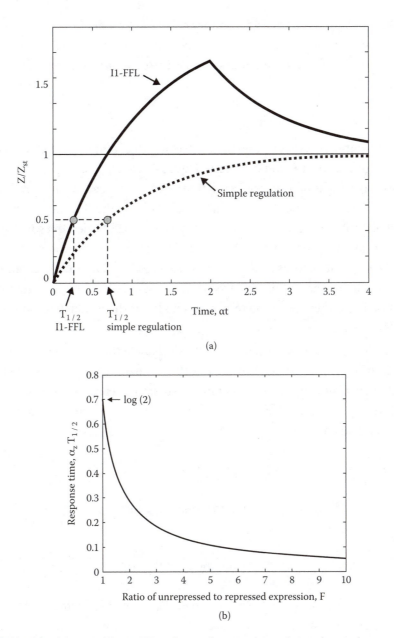

FIGURE 4.13 (a) Response time of the I1-FFL is shorter than simple regulation that reaches same the steady-state level. Simple regulation, dashed line; I1-FFL, full line. (b) Response time of the I1-FFL as a function of the repression coefficient F. F is the ratio of unrepressed to repressed Z expression. Also shown is the normalized response time of simple regulation, $\alpha_z T_{1/2} = \log(2)$.

several values of the **repression factor** F, defined as the ratio of the maximal and basal activity of the Z promoter, also equal to the ratio of the unrepressed and repressed steady-state levels of Z:

$$F = \beta_z / \beta'_z = Z_m / Z_{st} \tag{4.7.9}$$

When the repressor has a strong inhibitory effect on Z production, that is, when F >> 1, Z dynamics show a pulse-like shape. In the pulse, Z levels first increase and then decline to a low level. The I1-FFL can therefore act as a **pulse generator** (Mangan and Alon, 2003; Basu et al., 2004).

4.7.3   The I1-FFL Speeds the Response Time

In addition to pulse generation, the I1-FFL has another property: it can accelerate the response time of the system. You can see in Figure 4.13a that the response time of the I1-FFL is shorter than that of a simple-regulation circuit that reaches the same steady-state level of Z. The response time can be found graphically by the time at which the dynamics cross a horizontal line halfway to the steady-state level (dashed lines in Figure 4.13a). Note that one cannot speed the response time of the simple-regulation circuit by increasing its production rate, because such an increase would lead to an unwanted increase of the steady state level. The I1-FFL achieves its fast response time by initially using a high production rate, and then using the repressor Y to lower the production rate at a delay, to reach the desired steady-state level.

To analyze this speed-up quantitatively, let us calculate the response time $T_{1/2}$, the time to reach half of the steady-state level. In the I1-FFL, half steady state is reached during the initial fast stage of Z production, before Y crosses its repression threshold. Thus, the response time, $T_{1/2}$, is found by using Equation 4.7.4 by asking when the concentration of Z levels reaches halfway to Z_{st}:

$$Z_{1/2} = Z_{st}/2 = Z_m \left(1 - e^{-\alpha_z T_{1/2}}\right) \qquad (4.7.10)$$

which can be solved to give an expression that depends on the repression coefficient F = Z_m/Z_{st}:

$$T_{1/2} = 1/\alpha_z \log \left[2F/(2F - 1)\right] \qquad (4.7.11)$$

Note that, as shown in Figure 4.13b, the larger the repression coefficient F, the faster the response time becomes ($T_{1/2} \sim (2\alpha_z F)^{-1}$ at large F). In other words, the stronger the effect of Y in repressing production of Z, the faster the performance of the I1-FFL compared to an equivalent simple-regulation circuit X → Z made to reach the same steady-state level of Z. The response time becomes very small[1] when F >> 1, approaching $T_{1/2} = 0$. At the opposite extreme, the limit of no repression, F = 1, we find:

$$T_{1/2} = \log(2)/\alpha_z$$

---

[1] As mentioned previously, the response time cannot be smaller than the minimal time it takes proteins to be transcribed and translated, which is a few minutes in *E. coli* and often longer in human cells (see Table 2.1). Furthermore, the production rates β_z cannot be infinitely large — it is, for example, limited by the production capacity of the ribosomes, which is on the order of 10^6 proteins/cell generation in *E. coli*.

which is the same as the response time for simple regulation that we derived in Chapter 2. Indeed, when F = 1, the I1-FFL degenerates into a simple-regulation circuit because the repressor Y has no effect on Z, and the edge between Y and Z is nonfunctional.

4.7.4 Response Acceleration Is Sign Sensitive

In contrast to the accelerated response seen after ON steps, the response after the signal S_x is removed occurs with the same dynamics as for a simply regulated gene (no acceleration or delay). In both simple and I1-FFL circuits, OFF steps of S_x lead to an immediate shut-down in Z production. This immediate response to OFF steps in the I1-FFL is due to the AND logic of the Z promoter, in which unbinding of X* is sufficient to stop production (Figure 4.11b). After production stops, the concentration of protein Z decays exponentially according to its degradation/dilution rate. Hence, no speed-up is found in the OFF direction relative to simple regulation.

Thus, the I1-FFL is a *sign-sensitive response accelerator*. Sign-sensitive means that response acceleration occurs only for steps of S_x in one direction (ON) and not the other (OFF). I1-FFLs with OR gates have generally the same function as those with AND-gates, but accelerate OFF and not ON responses.

4.7.5 Experimental Study of the Dynamics of an I1-FFL

An experimental study of response dynamics of an I1-FFL is shown in Figure 4.14. This experiment employed a well-characterized system, which allows *E. coli* to grow on the sugar galactose as a carbon and energy source. As in other sugar systems, the genes in the galactose system are not highly expressed in the presence of glucose, a superior energy source. The galactose utilization genes are expressed at a low but significant level when both glucose is absent and galactose is absent, to allow the cell to grow rapidly on galactose should it appear in the environment. When galactose appears, the genes are fully expressed. The galactose genes are regulated by an I1-FFl, with the activator CRP and the repressor GalS. High resolution measurements show that the response of the output genes is accelerated upon glucose starvation (an ON step of S_x) compared to simply regulated genes (Figure 4.14). Removal of the repressor interaction in the I1-FFL abolishes this acceleration.

In addition to studying this network motif within a natural context, one can study it by making a synthetic I1-FFL made of well-characterized regulators. Weiss and colleagues constructed an I1-FFL using the activator LuxR as X, the repressor C1 of phage lambda as Y, and green fluorescent protein as the output gene Z (Basu et al., 2004). This "synthetic circuit" in *E. coli* showed pulse-like responses to steps of the input signal S_x (the inducer of LuxR). The synthetic construction of gene circuits is a promising approach for isolating and studying their properties.[1]

---

[1] Synthetic gene circuits are reviewed in (Hasty et al., 2002; Sprinzak and Elowitz, 2005). Examples include switches (Gardner et al., 2000; Becskei, 2001; Kramer et al., 2004; You et al., 2004), oscillators such as the repressilator (Elowitz and Leibler, 2000, Atkinson et al., 2003) and cascades (Rosenfeld and Alon, 2002; Yokobayashi et al., 2002; Hooshangi et al., 2005; Pedraza et al., 2005.)

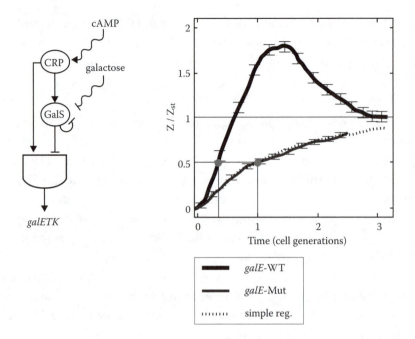

FIGURE 4.14 Dynamics of the I1-FFL in the galactose system of *E. coli*. The output genes, such as *galETK*, break down galactose for use as an energy and carbon source for the cells. The system is expressed fully upon glucose starvation, when the input signal cAMP is at high concentration in the cell. The second input signal is galactose, that inactivates the repressor GalS. Dynamics of the *galETK* promoter were measured by means of green fluorescent protein (GFP) expressed from the *galETK* promoter. Normalized GFP fluorescence is shown after an ON step of cAMP (heavy curve) in the absence of galactose. Also shown is an experiment on a bacterial strain in which the GalS site in the *galETK* promoter was deleted, removing the repression arrow in the I1-FFL (light curve). The dynamics of simply regulated promoters is shown in the dotted line. The I1-FFL accelerates the response time of the gal genes relative to simple regulation. (From Mangan et al., 2006.)

4.7.6   Three Ways to Speed Your Responses (An Interim Summary)

We have by now seen three different ways to speed the response time of transcription networks. The basic problem is that the response time of transcription regulation can be slow, on the order of the cell generation time for proteins that are not degraded. This is a drawback for networks that need to respond rapidly to external signals. The three ways to speed response times of gene regulation that we have discussed are:

1. *Increased degradation rate*: As we saw in Chapter 2, the response time of simple gene regulation is inversely proportional to the degradation/dilution rate, $T_{1/2} = \log(2)/\alpha$, where α is a sum of the rate of specific degradation of the protein and the rate of dilution by cell growth: $\alpha = \alpha_{\text{deg}} + \alpha_{\text{dil}}$. Therefore, increasing the degradation rate α_{deg} yields faster responses. There is a cost to this strategy: to maintain a given steady state, $X_{\text{st}} = \beta/\alpha$, one needs to increase the production rate β to balance the effects of increased degradation α. This creates a futile cycle, where the protein is rapidly produced and rapidly degraded. This cycle can be selected by evolution in some systems, despite the costs of increased production, due to the benefit of

speeding the response time. The increased speed applies both to turn-ON and turn-OFF of gene expression.

2. *Negative autoregulation*: As we saw in Chapter 3, negative autoregulation can speed responses by a large factor. This speed-up is due to the ability to use a strong promoter (large production rate β) to give rapid initial production, and then to turn production off by self-repression when the desired steady state is reached. Note that only turn-ON is speeded; turn-OFF is not, but rather has the same OFF response time as simple regulation. The negative autoregulation strategy works only for proteins that can repress themselves, namely, only for transcription factor proteins.

3. *Incoherent FFL*: The incoherent FFL can significantly speed up ON responses, as we saw in the previous section. This is due to initially rapid production that is later turned off by a delayed repressor, to achieve a desired steady state. This speed-up applies to the low-induction state in the presence of S_y. It can be used to speed the response time of any target protein, not only transcription factors.

Designs 2 and 3 can work together with 1: a large degradation rate can further speed the response of negative autoregulation and incoherent FFLs.

4.8   WHY ARE SOME FFL TYPES RARE?

We have so far examined the structure and function of the two most common FFL types. We will now ask why the other six FFL types are rare in transcription networks (Figure 4.4). To address this, we need to consider the steady-state computations performed by the FFLs. We will see that some of the rare forms have a functional disadvantage in these computations: they lack responsiveness to one of their two inputs.

4.8.1   Steady-State Logic of the I1-FFL: S_y Can Turn on High Expression

The FFL has two input signals, S_x and S_y. Up to now, we have considered changes only in one of the two inputs of the FFL, namely, S_x, and studied the dynamics in the presence of the second input signal, S_y. What happens in the I1-FFL if we remove S_y? The signal S_y causes the repressor Y to assume its active form, Y^*, and bind the promoter of gene Z to inhibit its expression. When S_y is removed, Y becomes inactive and unbinds from the promoter of gene Z. As a result, Z is not repressed and is expressed strongly (Figure 4.15).

The resulting *steady-state logic* of the I1-FFL with an AND gate is shown in Table 4.2. The second input S_y has a strong effect on the steady-state level of Z, modulating it by a factor $F = \beta_z / \beta_z'$.

4.8.2   I4-FFL, a Rarely Selected Circuit, Has Reduced Functionality

As mentioned above, not all FFL types are found in equal amounts in transcription networks. Among the incoherent FFLs, for example, the most common form is I1-FFL(about 30–40% of known FFLs), whereas the other forms are rare (I3-FFLs and I4-FFLs are in total less than 5% of known FFLs). Why?

To address this question, we will focus on two very similar structures, I1-FFL and I4-FFL. Both circuits have two activation arrows and one repression arrow (Figure 4.16). The

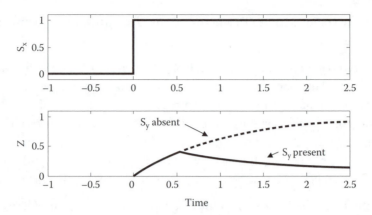

FIGURE 4.15 The effect of input signal S_y on the dynamics of the I1-FFL. When S_y is absent, Y is not active as a repressor, and the concentration of protein Z shows an increase to a high unrepressed steady state (dashed line).

only difference is that in the I1-FFL, X activates Y, which represses Z, whereas in the I4-FFL, X represses Y, which activates Z. The minus and plus edges in the indirect regulation path have simply changed position. How can this subtle change result in such a large difference in the appearance of the two circuits in transcription networks?

The structural difference between these two circuits means that in the I1-FFL, X activates both of its target genes, whereas in I4-FFL, X activates one target, Z, and represses the other, Y. Can a transcription factor be both an activator and a repressor? As mentioned in Section 2.3.1, the answer is yes: transcription factors such as the bacterial glucose starvation sensor CRP activate many target genes, but act as repressors for other genes. The molecular mechanism is often simple to understand (Collado-Vides et al., 1991; Ptashne and Gann, 2002). In bacteria, for example, an activator often binds a site that is close to the binding site of RNA polymerase (RNAp), helping RNAp to bind or to start transcription once it binds. If the activator binding site is moved so that it overlaps the space occupied by RNAp, binding of the activator protein interferes with binding of RNAp, and the activator acts as a repressor. Similar features, where a transcription factor can activate some targets and repress others, are commonly found in eukaryotic regulators, though the detailed mechanisms can vary. Thus, I4-FFL is a biologically feasible pattern.

What about dynamic behavior? Is I4-FFL a sign-sensitive accelerator and pulse generator as well? The answer, again, is yes. It is easy to see that upon an ON step of S_x, Z begins to be produced vigorously, activated by both Y and X. At the same time, since X represses

TABLE 4.2 Steady-State Response of the I1-FFL to Various Combinations of Input Signals

S_x	S_y	Z_{st}
0	0	0
0	1	0
1	0	1 High, β_z/α_z
1	1	0 Low, β'_z/α_z

Note: $S_x = 0$ means that S_x is below the activation threshold of transcription factor X, and $S_x = 1$ means saturating signal. Similar definitions apply to S_y.

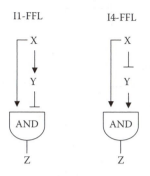

FIGURE 4.16 The incoherent type-1 FFL and type-4 FFL.

Y, the levels of Y begin to drop. When Y goes below its activation threshold for Z, the production rate of Z decreases and Z levels decline. This yields a pulse-like shape of the dynamics (Figure 4.17), just as in the I1-FFL, with a speed-up of the response time. The magnitude of the speed-up relative to simple regulation is the same as in I1-FFL. When S_x goes OFF, Z production stops at once (due to the AND gate), just as in the case of I1-FFL. Thus, I4-FFL is a sign-sensitive accelerator. It has all of the dynamical capabilities of I1-FFL in response to S_x signals. The same applies to I2-FFL and I3-FFL (except that they accelerate responses to OFF steps). What, then, might explain the difference in the occurrence of I1-FFLs and I4-FFLs in transcription networks?

The main difference between the two FFL forms is in their *steady-state logic*. We saw above that I1-FFL responds to both S_x and S_y. In contrast, the steady-state output of I4-FFL does not depend on S_y. To see this, note that when S_x is present, production of Y is repressed and its concentration declines. At steady-state, protein Y is not present at functional levels. Therefore, when S_x is present, S_y cannot affect Z production, because Y — the detector protein for S_y — is not present. When S_x is absent, on the other hand, Z is OFF regardless of S_y, due to the AND logic. Thus, S_y *does not affect the steady-state level of Z in I4-FFL*. The I4-FFL is not responsive to one of its two inputs (Table 4.3).

The lack of responsiveness to one of the two inputs may be one of the reasons why I4-FFL is selected less often than I1-FFL. The same reasoning applies also to I3-FFL.

Similar conclusions apply to the rare and common forms of coherent FFLs. Coherent type 3 and 4 FFLs have the same sign-sensitive delay properties as the much more common

TABLE 4.3 Steady-State Output in the I1-FFL and I4-FFL as a Function of the Input Signals S_x and S_y

S_x	S_y	Z_{st} in I1-FFL		Z_{st} in I4-FFL	
0	0	0		0	
0	1	0		0	
1	0	1	High, β_z/α_z	0	Low, β'_z/α_z
1	1	0	Low, β'_z/α_z	0	Low, β'_z/α_z

Note that the logic function of the entire I4-FFL circuit, $g(S_x, S_y)$, is *different* from the input function of the Z promoter, $f(X^*, Y^*) = X^* AND Y^*$.

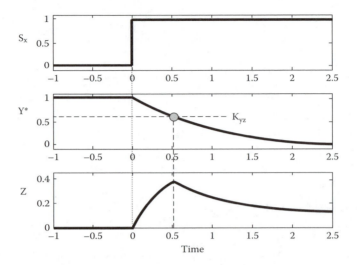

FIGURE 4.17 Dynamics of the I4-FFL following a step of S_x. In the presence of S_x, protein X is active and activates Z production. It also represses the production of Y. When Y levels decay below the activation coefficient K_{yz}, the concentration of Z begins to drop. Production and decay rates are $\beta_z = 1$, $\alpha_z = \alpha_y = 1$, and $F = 10$. The signal S_y is present throughout.

type-1 coherent FFL. However, these types with AND logic cannot respond to the input signal S_y, for the same reasons discussed above.

Note that we have analyzed here only AND gate input functions, and not OR gates. The discussion for OR gates is more complicated because they can have multiple intermediate states of Z. FFL types I3 and I4 with OR gates can in principle be responsive to both inputs. Similarly, the I2-FFL with AND logic is just as responsive as the I1-FFL. It is an interesting question why these circuits are not commonly found in transcription networks.

4.9 CONVERGENT EVOLUTION OF FFLs

How does evolutionary selection act on the three regulation edges in the FFL? It is reasonable that the most important function of the regulators X and Y is to respond to the signals S_x and S_y and accordingly regulate Z. Thus, the first-order selection is for the simple V-shaped structure where X and Y regulate Z (Figure 4.18a). It is the third edge, *the edge from X to Y, that needs special explanation*. Recall that it only takes one or a few mutations in the binding site of X in the Y promoter to abolish the edge X → Y. If it does not add a useful function (or if it actually destroys a function), this edge will rapidly be lost in evolution.

In the common FFL types, C1-FFL and I1-FFL, we have seen that the edge between X and Y can add a function, such as persistence detection or pulse generation and response acceleration. Presumably, such functions are useful enough in some systems to select this edge: Mutant organisms without the edge are lost from the population due to their decreased fitness. On the other hand, in I4-FFL, adding the edge between X and Y can cause a loss of functionality — the entire circuit no longer responds to S_y. This might cause such an edge to be lost during evolution.

How did FFLs evolve? The most common form of evolution for genes is conservative evolution, where two genes with similar function stem from a common ancestor gene. This is reflected in a significant degree of sequence similarity between the genes. Such genes are said to be **homologous**.

Did FFLs evolve in a similar way, where an ancestor FFL duplicated and gave rise to the present FFLs? It appears that the answer is no in most cases. For example, homologous genes Z and Z' in two organisms are often both regulated by FFLs in response to the same environmental stimuli. If the two FFLs had a common ancestor FFL, the regulators X and Y in the two FFLs would also be homologous. However, the regulators are usually not homologous in such FFL pairs (Figure 4.18b). The sequence of the regulators is often so dissimilar that they clearly belong to completely different transcription factor families. That is, evolution seems to have *converged independently* on the same regulation circuit in many cases (Conant and Wagner, 2003; Babu et al., 2004). Presumably, the FFL is rediscovered by evolution because it performs an important function in the different organisms.

More about gene circuit evolution and selection of FFLs will be discussed in Chapter 10.

4.10   SUMMARY

We have seen that sensory transcription networks have a measure of simplicity. Of the 13 possible three-gene regulation patterns, only one, the FFL, is a network motif. Furthermore, of eight possible FFL types, only two are commonly found.

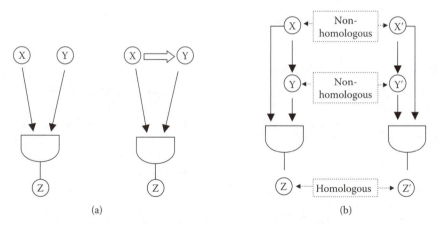

FIGURE 4.18 On the evolution of the FFLs. (a) The V-shaped pattern in which X and Y regulate Z is strongly selected because it allows regulation based on two inputs. The edge from X to Y (white arrow) can be selected based on an additional dynamical function (e.g., sign-sensitive delay, acceleration, pulse generation). (b) In many cases homologous genes Z and Z' in different organisms are both regulated by a FFL in response to the same stimuli, but the two regulators X and Y in the FFL of Z are not homologous to the regulators X' and Y' in the FFL of Z'. This suggests that rather than the duplication of an ancestral FFL, Z and Z' acquired FFL regulation by convergent evolution: the FFL was rediscovered in each system. Homology means sufficient similarity in the genes sequence to indicate that the genes have a common ancestral gene.

The two common FFL types can carry out specific dynamical functions. The coherent type-1 FFL (C1-FFL) can act as a sign-sensitive delay element. Thus, it can function as a persistence detector, filtering away brief fluctuations in the input signal. With AND logic, brief ON pulses of the input signal are filtered out, whereas with OR logic, brief OFF pulses are filtered out. This function can help protect gene expression in environments with fluctuating stimuli.

The second common FFL type, the incoherent type-1 FFL (I1-FFL), can act as a pulse generator and a response accelerator. This acceleration can be used in conjunction with the other mechanisms of acceleration, such as increased degradation and negative autoregulation.

Some types of FFL have reduced functionality relative to other types. In particular, some of the FFL types cannot respond to one of their two inputs. This reduced functionality may explain, at least partly, why these FFL types are relatively rare in transcription networks.

This chapter did not exhaust all of the possible dynamical functions of the FFLs. These circuits may carry out additional functions (Wang and Purisima, 2005; Ghosh et al., 2005; Hayot and Jayaprakash, 2005; and Ishihara et al., 2005), some of which are discussed in the exercises.

Evolution seems to have converged again and again on the FFLs in different gene systems and in different organisms. Thus, this recurring network motif is an example of a pattern that may have been selected for its specific dynamical functions. As we will see in Chapter 6, the FFL is also a network motif in several other types of biological networks.

FURTHER READING

Feed-Forward Loop Network Motif

Mangan, S. and Alon, U. (2003). Structure and function of the feed-forward loop network motif. *Proc. Natl. Acad. Sci. U.S.A.*, 100: 11980–11985.

Shen-Orr, S., Milo, R., Mangan, S., and Alon, U. (2002). Network motifs in the transcriptional regulation network of *Escherichia coli. Nat. Genet.*, 31: 64–68.

Experimental Study of the Dynamics of the Coherent FFL

Kalir, S., Mangan, S., and Alon, U. (2005). A coherent feed-forward loop with a SUM input function protects flagella production in *Escherichia coli. Mol. Syst. Biol.*, msb4100010:E1-E6.

Mangan, S., Zaslaver, A., and Alon, U. (2003). The coherent feed-forward loop serves as a sign-sensitive delay element in transcription networks. *J. Mol. Biol.*, 334: 197–204.

Experimental Study of the Dynamics of Incoherent FFL

Basu, S., Mehreja, R., Thiberge, S., Chen, M.T., and Weiss, R. (2004). Spatiotemporal control of gene expression with pulse-generating networks. *Proc. Natl. Acad. Sci. U.S.A.*, 101: 6355–6360.

Mangan, S., Itzkovitz, S., Zaslaver, A., Alon, U. (2006). The incoherent feed-forward loop accelerates the response time of the Gal system of *Escherichia coli. J. Mol. Biol.*, 356: 1073–1083.

Convergent Evolution of Network Motifs

Babu, M.M., Luscombe, N.M., Aravind L., Gerstein M., and Teichmann, S.A. (2004). Structure and evolution of transcriptional regulation networks. *Curr. Opin. Struct. Biol.* 14: 283–291.

Conant, G.C. and Wagner, A. (2003). Convergent evolution of gene circuits. *Nat. Genet.*, 34: 264–266.

EXERCISES

4.1. *The second input.* What is the effect of steps of S_y on the expression dynamics of Z in the C1-FFL with AND logic? Are there delays in Z expression for ON or OFF steps of S_y? What is the response time of Z for such steps? Assume that S_x is present throughout.

4.2. *OR gate logic.* Analyze the C1-FFL with OR logic at the Z promoter. Are there delays following ON or OFF steps of S_x? What could be the biological use of such a design?

Solution:

After an ON step of S_x, X becomes active X*. On a rapid timescale it binds the Z promoter. Since Z is regulated by OR logic, X* alone can activate transcription without need of Y. Therefore, there are no delays following an ON step of S_x.

After an OFF step of S_x, X* rapidly becomes inactive, X. However, protein Y is still present in the cell, and if S_y is present, Y is active. Since the Z input function is an OR gate, Y* can continue to activate transcription of Z even in the absence of X*. Therefore, Z production persists until Y degrades/dilutes below its activation threshold for Z. The dynamics of Y are given by $dY/dt = -\alpha Y$, (there is no production term because X is inactive following the removal of S_x), so that Y $= Y_m e^{-\alpha t}$, where Y_m is the level of Y at time t = 0. The OFF delay is given by the time it takes Y to reach its activation threshold for Z, K_y: solving for this time, $Y(T_{OFF}) = Y_m e^{-\alpha T_{off}} = K_y$, yields

$$T_{OFF} = 1/\alpha \log (Y_m/K_y)$$

In summary, the OR gate C1-FFL shows sign-sensitive delays. It has a delay following OFF but not ON steps of S_x. The delay depends on the presence of S_y. This behavior is opposite that of the C1-FFL with an AND gate, which shows delay upon ON but not OFF steps.

The OR gate C1-FFL could be useful in systems that need to be protected from sudden loss of activity of their master regulator X. The OR gate FFL can provide continued production during brief fluctuations in which X activity is lost. This protection works for OFF pulses shorter than T_{OFF}. Note that T_{OFF} can be tuned by evolutionary selection by adjusting the biochemical parameters of protein Y, such as its expression level, Y_m, and its activation threshold, K_y.

4.3. *A decoration on the FFL.* The regulator Y in C1-FFLs in transcription networks is often negatively autoregulated. How does this affect the dynamics of the circuit, assuming that it has an AND input function at the Z promoter? How does it affect the delay times? The Y regulator in an OR gate C1-FFL is often positively autoregulated. How does this affect the dynamics of the circuit? How does it affect the delay times?

4.4. *The diamond.* The four-node diamond pattern occurs when X regulates Y and Z, and both Y and Z regulate gene W.

a. How does the mean number of diamonds scale with network size in random ER networks?

b. What are the distinct types of sign combinations of the diamond (where each edge is either activation + or repression –)? How many of these are coherent? (Answer: 10 types, of which 6 are coherent).

c. Consider a diamond with four activation edges. Assign activation thresholds to all edges. Analyze the dynamics of W following a step of S_x, for both AND and OR logic at the W promoter. Are there sign-sensitive delays?

Solution (partial):

a. The diamond has $n = 4$ nodes and $g = 4$ edges. The number of diamonds therefore scales in ER networks as $N^{n-g} = N^0$. Hence, the number of diamonds does not depend on the ER network size.

c. A diamond generally has unequal response times for the arm through Y and the arm through Z. For example, suppose that Y and Z have the same production and degradation rates, but that their thresholds to activate W are different. Without loss of generality, suppose that Z has a lower threshold, $K_{zw} < K_{yw}$. We will solve for an AND gate at the W promoter. Following an ON step of S_x, both Y and Z must accumulate and cross their thresholds to activate W. The response is therefore governed by the higher of the two thresholds, since both Y and Z must cross their thresholds. Hence:

$$T_{ON} = 1/\alpha \log [Y_{st}/(Y_{st} - K_{yw})]$$

In contrast, after an OFF step, only one of the two regulators must go below its threshold to deactivate the AND gate at the W promoter. Again, the OFF time corresponds to the higher of the two thresholds because it is crossed first:

$$T_{OFF} = 1/\alpha \log (Y_{st}/K_{yw})$$

The delay is asymmetric ($T_{ON} \neq T_{OFF}$) unless K_{yw} is equal to $Y_{st}/2$. Note that Z does not affect the dynamics at all in this circuit (assuming logic input functions).

4.5. *Type three.* Solve the dynamics of the type-3 coherent FFL (Figure 4.3) with AND logic at the Z promoter in response to steps of S_x. Here, AND logic means that Z is produced if both X* and Y* do not bind the promoter. Are there delays? What is the steady-state logic carried out by this circuit? Compare to the other coherent FFL types.

4.6. *Shaping the pulse.* Consider a situation where X in an I1-FFL begins to be produced at time $t = 0$, so that the level of protein X gradually increases. The input signals S_x and S_y are present throughout. How does the pulse shape generated by the I1-FFL depend on the thresholds K_{xz}, K_{xy}, and K_{yz}, and on β, the production rate of protein X? Analyze a set of genes Z_1, Z_2, ..., Z_n, all regulated by the same X and Y in

I1-FFLs. Design thresholds such that the genes are turned ON in the rising phase of the pulse in a certain temporal order and turned OFF in the declining phase of the pulse with the same order. Design thresholds such that the turn-OFF order is opposite to the turn-ON order. Plot the resulting dynamics.

4.7. *Amplifying intermediate stimuli.* This problem highlights an additional possible function of incoherent type-1 FFLs for subsaturating stimuli S_x. Consider an I1-FFL, such that the activation threshold of Z by X, K_{zx}, is smaller than the activation threshold of Y by X, K_{yx}. That is, Z is activated when $X^* > K_{zx}$, but it is repressed by Y when $X^* > K_{yx}$. Schematically plot the steady-state concentration of Z as a function of X^*. Note that intermediate values of X^* lead to the highest Z expression.

4.8. *The diamond again.* The diamond pattern occurs when X regulates Y and Z, and both Y and Z regulate gene W. Analyze the 10 types of diamond structures (where each edge is either activation + or repression –) with respect to their steady-state responses to the inputs S_x, S_y, and S_z. Use an AND input function at the W promoter. Do any diamond types lack responsiveness to any input? To all three inputs?

4.9. *Repressilator.* Three repressors are hooked up in a cycle X ⊣ Y ⊣ Z and Z ⊣ X. What are the resulting dynamics? Use initial conditions in which X is high and Y = Z = 0. Solve graphically using logic input functions. This circuit was constructed in bacteria using three well-studied repressors, one of which was also made to repress the gene for green fluorescent protein (Elowitz and Leibler, 2000). What would the resulting bacteria look like under a microscope that dynamically records green fluorescence?

4.10. *Interconnected FFLs.* Consider a coherent type-1 FFL with nodes X, Y, and Z, which is linked to another coherent type-1 FFL in which Y activates Y_1, which activates Z.

a. Sketch the dynamics of Z expression in response to steps of the signals S_x, S_y, and S_{y1} (Steps in which one of the signals goes ON or OFF in the presence of the other signals). Can the dynamics of the interconnected circuit be understood based on the qualitative behavior of each FFL in isolation?

b. Repeat for the case where Y represses Z, so that the X, Y, Z FFL is an incoherent type-1 FFL. Assume that Y_1 binding to the Z promoter can alleviate the repressing effect of Y.

CHAPTER 5

Temporal Programs and the Global Structure of Transcription Networks

5.1   INTRODUCTION

We have seen that transcription networks contain recurring network motifs that can perform specific dynamical functions. We examined two of these motifs in detail, autoregulation and the feed-forward loop. In this chapter we will complete our survey of motifs in sensory transcriptional networks. We will see that sensory transcription networks are largely made of just four families of network motifs: the two we have studied, including feed-forward loops with multiple outputs, and two families of larger motifs. The larger motif families are called single-input module (SIM) and dense overlapping regulons (DORs).

The SIM network motif is a simple pattern in which one regulator controls a group of genes. Despite its simple structure, the SIM has an interesting dynamical function: it can generate temporal programs of expression, in which genes are turned on one by one in a defined order. In *Escherichia coli*, these temporal programs are found to match the functional order of the gene products. This is a "just-when-needed" production strategy, not making a protein before it is needed. Such a strategy is optimal for rapid production of a system made of different types of proteins, under constraints of limited resources for producing these proteins.

More detailed temporal programs can be generated by a generalized network motif related to the feed-forward loops (FFLs) we have examined in the previous chapter. This generalized motif is an FFL with multiple output genes. We will see that the multi-output FFL can generate temporal programs with different orders of activation and inactivation of the genes.

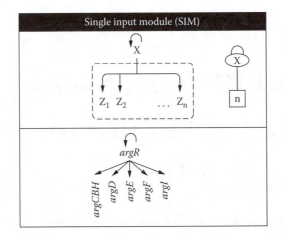

FIGURE 5.1 The single-input module (SIM) network motif. Transcription factor X regulates a group of genes that have no additional transcription factor inputs. X usually also regulates itself. An example of a SIM, the arginine biosynthesis pathway (in the arginine system, all regulations are repression.)

The last motif family, called DORs (for densely overlapping regulons), is a dense array of regulators that combinatorially control output genes. The DORs can carry out decision-making calculations, based on the input functions of each gene.

Finally, we will discuss how the network motifs fit together to build the global structure of the transcription network. These four motif families appear to account for virtually all of the interactions in sensory transcription networks.

5.2   THE SINGLE-INPUT MODULE (SIM) NETWORK MOTIF

The network motifs we have studied so far all had a defined number of nodes (one node in the autoregulation motif, three nodes in FFLs). We will now look for larger motifs. Each of these larger motifs corresponds to a family of patterns that share a common architectural theme.[1] The first such motif family found in transcription networks is called the **single-input module** (Figure 5.1), or **SIM** for short (Shen-Orr et al., 2002).

In the SIM network motif, a master transcription factor X controls a group of target genes, $Z_1, Z_2, ..., Z_n$ (Figure 5.1). Each of the target genes in the SIM has only one input: no other transcription factor regulates any of the genes. In addition, the regulation signs (activation/repression) are the same for all genes in the SIM. The last feature of the SIM is that the master transcription factor X is usually autoregulatory.

---

[1] Detection of large network motifs, and more generally, counting of large subgraphs, poses interesting computational problems due to the huge number of possible subgraphs. An efficient algorithm for counting subgraphs and detecting motifs avoids part of the problem by random sampling of subgraphs from the network. (From Kashtan et al., 2004a; Wernicke and Rasche, 2006; Berg and Lassig, 2006.)

The SIMs are a family of structures with a free parameter, the number of target genes n. They are a strong network motif when compared to random networks.[1] This is because it is rare to find in random networks a node regulating, say, 14 other nodes with no other edge going into any of these nodes. Despite their simple structure, we will see that SIMs turn out to have interesting dynamics.

What is the function of SIMs? The most important task of a SIM is to control a group of genes according to the signal sensed by the master regulator. The genes in a SIM always have a common biological function. For example, SIMs often regulate genes that participate in a specific metabolic pathway (Figure 5.2). These genes work sequentially to assemble a desired molecule atom by atom, in a kind of molecular assembly line.[2]

Other SIMs control groups of genes that respond to a specific stress (DNA damage, heat shock, etc.). These genes produce proteins that repair the different forms of damage caused by the stress. Such stress response systems usually have subgroups of genes that specialize in certain aspects of the response. Finally, SIMs can control groups of genes that together make up a protein machine (such as a ribosome). The gene products assemble into a functional complex made of many subunits.

5.3   SIMS CAN GENERATE TEMPORAL EXPRESSION PROGRAMS

In addition to controlling a gene module in a coordinated fashion, the SIM has a more subtle dynamical function. The SIM can generate **temporal programs of expression**, in which genes are activated one by one in a defined order.

A simple mechanism for this temporal order is based on different thresholds of X for each of the target genes Z_i (Figure 5.3). The threshold of each promoter depends on the specific binding sites of X in the promoter. These sites can be slightly different in sequence and position, resulting in different activation thresholds for each gene. When X activity

---

[1] Even small SIMS are significant when compared to ER random networks (ER networks were discussed in Chapters 3 and 4). ER networks have a degree sequence (distribution of edges per node) that is Poisson, so that there are exponentially few nodes that have many more edges than the mean connectivity λ. Thus, ER networks have very few large SIMs. In contrast, real transcription networks have degree distributions with long tails: they show a few nodes with many more outgoing edges than the mean (Appendix C). These nodes correspond to transcription factors with many target genes, known as global regulators. To control for this, one may use randomized networks that preserve the degree sequence of the real network, called degree-preserving random networks (DPRNs), mentioned in Chapter 4. In *E. coli*, SIMS with n > 12 are significantly overrepresented relative to DPRN. What about small SIMs that appear about as often as in DPRN? Statistical significance is a powerful tool for detecting interesting structures, but it is important to remember that even structures that are not significant in comparison to strict random ensembles can turn out to be biologically interesting.

[2] Note that often the final product of the pathway in a SIM is the signal sensed by the master regulator X (Figure 5.2). For example, the repressor ArgR regulates several genes that make up the pathway for biosynthesis of the amino acid arginine, one of the 20 amino acids of which proteins are made (Figure 5.1). Arginine binds the repressor ArgR and causes it to become active, and hence to bind and repress the promoters of the biosynthesis genes. Thus, the more arginine is present, the more the genes are repressed. This is a negative feedback loop. One arm of this feedback (diffusion and binding of arginine to ArgR on a 10-msec timescale) is much faster than the other arm (transcription regulation of arg genes on a timescale of many minutes). This composite feedback loop made of two interactions with different timescales is a common network motif in biology, as mentioned in Chapter 6.

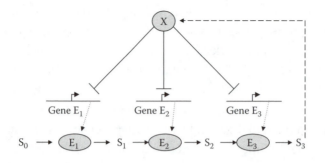

FIGURE 5.2 A single-input module (SIM) regulating a three-step metabolic pathway. The master repressor X represses a group of genes that encode for enzymes E_1, E_2, and E_3 (each on a different operon). These enzymes catalyze the conversion of substrate S_0 to S_1 to S_2, culminating in the product S_3. The product S_3 is the input signal of X: It binds to X and increases the probability that X is in its active state, X*, in which it binds the promoters to repress the production of enzymes. This closes a negative feedback loop, where high levels of S_3 lead to a reduction in its rate of production.

changes gradually in time, it crosses these thresholds, K_i, at different times, and the genes are turned ON or OFF in a specified order (Figure 5.3).[1]

When the activity of X increases gradually, it first activates the gene with the lowest threshold. Then it activates the gene with the next lowest threshold, etc. (Figure 5.3). However, when X activity goes down, the genes are affected in reverse order. Hence, the first gene activated is the last one to be deactivated (Figure 5.3). This type of program is called a **last-in-first-out (LIFO)** order.

The faster the changes in the activity of X, the more rapidly it crosses the different thresholds, and the smaller the delay between the genes. In many systems, there is an asymmetry in the naturally occurring dynamics of the regulator activity. For example, sometimes turn-ON is fast, but turn-OFF is gradual. In such cases, temporal order will be more pronounced in the slow phase of the transcription factor activity profile.[2]

Experimentally, temporal order is found in a wide variety of systems in *E. coli* with SIM architecture. This includes metabolic pathways (Zaslaver et al., 2004) such as the arginine system (Figure 5.4) (See color insert following page 112) and damage repair systems such as the SOS DNA repair system (Ronen et al., 2002). The genes in these systems are expressed in a defined order, with delays on the order of 0.1 generation between genes (about 5 to 10 min).

---

[1] In the previous chapter, we considered changes in activity X* that are much faster than the response time of the network. Here we consider cases where X* changes gradually. For example, transcription of gene X may be itself controlled by another transcription factor. This transcription factor can activate transcription of X at time t = 0, resulting in a gradual increase in the concentration of protein X with time. A second common example occurs if X* is governed by a feedback as a result of the action of the downstream genes, as in Figure 5.2. For example, the signal that activates X can be the metabolic product of a metabolic pathway regulated by X. In this case, the level of the metabolic product can change slowly as enzymes in the pathway are made, and the corresponding changes in X activity will occur on a relatively slow timescale.

[2] For example in the case of metabolites there is often an asymmetry in the timescales of regulation. Intracellular levels of externally supplied metabolites such as amino acids drop within a minute or less once the amino acid stops being externally supplied. It then takes many minutes until the biosynthetic enzymes are produced and endogenous production of the amino acid can commence.

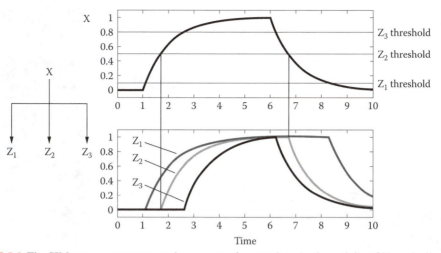

FIGURE 5.3 The SIM can generate temporal programs of expression. As the activity of X gradually rises, it crosses the different thresholds for each target promoter in a defined order. The gene with the lowest threshold, Z_1, is turned ON first, whereas the gene with the highest threshold, Z_3 is turned ON last. When X activity declines, it crosses the thresholds in reverse order (last-in-first-out, or LIFO order).

Is there any meaning to the temporal order found in the SIMs? In *E. coli* metabolic pathways, such as arginine biosynthesis, the following principle unifies the experimental findings: *the earlier the protein functions in the pathway, the earlier its gene is activated.*

Thus, the temporal order of the genes matches their functional order. This is an economical design, because proteins are not produced before they are needed. Such a just-when-needed production strategy can be shown, using simplified mathematical models, to be optimal for rapidly reaching a desired flux of a metabolic product under constraints of producing a minimal total number of protein enzymes (Klipp et al., 2002; Zaslaver et al., 2004).[1]

The precise temporal order generated by a SIM can be varied by mutations that change the relative order of the thresholds of the genes. For example, mutations in the binding site of X in the promoter of a gene can change the affinity of X to the site, changing the threshold (Kalir and Alon, 2004). This suggests that the observed temporal order is maintained against mutations due to the selective advantage afforded by just-when-needed production strategies.

Temporal order is found also in damage repair systems controlled by SIMs. In damage repair systems, turn-ON is usually fast, because the regulator needs to be activated rapidly in order to rapidly mobilize all repair processes. As damage is repaired, the input signal of the regulator declines and the genes get turned off gradually, reaching 50% of their maximal promoter activity at different times. In the systems that have been studied, the genes responsible for the mildest form of repair are turned off first, and those responsible for more severe damage repair are turned off later (Ronen et al., 2002).

---

[1] In addition, in many systems, a second principle is found: the earlier the protein functions in the pathway, the higher its maximal promoter activity (Heinrich and Klipp, 1996; Zaslaver et al., 2004).

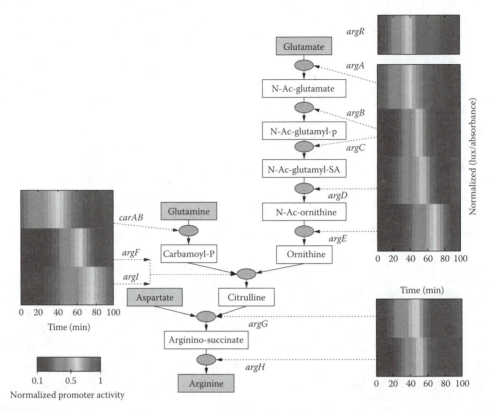

FIGURE 5.4 (See color insert following page 112) Temporal order in the arginine biosynthesis system. The promoters are activated in a defined order with delays of minutes between promoters. Color bars show expression from the promoters of the different operons in the system, measured by means of a luminescent reporter gene. The position of each gene product in the pathways that produce arginine is shown. Metabolites are in rectangles and enzymes in ellipses. (From Zaslaver et al., 2004.)

Temporal order also characterizes a large number of other global cellular responses. Examples include genes timed throughout the cell cycle in bacteria (Laub et al., 2000; McAdams and Shapiro, 2003) and yeast (Spellman et al., 1998), genes regulated by different phases of the circadian clock that keeps track of the time of day (Young, 2000; Duffield et al., 2002), as well as genes in developmental processes (Dubrulle and Pourquie, 2002; Kmita and Doboule, 2003).

In these global well-timed responses, genes are often regulated by a master regulator and also coregulated by additional regulators responsible for smaller subsystems. Temporal order may be generated by the action of a master coordinating regulator even if the network pattern is not strictly a SIM, in the sense that it has more than one regulator. The present analysis of temporal order applies also to circuits with multiple regulators, as long as all regulators except one have a constant activity during the interval of interest.

How did SIMs evolve? There are many examples of SIMs that regulate the same gene systems in different organisms. However, the master regulator in the SIM is often different in each organism, despite the fact that the target genes are highly homologous (Ihmels et al., 2005; Tanay et al., 2005). This means that rather than duplication of an ancestral

SIM together with the regulator to create the modern SIMs, evolution has converged on the same regulation pattern in the different organisms, just as we saw that evolution can converge on the FFL network motif (see Section 4.9). Presumably, the SIM regulatory pattern is rediscovered and preserved against mutations because it is useful enough to be selected.

In short, the SIM can generate just-when-needed temporal programs. As was mentioned above, the SIM circuit generates LIFO order: the activation order of the genes is *reversed* with respect to the deactivation order (Figure 5.3). However, in many cases, it seems more desirable to have an activation order that is the *same* as the deactivation order: the first promoter turned on is also the first turned off (**first-in-first-out (FIFO)** order). FIFO order is desirable for assembly processes that require parts in a defined order, some early and some late. In this case, when the process is de-activated, it is better for the early genes to be turned OFF before the late genes. This FIFO order prevents waste from needlessly producing early genes proteins after late ones are OFF. We next describe circuitry that can achieve FIFO order. To describe this circuit, we first discuss generalizations of network motifs and an additional important motif family in sensory transcription networks, the multi-output feedforward loop.

5.4 TOPOLOGICAL GENERALIZATIONS OF NETWORK MOTIFS

We have so far discussed relatively simple network motifs. When considering larger and more complex subgraphs, one is faced with a large number of possible patterns. For example, there are 199 possible four-node directed patterns (Figure 5.5) and over 9000 five-node patterns. There are millions of distinct seven-node patterns. In order to try to group these patterns into families that share a functional theme, one can define **topological generalizations of motifs** (Kashtan et al., 2004b).

To describe topological motif generalizations, consider the familiar feed-forward loop (FFL). The FFL is a three-node pattern with nodes X, Y, and Z (Figure 5.6a). The simplest form of topological generalization is obtained by choosing a node, say X, and duplicating it along with all of its edges (Figure 5.6b). The resulting pattern is a double-input FFL. This can be repeated to obtain multi-input FFL generalizations. There are two other simple topological generalizations of the FFL obtained by replicating the appropriate nodes, called multi-Y and multi-output FFLs (Figure 5.6b and c).

In principle, since the FFL is a network motif, any of these three patterns could also be network motifs; that is, any of them could occur significantly more often than in randomized networks. However, *only one of these generalizations is actually a motif* in transcription networks. The chosen generalization is the **multi-output FFL**.

We can now ask why. What might be the function of the multi-output FFL? To address this question, we will consider a well-characterized case of the multi-output FFL and see that it can generate a FIFO temporal program, in contrast to the LIFO order generated by SIMs.

FIGURE 5.5 The 199 four-node connected subgraphs.

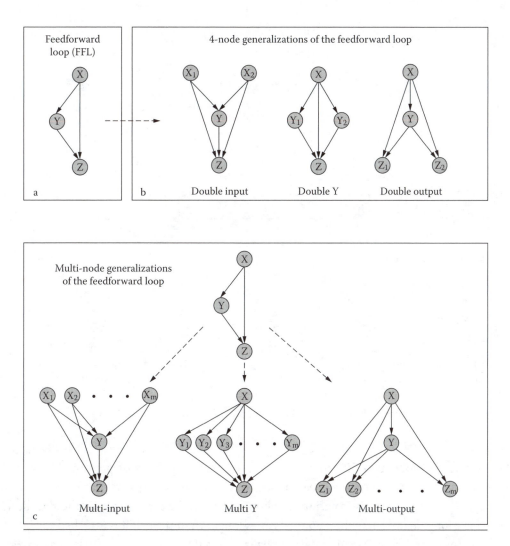

FIGURE 5.6 Simple topological generalizations of the FFL. Each topological generalization corresponds to a duplication of one of the nodes of the FFL and all of its edges. (a) The FFL. (b) Generalizations based on duplicating one node. (c) Multi-node generalizations. (From Kashtan et al., 2004.)

5.5   THE MULTI-OUTPUT FFL CAN GENERATE FIFO TEMPORAL ORDER

To study the multi-output FFL, we shall examine this network motif in the gene system that controls the production of flagella, *E. coli*'s outboard motors. When *E. coli* is in a comfortable environment with abundant nutrients, it divides happily and does not try to move. When conditions become worse, *E. coli* makes a decision to grow several nanometer-size motors attached to helical flagella (propellers), which allow it to swim. Its also generates a navigation system that tells it where to go in search of a better life. Chapter 7 will tell in detail about this navigation (chemotaxis) system. We will now consider the genes that make the parts of the flagella motor.

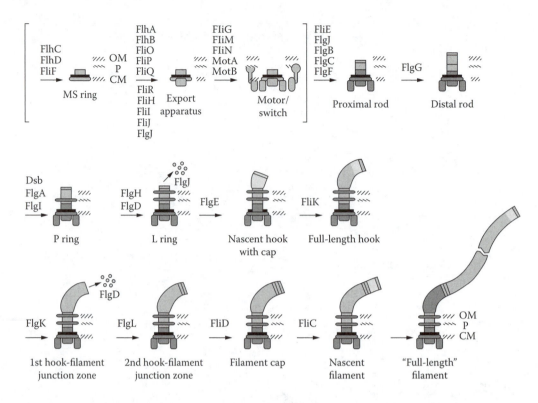

FIGURE 5.7 The flagellar motor of *E. coli* and its assembly steps. (From Macnab, 2003.)

The flagella motor is a 50-nm device built of about 30 types of protein (Figure 5.7; see also Figure 7.3). The motor is electrical, converting the energy of protons moving in through the motor to drive rotation[1] at about 100 Hz. The motor rotates the flagellum, which is a long helical filament, about 10 times longer than the cell it is attached to (*E. coli* is about 1 micron long). Flagella rotation pushes the cell forward at speeds that can exceed 30 microns/sec.

The motor is put together in stages (Figure 5.7). This is an amazing example of biological self-assembly, like throwing Lego blocks in the air and having them assemble into a house. The motor and flagellum have a hollow central tube through which the proteins move to assemble each stage. Thus, each stage of the motor acts as a transport device for the proteins in the next stage.

We will focus on the transcription network that controls the production of the motor proteins. The proteins that build up the flagella motor are encoded by genes arranged in six operons (an operon is a group of genes transcribed on the same piece of mRNA). The flagella motor operons are regulated by two transcription factors, both activators, X and Y. The master regulator X activates Y, and both jointly activate each of the six operons, Z_1, Z_2, ..., Z_6. This regulatory pattern is a **multi-output FFL** (Figure 5.8).

---

[1] Cells maintain a proton gradient across their membrane by continually pumping out protons at the expenditure of ATP. Thus, the motors are effectively powered by this "proton-motive force."

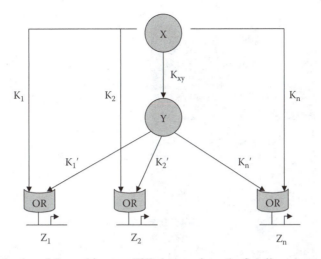

FIGURE 5.8 Schematic plan of the multi-output FFL that regulates the flagella motor genes. Shown are the logic input functions at each promoter and the activation thresholds. X = flhDC, Y = fliA, Z_1 = fliL, Z_2 = fliE, etc.

In this multi-output FFL, each operon can be activated by X in the absence of Y, and by Y in the absence of X. Thus, the input functions are similar to OR gates.[1]

Experiments using high-resolution gene expression measurements found that the six flagella operons show a defined temporal order of expression (Kalir et al., 2001; Kalir and Alon, 2004) (Figure 5.9) (See color insert following page 112). When the bacteria sense the proper conditions, they activate production of protein X.[2] The concentration of X gradually increases, and as a result, the Z genes get turned ON one by one, with about 0.1 cell generations between them. The order in which the operons are turned on is about the same as the order in which the proteins they encode participate in the assembly of the motor: first a ring in the inner membrane, then a rod, a second ring, etc. This is the principle of just-when-needed production that we discussed in the single-input module (SIM) network motif: the temporal order matches the functional order of the gene products.

The SIM architecture, however, has a limitation, as mentioned before: the turn-OFF order is *reversed* with respect to the turn-OFF order (last-in-first-out, or LIFO order) (Figure 5.3). In contrast to the SIM, the flagella turn-OFF order is the *same* as the turn-ON order: the first promoter turned on is also the first turned off when flagella are no longer needed. In other words, the genes show a first-in-first-out (FIFO) order.

How can FIFO order be generated by the multi-output FFL? The mechanism is easy to understand (Figure 5.10). Recall that in the flagella system, X and Y effectively function in OR gate logic at the Z promoters. Thus, X alone is sufficient to turn the genes

---

[1] While we use a logical OR gate for clarity, the input functions in this system are actually additive (SUM input functions) (Kalir and Alon, 2004). This does not change the conclusions of the present discussion.

[2] The promoter of the gene that encodes X is controlled by multiple transcription factors that respond to signals such as glucose starvation, osmotic pressure, temperature, and cell density. Cell density is sensed via a quorum-sensing signaling pathway. In quorum-sensing, a small signal molecule is secreted and also sensed by the cells. When the culture is dense enough, the signal molecule exceeds a threshold concentration and activates the quorum-sensing signal transduction pathway in the cells.

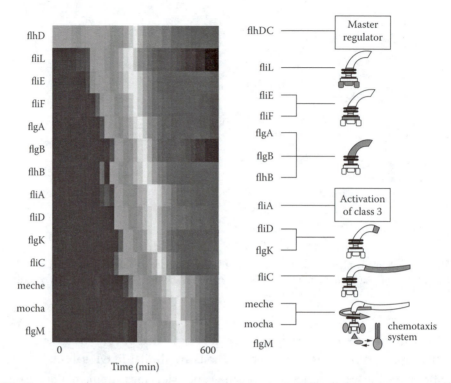

FIGURE 5.9 (See color insert following page 112) Temporal order in the flagella system of *E. coli*. Colored bars are the normalized expression of each promoter, where blue is low and red is high expression. Activity of each promoter was measured by means of a green fluorescent (GFP) reporter. The flagellum is shown schematically on the right. The position of the gene products within the flagella are shown in shaded tones. The temporal order matches the assembly order of the flagella, in which proteins are added going from the intracellular to the extracellular sides. (From Kalir et al., 2001.)

on. Therefore, the turn-ON order is determined by the times when X crosses the activation thresholds for the promoters of Z_1, Z_2, etc. These thresholds are K_1, K_2, ..., K_n. The promoter with the lowest K is turned on first, and the one with the highest K is turned on last (Figure 5.10). If this were all, genes would be turned off in the reverse order once X levels decline, resulting in LIFO order, just like in the SIM. But here Y comes to the rescue. When production of X stops so that protein X decays away, Y is still around. Its production only stops when X decays below the threshold for activation of the Y promoter, K_{xy}. Thereafter, the levels of protein Y gradually decay by degradation/dilution. Since Y is still present after X levels have decayed, the turn-OFF order (in a properly designed FFL) is governed by Y, which has its own thresholds for each of the genes, K_1', K_2', ..., K_n'. A FIFO order is achieved if the order of the thresholds of Y is *reversed* compared to that of X. That is, if the X thresholds are such that $K_1 < K_2$, so that promoter 1 is turned on before 2, the Y thresholds are arranged so that $K_1' > K_2'$, so that promoter 1 is turned off before 2 (Figure 5.10). This is the design that is experimentally found in the flagella system (Kalir and Alon, 2004). The temporal order in this system was shown to change by mutations that affected these activation thresholds.

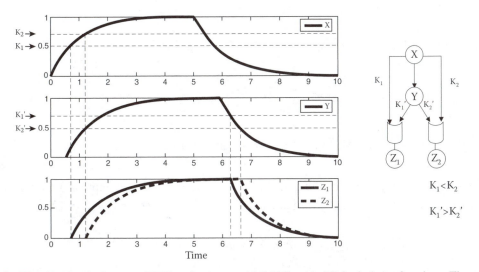

FIGURE 5.10 First-in-first-out (FIFO) order in a multi-Z FFL with OR logic input functions. The output genes Z_1 and Z_2 are turned on when X crosses activation thresholds K_1 and K_2 (dashed lines). The genes are turned off when Y decays below activation thresholds K_1' and K_2'. When the order of K_1 and K_2 is opposite of that of K_1' and $K_2',$ FIFO order is obtained.

5.5.1 The Multi-Output FFL Can Also Act as a Persistence Detector for Each Output

In addition to generating a FIFO temporal order, the multi-output FFL also conveys all of the functions of the feed-forward loop that we discussed in Chapter 4. In particular, each of the output nodes benefits from the sign-sensitive filter property of the FFL. For example, in the flagella system, the FFL functions as a device that delays the deactivation of the Z genes following the loss of X activity (as described in Section 4.6.6). In other words, this FFL filters away brief OFF pulses of X, allowing deactivation only when X activity is gone for a persistent length of time.

The OR gate FFL can therefore provide an uninterrupted input source. It allows expression even if the activity of X is briefly lost. As mentioned in Chapter 4, the delay afforded by Y in the flagella system was found to be on the order of a cell generation, which is similar to the time needed to complete the assembly of a flagellar motor (Kalir et al., 2005). Telologically speaking, such a protection function could be useful for flagella production, because the swimming cell is likely to encounter fluctuating environments. This can cause the production of X to fluctuate over time, because multiple environmental factors regulate the promoter of the gene that encodes X. The OR FFL can ensure that flagella production stops only when the appropriate conditions have been sensed for a persistent length of time. A transient deactivation of the X promoter would not be sufficient to turn off flagella production.

To summarize, the *multi-output* FFL is the generalization of the FFL that occurs most frequently in sensory transcription networks. We will see in the next chapter that other biological networks can display different FFL generalizations. The multi-output FFL confers to each of the output genes all of the dynamic functions of the FFL that we discussed

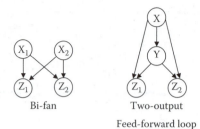

FIGURE 5.11 The four-node network motifs in sensory transcription networks.

in Chapter 4. In addition, it can generate a FIFO temporal order of expression, by means of opposing hierarchies of activation coefficients for the two regulators X and Y.

5.6 SIGNAL INTEGRATION AND COMBINATORIAL CONTROL: BI-FANS AND DENSE OVERLAPPING REGULONS

We now complete our survey of network motifs in sensory transcription networks. We have so far considered three network motif families: autoregulation, FFLs (with single or multiple outputs), and SIMs. Our fourth and final network motif family stems from the analysis of four-node patterns.

We have mentioned that there are 199 possible four-node patterns with directed edges (Figure 5.5). Of these, *only two* are significant motifs in the known sensory transcription networks. Again, the networks appear to show a striking simplicity because they contain only a tiny fraction of the possible types of subgraphs. The significant pair of network motifs is the two-output FFL, which belongs to the family of multi-output FFLs we have just examined, and an overlapping regulation pattern termed the **bi-fan** (Figure 5.11)[1]. In the bi-fan, two input transcription factors, X_1 and X_2, jointly regulate two output genes, Z_1 and Z_2.

The bi-fan gives rise to a family of motif generalizations, shaped as a layer of inputs with multiple overlapping connections to a layer of outputs (Figure 5.12). This family of patterns is our last network motif, called **dense overlapping regulons** (a regulon is the set of genes regulated by a given transcription factor), or **DORs** for short[2] (Figure 5.13) (Shen-Orr et al., 2002).

The DOR is a row of input transcription factors that regulate a set of output genes in a densely overlapping way. The DORs are usually not fully wired; that is, not every input regulates every output. However, the wiring is much denser than in the patterns found in randomized networks. To understand the function of the DOR requires knowledge of the multi-dimensional input function that integrates the inputs at the promoter of each gene, described in Section 2.3.5. The DOR can be thought of as a combinatorial decision-

---

[1] Several four-node patterns are also found that correspond to FFLs with a single dangling edge going into or out of one of the FFL nodes. These patterns are also network motifs in the sense that they occur more often than in randomized networks. However, they do not appear to have additional special functionality beyond what we have already described.

[2] The DOR can be detected by algorithms that search for dense layers of connected nodes (Shen-Orr et al., 2002; Sprin and Mirny, 2003).

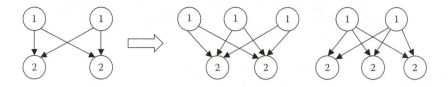

FIGURE 5.12 The bi-fan generalizes to larger patterns with a row of inputs and a row of outputs.

making device. It functions as an array of gates (input functions) that integrate multiple inputs to compute the regulation of each output gene.

Transcription networks, such as those of *E. coli* and yeast, show several large DORs, each controlling tens to hundreds of genes. The genes in each DOR have a shared global function, such as stress response, nutrient metabolism, or biosynthesis of key classes of cellular components. Often, a global regulator governs many of the genes, supplemented by numerous regulators that regulate subsets of the genes. For example, in *E.coli*, the global regulator CRP senses glucose starvation and, together with multiple transcription factors that each sense a different sugar, determines which sugar utilization genes are activated in response to the sugars in the environment. The DORs form the backbone of the network's global structure, as we will see next.

5.7 NETWORK MOTIFS AND THE GLOBAL STRUCTURE OF SENSORY TRANSCRIPTION NETWORKS

We have described the four main network motif families in sensory transcription networks: autoregulation, feed-forward loops, SIMs, and DORs. How are these motifs positioned with respect to each other in the network? How much of the network do they cover?

To answer these questions, one needs to view an image of the network. It is difficult to draw a complex network in an understandable way (Figure 2.3). However, network motifs can help to produce a slightly less complicated image, based on the following coarse-graining procedure. To draw the network, replace each occurrence of a SIM that regulates n genes by a square marked with the number n. Replace each multi-output FFL by a triangle marked with the number of output genes. Replace each DOR with a rectangular box that groups its inputs, outputs, and connections. The result is still an intricate picture, but one that can help us to understand the global network structure (Figure 5.14) (See color insert following page 112).[1]

The coarse-grained network obtained by this procedure, such as that shown in Figure 5.14, shows that sensory transcription networks such as those of *E. coli* and yeast are made of a layer of DORs. The DORs form a single layer—they do not form cascades. That is,

---

[1] Network motifs can be used in some networks to systematically form a coarse-grained network, in which each node corresponds to an entire pattern in the original network. This leads to a simplified network description. In some cases, the coarse-graining process can be repeated, yielding several levels of simplification. One typically finds that the network motifs at each level are different: in this sense, real-world networks are self-dissimilar. For details, see Itzkovitz et al., 2005.

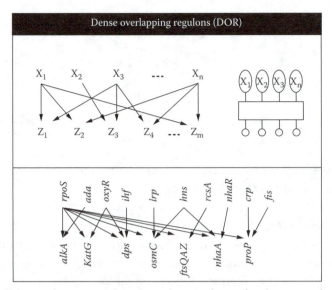

FIGURE 5.13 The dense overlapping regulons (DORs) network motif and an example in the *E. coli* stress response and stationary phase system (Baumberg, 1998). (From Shen-Orr, 2002.)

there is no DOR at the output of another DOR. Thus, most of the computation done by the network is done at a cortex of promoters within the DORs.

The layer of DORs also contains most of the other motifs. The FFLs and SIMs are integrated within the DORs. Many of the FFLs are multi-output, with the same X and Y regulating several output genes. Negative autoregulation is often integrated within FFLs and also decorates the master regulators of SIMs. Overall, the rather simple way in which the network motifs are integrated makes it possible to understand the dynamics of each motif separately, even when it is embedded within larger patterns.

Virtually all of the genes are covered by these four network motifs in the organisms studied so far, including the known part of the sensory networks of bacteria, yeast, worms, fruit flies, and humans (Harbison et al., 2004; Odom et al., 2004; Penn et al., 2004; Boyer et al., 2005). Thus, these network motifs represent the major types of patterns that occur in sensory transcription networks.

A striking feature of the global organization of sensory transcription networks is the relative absence of long cascades of transcription interactions, X → Y → Z →, etc. Most genes are regulated just one step away from their transcription factor inputs.

Why are long cascades relatively rare? One possible reason is response time constraints. We have seen in Chapter 2 that information transmission down transcription cascades is slow: protein Y needs to accumulate to cross the threshold for regulation of gene Z. This accumulation time is on the order of the lifetime of protein Y, usually on the order of many minutes to hours. Thus, long cascades would typically be far too slow for sensory transcription networks that need to respond quickly to environmental stresses and nutrients. Sensory transcription networks are "rate limited," with components that are often

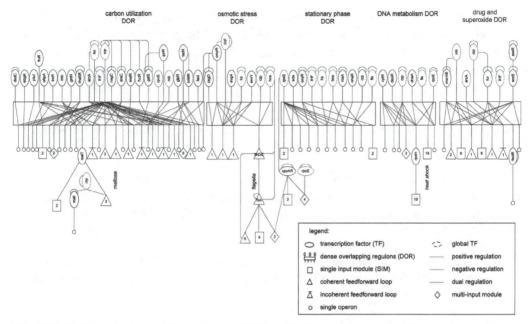

FIGURE 5.14 (See color insert following page 112) The global structure of part of the *E. coli* transcription network. Ellipses represent transcription factors that read the signals from the environment. Circles are output genes and operons. Rectangles are DORs. Triangles are outputs of single- or multi-output FFLs. Squares are outputs of SIMs. Blue and red lines correspond to activation and repression interactions. (From Shen-Orr et al., 2002.)

slower than the timescales on which the network needs to respond. Rate-limited networks tend not to employ long cascades (Rosenfeld and Alon, 2003).

Cascades are relatively rare in sensory transcription networks, but other biological networks do have long cascades, as we will see in the next chapter. Cascades are found in networks whose interactions are rapid with respect to the timescale on which the network needs to function. This includes developmental transcription networks that control slow developmental processes and signal transduction networks whose components are fast. In these non-rate-limited networks, long cascades are prominent network motifs.

Experiments on the global transcription of cells allow one to glimpse which parts of the network are active under which stimuli. Such studies in yeast suggest that SIMs and DORs tend to operate in systems that require fast signal propagation and coordinated expression, whereas FFLs occur in systems that guide progression in stages such as the cell cycle (Yu et al., 2003; Luscombe et al., 2004). Future work in this direction could help refine our understanding of motif dynamics and help to study the interactions between networks motifs.

In summary, sensory transcription networks across organisms appear to be built of four motif families: autoregulation, FFLs, SIMs, and DORs. These network motifs and their main functions are summarized in Figure 5.15. Almost all of the genes participate in these motifs. Each network motif carries out a defined dynamical function, such as speeding network response, temporal program generation, sign-sensitive filtering, or combinatorial decision making. The networks have a shallow architecture in which one-step cascades are far more common than long cascades. Most other patterns, such as three-node feedback loops, are conspicuously absent.[1] Hence, the subgraph content of these networks is much simpler than it could have been. They seem to be built of a small set of elementary circuit patterns, the network motifs.

FURTHER READING

Shen-Orr, S., Milo, R., Mangan, S., and Alon, U. (2002). Network motifs in the transcriptional regulation network of *Escherichia coli*. *Nat. Genet.*, 31: 64–68.

Single-Input Module and Temporal Order

Laub M.T., McAdams, H.H., Feldblyum, T., Fraser, C.M., and Shapiro, L. (2000). Global analysis of the genetic network controlling a bacterial cell cycle. *Science*, 290: 2144–2148.
McAdams, H.H. and Shapiro, L. (2003). A bacterial cell-cycle regulatory network operating in space and time. *Science*, 301: 1874–7.
Zaslaver, A., Mayo, A.E., Rosenberg, R., Bashkin, P., Sberro, H., Tsalyuk, M., Surette, M.G., and Alon, U. (2004). Just-in-time transcription program in metabolic pathways. *Nat. Genet.*, 36: 486–491.

Generalizations of Network Motifs

Berg, J. and Lassig, M. (2004) Local graph alignment and motif search in biological networks. *Proc. Nati. Acad. Sci. U.S.A.*, 101: 14689–94.
Kashtan, N., Itzkovitz, S., Milo, R., and Alon, U. (2004). Topological generalizations of network motifs. *Phys. Rev. E*, 70: 031909.

Multi-Output FFLs and Temporal Order

Kalir, S. and Alon, U. (2004). Using a quantitative blueprint to reprogram the dynamics of the flagella gene network. *Cell*, 117: 713–720.

Cascade Structure of Transcription Networks

Hooshangi, S., Thiberge S., and Weiss, R. (2005). Ultra-sensitivity and noise propagation in a synthetic transcriptional cascade. *Proc. Natl. Acad. Sci. U.S.A.*, 102: 35886.
Pedraza, J.M. and van Ovdenaarden, A. (2005). Noise propagation in gene networks. *Science*, 307: 1965–9.
Rosenfeld, N. and Alon, U. (2003). Response delays and the structure of transcription networks. *J. Mol. Biol.*, 329: 645–654.

---

[1] Three-node feedback loops are often unstable (a common property of negative loops) or multistable (a common property of positive loops), both undesirable properties for sensory networks that need to perform reversible and reliable responses (Prill et al., 2005). Sensory transcription networks belong to the family of directed acyclic graphs (DAGs). However, they have even fewer subgraph types than generic DAGs.

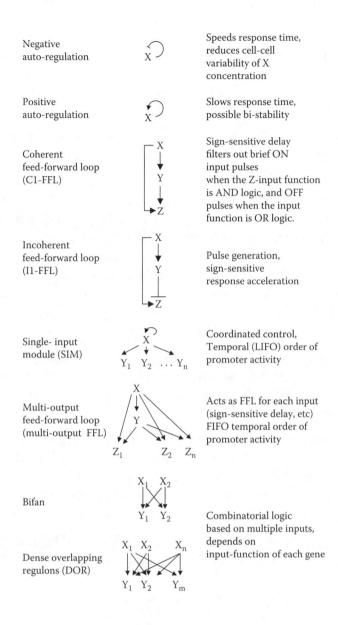

Negative auto-regulation		Speeds response time, reduces cell-cell variability of X concentration
Positive auto-regulation		Slows response time, possible bi-stability
Coherent feed-forward loop (C1-FFL)		Sign-sensitive delay filters out brief ON input pulses when the Z-input function is AND logic, and OFF pulses when the input function is OR logic.
Incoherent feed-forward loop (I1-FFL)		Pulse generation, sign-sensitive response acceleration
Single- input module (SIM)		Coordinated control, Temporal (LIFO) order of promoter activity
Multi-output feed-forward loop (multi-output FFL)		Acts as FFL for each input (sign-sensitive delay, etc) FIFO temporal order of promoter activity
Bifan		Combinatorial logic based on multiple inputs, depends on input-function of each gene
Dense overlapping regulons (DOR)		

FIGURE 5.15 Network motifs in sensory transcription networks.

EXERCISES

5.1. *Equal timing.* Consider a SIM controlled by regulator X that activates downstream genes Z_i, $i = 1, ..., n$, with thresholds K_i. At time $t = 0$, X begins to be produced at a constant rate β. Design thresholds such that the genes are turned on one after the other at equal intervals (use logic input functions).

5.2.   *Robust timing.*

  a. For the system of exercise 5.1, are there biological reasons that favor placing the thresholds K_i much smaller than the maximal level of X? Consider the case in which X is an activator that begins to be produced at time t = 0, and consider the effects of cell–cell variations in the production rate of X.

  b. Would a design in which X is a repressor whose production stops at time t = 0 provide more robust temporal programs? Explain.

Solution:

  a. Imagine that K_i is close to the maximal level of X, $X_{st} = \beta/\alpha$. Since the production rate of X, β, varies from cell to cell, some cells have higher β than others over their entire generation time (see Appendix D). Hence, there are cell–cell variations in X_{st}. In cells in which production is low, we might have $X_{st} < K_i$: in these cells X* does not cross K_i. Therefore, the downstream gene Z_i is not expressed, and the cell is at a disadvantage. Thus, designs that provide the required timing and in which K_i is much smaller than the mean X_{st} (smaller than the lowest expected X_{st} given the variability in β) have an advantage.

  Let us consider the case in which K_i are much lower than X_{st}. In this case, X crosses these thresholds at early times and we can use the approximation of linear production $X(t) \sim \beta t$ (Equation 2.4.7). Thus, the activation times of the genes, found by asking when $X(t) = K_i$, are $t_i = K_i/\beta$.

  Low thresholds thus ensure that all genes can be activated despite the noise in production rate. One can tune K_i and β to achieve the required timing. Note that the activation times can vary from cell to cell if β varies. A factor 2 change in β would lead to a factor 2 change in t_i. The relative *order* of the turn-ON events of different genes in a SIM in the same cell would, however, not change.

  b. When X production stops, it decays $X(t) = (\beta/\alpha)\,e^{-\alpha t}$. The turn-ON time of gene i is the time when the level of repressor X goes below its threshold: $X(t_i) = K_i$, so that:

$$t_i = \alpha^{-1}\log\frac{\beta}{\alpha K_i}$$

  Note that β appears only in the logarithm in this expression, so that the activation time t_i is quite robust with respect to fluctuations in production rate, more so than for an activator whose concentration increases with time (where, as we saw above, $t_i \sim K_i/\beta$).

  This increased robustness might be one reason that repressor cascades are often used in developmental transcription networks (Chapter 6) (Rappaport et al., 2005).

5.3. *The multi-output OR FFL.* In a multi-output C1-FFL with OR gate logic at the Z promoters, transcription factor X begins to be produced at a constant rate β at time $t = 0$. At time $t = T$, the production rate β suddenly becomes equal to zero. Calculate the dynamics of the downstream genes Z_i. What are the delays between genes? (Use logic input functions).

5.4. What are the topological generalizations of the diamond pattern $(X \to Y_1, X \to Y_2, Y_1 \to Z, Y_2 \to Z)$ based on duplication of a single node and all of its edges? How are these different from DORs? What are the topological generalizations of the bi-fan $(X_1 \to Y_1, X_2 \to Y_1, X_1 \to Y_2, X_2 \to Y_2)$? Most of these five-node generalizations of the diamond and bi-fan are network motifs in the neuronal network of *C. elegans*, as we will see in the next chapter.

5.5. SIM with auto-regulation: what is the effect of autoregulation on the master transcription factor X in a SIM? Plot schematically and compare the dynamics of the output genes in a given SIM with positive autoregulation of X, negative autoregulation of X, and no autoregulation of X. Explain when each design might be useful.

5.6. Bi-fan dynamics: Consider a bi-fan in which activators X_1 and X_2 regulate genes Z_1 and Z_2. The input signal of X_1, S_{x1}, appears at time $t=0$ and vanishes at time $t=D$. The input signal of X_2, S_{x2}, appears at time $t=D/2$ and vanishes at $t=2D$. Plot the dynamics of the promoter activity of Z_1 and Z_2, given that the input functions of Z_1 and Z_2 are AND and OR logic, respectively.

CHAPTER 6

Network Motifs in Developmental, Signal Transduction, and Neuronal Networks

6.1   INTRODUCTION

In the previous chapters we considered network motifs in sensory transcription networks. We saw that these networks are made of a small set of network motifs, each with a defined function. We will now ask whether network motifs appear also in other kinds of biological networks.

The sensory transcription networks we have studied are designed to rapidly respond to changes in the environment. In this chapter, we will first examine a different type of transcription network, one that governs the fates of cells as an egg develops into a multi-cellular organism, or more generally as a cell differentiates into a different type of cell. This type of network is called a **developmental transcription network**.

The main difference between sensory and developmental transcription networks is the timescale on which they need to operate and the reversibility of their action. Sensory transcription networks need to make rapid and reversible decisions on timescales that are usually shorter than a cell generation time. In contrast, developmental networks often need to make irreversible decisions on the slow timescale of one or more cell generations. We will see that these differences lead to new network motifs that appear in developmental, but not in sensory, networks.

In addition to transcription networks, the cell uses several other networks of interactions, such as protein–protein interaction networks, signal transduction networks, and metabolic networks. Each network corresponds to a different mode of interaction between biomolecules. Thus, one can think of the cell as made of several superimposed networks, with different colors of edges. Transcription networks correspond to one color of edges, protein–protein interactions to a different color, etc.

There is a separation of timescales between these different network layers. Transcription networks are among the slowest of these networks. As we have discussed, they often show dynamics on the scale of hours. Other networks in the cell function on much faster timescales. For example, **signal transduction networks**, which process information using interactions between signaling proteins, can function on the timescale of seconds to minutes.

We will describe some of the network motifs that appear in signal transduction networks, such as structures called multi-layer perceptrons. Since we currently lack many of the precise biochemical details, we will use toy models to understand the broad types of computations made possible by these motifs. We will also examine composite motifs made of two kinds of interactions (two different colors of edges).

In addition to these molecular networks, biological networks can also be defined on larger scales. For example, one may consider networks of interactions between cells. One important cellular network is the network of synaptic connections between neurons. We will examine the best-characterized **neuronal network**, from the worm *Caenorhabditis elegans*. This network contains several strong network motifs. Notably, neuronal wiring shares some of the same network motifs that are found in molecular interaction networks. We shall discuss the possibility that these motifs perform the same basic information processing functions in both types of networks.

6.2   NETWORK MOTIFS IN DEVELOPMENTAL TRANSCRIPTION NETWORKS

The transcription networks we have studied so far are built to sense and respond to external changes. This type of network is termed a **sensory transcription network**. Sensory transcription networks are found in almost all cells.

However, organisms also have another type of transcription network. In all multi-cellular organisms and in many microorganisms, cells undergo differentiation processes: they can change into other cell types. An important example is the development of a multicelled organism. Multicelled organisms begin life as a single celled egg, which divides to form the diverse cell types of the body. As the cells divide, they differentiate into different tissues. To become part of a new tissue, the cell needs to express a specific set of proteins. The specific set of proteins expressed by the cell determines whether it will become, say, nerve or muscle. These differentiation processes are governed by transcription networks, known as **developmental transcription networks** (Davidson, 2001; Davidson et al., 2002; Albert and Othmer, 2003; Sanchez and Thieffry, 2003; Gunsalus et al, 2005; Levine and Davidson, 2005; Stathopoulos and Levine, 2005).

Developmental transcription networks of well-studied organisms such as fruit flies, worms, sea urchins, and humans show several strong network motifs. They display most of the network motifs that we have described in sensory networks. For example, as in sensory networks, the feed-forward loop (FFL) is a strong network motif (Milo et al., 2004; Odom, 2004; Penn et al., 2004; Boyer et al., 2005). The most common FFL types in developmental networks appear to be the type-1 coherent and type-1 incoherent FFLs, just as

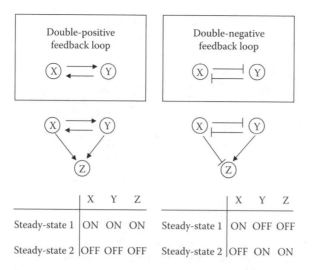

FIGURE 6.1 Positive transcriptional feedback loops with two nodes. The double-positive loop has two activation interactions, and the double-negative loop is made of two repression interactions. An output gene Z is regulated as shown. Each of the feedback loops has two steady states: both X and Y genes ON or both OFF in the double-positive loop, and one ON and the other OFF in the double-negative loop.

in sensory networks (Chapter 4, Figure 4.4). Developmental networks also display prominent autoregulation motifs and SIMs.

In addition to these motifs, developmental networks display a few additional network motifs that are not commonly found in sensory transcription networks. We will now describe these network motifs and their functions.

6.2.1 Two-Node Positive Feedback Loops for Decision Making

Developmental networks display a network motif in which two transcription factors regulate each other. This mutual regulation forms a feedback loop. In developmental networks, the regulation signs of the two interactions usually lead to *positive* feedback loops (Figure 6.1).

There are two types of positive feedback loops made of two transcription factors. Both types commonly appear in developmental networks. The first type of positive feedback loop is made of two positive interactions, so that the two transcription-factors activate each other. The second type has two negative interactions, where the two transcription factors repress each other.

The double-positive feedback loop has two stable steady states (Thomas and D'Ari, 1990): In one stable state, genes X and Y are both ON. The two transcription factors enhance each others' production. In the other stable state, X and Y are both OFF. A signal that causes protein X or Y to be produced can irreversibly lock the system into a state where both X and Y are ON and activate each other. This type of bi-stable switch is called a lock-on mechanism (Davidson et al., 2002).[1] Since X and Y are both ON or both OFF,

---

[1] Recall that positive autoregulation can also lock into a state of high expression (Section 3.5.1). Why, then, do two-node feedback loops appear if one-node loops are sufficient? One possible reason is that the double-positive feedback loop only locks on after an appreciable delay, and hence can filter out transient input signals. This is conceptually similar to the filtering function of the coherent FFL (Chapter 4).

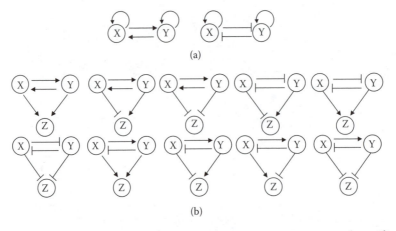

(a)

(b)

FIGURE 6.2 (a) Two-node feedback loops with autoregulation are a common network motif in developmental transcription networks. (b) The 10 distinct types of regulating feedback motifs, each corresponding to a different combination of regulation signs.

the double-positive feedback loop is most useful when genes regulated by X and genes regulated by Y encode proteins that belong to the same tissue.

The other type of positive feedback loop is made of two negative interactions. This double-negative loop also has two stable steady-states. In one stable state X is ON and Y is OFF, so that protein X represses Y expression and prevents it from being produced. The other stable state is the reverse: X is OFF and Y is ON. Thus, unlike the double positive feedback loop that can express both X and Y (or neither), the double-negative loop expresses *either* X or Y. This is useful when genes regulated by X belong to different cell fates than the genes regulated by Y.[1]

Often, the transcription factors in the two-node feedback loop also each have positive autoregulation (Figure 6.2a). Each positive autoregulation loop acts to enhance the production of the transcription factor once it is present in sufficient levels. This further stabilizes the ON steady states of the transcription factors.

The bi-stable nature of these motifs allows cells to make irreversible decisions and assume different fates in which specific sets of genes are expressed and others are silent (Demongeot et al., 2000).[2]

---

[1] A classic example of a double-negative feedback loop appears in phage lambda, a virus that infects *E. coli*. Here is a simplified description. The phage is composed of a protein container that houses a short DNA genome, which the phage can inject into the bacterium. This phage has two modes of existence. It uses two transcription factors, X and Y (called cro and C1), to control these two modes. In one mode, the lytic mode, it expresses many genes, including X, to produce about a hundred new phages, killing the bacterium. The genes expressed in the lytic mode are activated by X. In the other mode, called the lysogenic mode, the phage DNA integrates into the bacterial DNA and sits quiet: the lethal phage genes of the lytic mode are repressed by a single phage protein, Y. The two phage regulators X and Y form a double-negative feedback loop, to affect an either-or decision between these two modes of existence. Both regulators also show positive autoregulation. More details can be found in (Ptashne and Gann, 2002). For an example with mutually repressing micro-RNAs instead of transcription factors see (Johnston et al., 2005).

[2] The two-node positive feedback loop leads to bi-stability if the interactions between X and Y are strong and steep enough. Weaker interactions do not lead to bi-stability, but can lead to interesting effects, such as increasing the sensitivity of the two nodes to weak input signals.

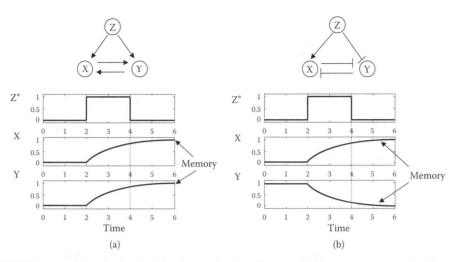

FIGURE 6.3 The regulated-feedback network motif in developmental transcription networks. (a) Double-positive feedback loop. When Z is activated, X and Y begin to be produced. They can remain locked ON even when Z is deactivated (at times after the vertical dashed line). (b) Double-negative feedback loop. Here Z acts to switch the steady states. Initially Y is high and represses X. After Z is activated, X is produced and Y is repressed. This state can persist even after Z is deactivated. Thus, in both (a) and (b), the feedback loop effectively stores a memory.

6.2.2 Regulating Feedback and Regulated Feedback

Two-node feedback loops can appear within larger motifs in developmental networks. There are two main three-node motifs that contain feedback loops (Milo et al., 2004). The first is a triangle pattern in which the mutually regulating nodes X and Y both regulate gene Z (Figure 6.2b), called regulating feedback.

The regulating-feedback network motif has 10 possible sign combinations (combinations of positive and negative interactions on its four edges; Figure 6.2b). In the simplest case, X and Y, which activate each other in a double-positive loop, have the same regulation sign on the target gene Z (both positive or both negative). In contrast, a double-negative feedback loop will often have opposing regulation signs for Z (Figure 6.1). The two sign combinations shown in Figure 6.1 are coherent, in the sense that any two paths between two nodes have the same overall sign.

In addition to the regulating-feedback motif, developmental networks show a network motif in which a two-node feedback loop is regulated by an upstream transcription factor (Figure 6.3). This motif is called regulated feedback. Again, several coherent sign combinations are commonly found. For example, the input transcription factor can be an activating regulator that locks the system ON in the case of a double-positive loop (Davidson et al., 2002). In the case of a double-negative loop, the regulator can have different signs for the two feedback nodes and act to switch the system from one steady state to the other (Gardner et al., 2000).

The regulated feedback motif can be considered as a memory element: the regulator Z can switch the feedback loop from one state to another, such that the state persists even after Z is deactivated (Figure 6.3). Hence, the circuit can remember whether Z was active

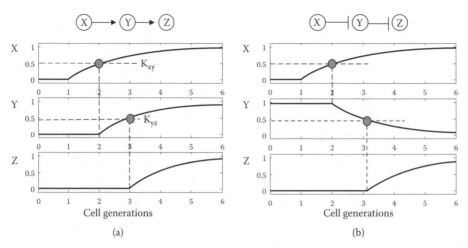

FIGURE 6.4 Transcription cascades can generate delays on the order of the cell generation time (in the case of stable proteins). Each step in the cascade activates or represses the next step when it crosses its threshold (dashed lines). Shown are a cascade of activators and a cascade of repressors.

in the past. This memory is a well-known feature of positive feedback loops (Demongeot et al., 2000; Smolen et al., 2000; Xiong and Ferrell, 2003). It can help cells to maintain their fate even after the original developmental signals that determined the fate have vanished.

6.2.3 Long Transcription Cascades and Developmental Timing

An additional important family of network motifs in developmental networks that is rare in sensory networks is long transcriptional cascades. Transcriptional cascades are chains of interactions in which transcription factor X regulates Y, which in turn regulates Z, and so on (Figure 6.4).

As we have seen in Chapter 2, the response time of each stage in the cascade is governed by the degradation/dilution rate of the protein at that stage of the cascade: $T_{1/2} = \log(2)/\alpha$. Recall that for stable proteins, this response time is on the order of a cell generation time (Section 2.4.1). Interestingly, developmental networks work on precisely this timescale, the scale of one or a few cell generations. This is because cell fates are assigned with each cell division (or several divisions) as the cells divide to form the tissues of the embryo. Hence, the timescale of transcription cascades is well suited to guide developmental processes. Development often employs cascades of repressors (Figure 6.4b), whose timing properties may be more robust with respect to fluctuations in protein production rates than cascades of activators (Rappaport et al., 2005) (see Exercises 5.2 and 8.4).

6.2.4 Interlocked Feed-Forward Loops in the *B. subtilis* Sporulation Network

The feed-forward loop (FFL) is another strong network motif in developmental networks. In developmental networks, the FFLs often form parts of larger and more complex circuits than in sensory transcription networks. Can we still understand the dynamics of such large circuits based on the behavior of the individual FFLs?

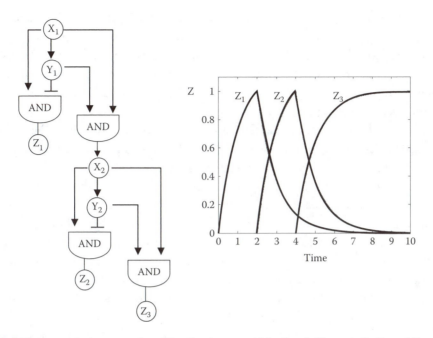

FIGURE 6.5 The transcription network guiding development of the *B. subtilis* spore. Z_1, Z_2, and Z_3 represent groups of tens to hundreds of genes. This network is made of two type-1 incoherent FFLs, which generate pulses of Z_1 and Z_2, and two type-1 coherent FFLs, one of which generates a delayed step of Z_3. (Based on Eichenberger et al., 2004.)

To address this question, we will discuss a well-mapped developmental network made of interlocking FFLs that governs differentiation in a single-celled organism, the bacterium *Bacillus subtilis*.

When starved, *B. subtilis* cells stop dividing and differentiate into durable spores. The spore contains many proteins that are not found in the growing bacterium. It is a resting cell, almost completely dehydrated. It can survive for a long time in a dormant state. When placed in the right conditions, the spore converts itself again into a normal bacterium.

When *B. subtilis* makes a spore, it must switch from making one subset of proteins to making another subset. This process, termed sporulation, involves hundreds of genes. These genes are turned ON and OFF in a series of temporal waves, each carrying out specific stages in the formation of the spore. The network that regulates sporulation (Eichenberger et al., 2004) is made of several transcription factors arranged in linked coherent and incoherent type-1 FFLs (Figure 6.5).

To initiate the sporulation process, a starvation signal S_x activates X_1. This transcription factor acts in an incoherent type-1 FFL (I1-FFL) to control a set of genes Z_1. In this I1-FFL, X_1 directly activates Z_1 and also activates Y_1, which represses Z_1. The I1-FFL generates a pulse of Z_1 expression (as described in Section 4.7). A second FFL is formed by Y_1 and X_1, which are both needed to activate X_2, resulting in a coherent type-1 FFL (C1-FFL) with AND logic. The C1-FFL ensures that X_2 is not activated unless the S_x signal is persistent (as was discussed in Section 4.6). Next, X_2 acts in an I1-FFL, where it activates genes Z_2 as well as their repressor Y_2. This results in a pulse of Z_2 genes, timed at a delay relative to the first pulse. Finally, Y_2 and X_2 together join in an AND gate C1-FFL to activate genes Z_3, which are turned on last. The result is a three-wave temporal pattern: first a pulse of Z_1 expression, followed by a pulse of Z_2 expression, followed by expression of the late genes Z_3.

Hence, the FFLs in this network are combined in a way that utilizes their delay and pulse-generating features to generate a temporal program of gene expression. The cascaded FFLs generate two pulses of genes followed by a third wave of late genes (Figure 6.5). The FFLs controlling Z_1, Z_2, and Z_3 are actually multi-output FFLs because Z_1, Z_2, and Z_3 each represent large groups of genes. This design can generate finer temporal programs within each group of genes, as described in Section 5.5.

We see that the FFLs in this network are linked such that the dynamics of the network can be easily understood based on the dynamics of each FFL. A similar situation was found in sensory transcription networks, in which FFLs are combined in particular fashions, to form multi-output FFLs. The dynamics of multi-output FFLs can also be understood based on the dynamics of each of the constituent three-node FFLs. It is important to note that there are, in principle, many other ways of linking FFLs. Most combinations of linked FFLs do not lend themselves to easy interpretation (Can you draw a few of these possible configurations?). In short, the FFLs in the *B. subtilis* sporulation network and in sensory networks seem to be linked in ways that allow easy interpretation based on the dynamics of each FFL in isolation. This appears to be the case also for network motifs in many other developmental networks.

Such **understandability** of circuit patterns in terms of simpler subcircuits could not have evolved to make life easier for biologists. Understandability is a central feature of engineering, because engineers build complex systems out of simpler subsystems that are well understood. These subsystems are connected so that each subsystem retains its behavior and works reliably. It is an interesting question whether understandability might be a common feature of networks that evolve to function.

6.3   NETWORK MOTIFS IN SIGNAL TRANSDUCTION NETWORKS

We have discussed transcription networks that operate slowly, on a timescale that can be as slow as the cell's generation time. To elicit rapid responses, the cell also contains much faster information processing networks, called **signal transduction networks**.

Signal transduction networks are composed of interactions between signaling proteins. Their function is to sense information from the environment, process this information, and accordingly regulate the activity of transcription factors or other effector proteins. For example, cells in animals usually do not divide unless they are stimulated by hormone proteins called growth factors. Specific growth factors can be sensed by cells, triggering a signal transduction pathway that culminates in the activation of genes that lead to cell division.

The inputs to signal transduction networks are typically detected by **receptor** proteins (Figure 6.6). Receptors usually have one end outside of the cell's membrane and the other end within the cell's cytoplasm. Their extracellular side can detect specific molecules called **ligands**. Binding of the ligand molecules causes a conformational change in the receptor, causing its intracellular side to become active and catalyze a specific chemical

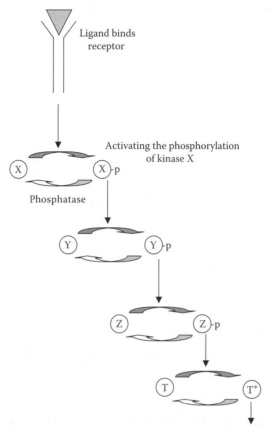

FIGURE 6.6 Protein kinase cascade: ligand binds the receptor, which leads, usually through adaptor proteins, to phosphorylation of kinase X. Kinase X is active when phosphorylated, X-p. The active kinase X-p phosphorylates its target kinase, Y. The active kinase, Y-p, in turn, phosphorylates Z. The last kinase in the cascade, Z-p, phosphorylates transcription factor T, making it active, T*. Finally, T* enters the nucleus and activates (or represses) transcription of genes. Phosphatases remove the phosphoryl groups (light arrows).

modification to a diffusible messenger protein within the cell. This modification can be thought of as passing one bit of information from the receptor to the messenger.

Once modified, the messenger protein can itself modify a second type of messenger protein, and so on. This network of modification interactions often functions on the timescale of seconds to minutes. It often culminates in modification of specific transcription factors, causing them to become active and control the expression of target genes.

The structure of signaling networks is a subject of current research, and many interactions are as yet unknown. Furthermore, as we will see in Chapters 7 to 9, the precise function of these networks can depend on subtle biochemical details. As a result, we can at present only draw tentative conclusions about the structure and function of signaling networks. However, the available data already show several intriguing network motifs. Here we will focus on a motif made of interacting signaling pathways that appears throughout diverse signal transduction networks, but does not appear in transcription networks.

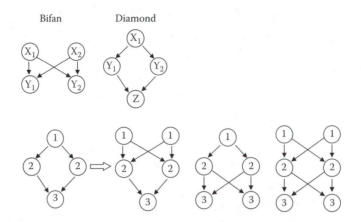

FIGURE 6.7 Network motifs in signal transduction networks. The main four-node motifs are the diamond and the bi-fan. Generalizations of the diamond are obtained by duplicating one of the four nodes and all of its edges. These generalizations are all also network motifs in signal transduction networks.

6.4   INFORMATION PROCESSING USING MULTI-LAYER PERCEPTRONS

In signal transduction networks, the nodes represent signaling proteins and the edges are directed interactions, such as covalent modification of one protein by another. Signaling networks show two strong four-node motifs, the bi-fan and the diamond (Figure 6.7) (Itzkovitz et al., 2005; Ma'ayan et al., 2005b). The bi-fan is also found in transcription networks. We saw that the bi-fan in transcription networks generalized to single-layer patterns called dense overlapping regulons, or DORs (Chapter 5, Figure 5.12 and Figure 5.13). The diamond, however, is a new network motif that is not commonly found in transcription networks.

The diamond network motif in signaling networks generalizes to form multi-layer patterns (Figure 6.7). These patterns resemble DOR-like structures arranged in cascades, with each DOR receiving inputs from an upstream DOR. In contrast, in transcription networks, as described in Chapter 5, DOR patterns do not occur in cascades; that is, a DOR is not normally found at the output of another DOR.

The multi-layer patterns in signaling networks usually show connections mainly from one layer to the next, and not, say, connections to nodes that are two layers down. Such structures are similar to patterns studied in the fields of artificial intelligence and artificial neural networks, called **multi-layer perceptrons** (Hertz et al., 1991; Bray, 1995).

6.4.1   Toy Model for Protein Kinase Perceptrons

Let us analyze the information processing capabilities of these multi-layer perceptrons. We will use a toy model of a common signal transduction module, **protein kinase cascades** (Figure 6.6) (Wiley et al., 2003; Hornberg et al., 2005; Ma'ayan et al., 2005b; Kolch et al., 2005). Protein kinase cascades are information processing pathways found in most

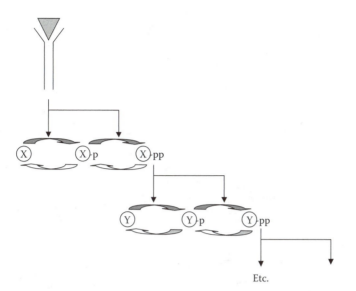

Etc.

FIGURE 6.8 Double phosphorylation in protein kinase cascades: protein kinases X, Y, and Z are usually phosphorylated on two sites and often require both phosphorylations for full activity.

eukaryotic organisms.[1] These cascades are made of kinases, proteins that catalyze the phosphorylation of specific target proteins (phosphorylation is the addition of a charged PO_4 group to a specific site on the target protein). The cascade is activated when a receptor binds a ligand and activates the first kinase in the cascade, X. Kinase X, when activated by the receptor, phosphorylates kinase Y on two specific sites (Figure 6.8). Kinase Y, when doubly phosphorylated, goes on to phosphorylate kinase Z. When kinase Z is doubly phosphorylated, it phosphorylates a transcription factor, leading to gene expression. Specific protein enzymes called phosphatases continually dephosphorylate the kinases (by removing the phosphoryl groups). Therefore, active protein kinase cascades display a cycle of phosphorylation and dephosphorylation.

Protein kinase cascades often use scaffold proteins to hold kinases in close proximity, (Levchenko et al., 2000; Park et al., 2003). Adaptor proteins can connect a given cascade to different input receptors in different cell types (Pawson and Scott, 1997). Thus, these cascades act as reusable modules (Schaeffer and Weber, 1999; Wilkins, 2001); the same cascade transduces a different signal in different tissues in an organism.

Protein kinase cascades are usually made of layers (Figure 6.9), often three layers. In the first layer, several related kinases X_1, X_2, …, can activate the next layer of kinases Y_1, Y_2. These, in turn, can activate the third layer of kinases Z_1, Z_2, …. This forms a multilayer perceptron that can integrate inputs from several receptors (Figure 6.9).

---

[1] Bacteria also use protein phosphorylation for signal transduction. However, the signal transduction networks in bacteria are usually much simpler than the protein kinase cascades of eukaryotes. Bacteria use two-component systems, with a receptor that transfers a phosphoryl group onto a diffusible response regulator protein (for example, CheA and CheY in the chemotaxis system described in Chapter 7). Sometimes several signaling proteins are linked in a phospho-relay architecture. It is interesting to ask why bacterial signaling is so much simpler than eukaryotic signaling networks.

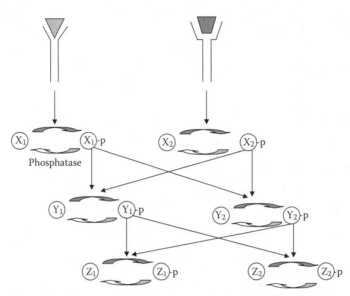

FIGURE 6.9 Multi-layer perceptrons in protein kinase cascades. Several different receptors in the same cell can activate specific top-layer kinases in response to their ligands. Each layer in the cascade often has multiple kinases, each of which can phosphorylate many of the kinases in the next layer.

To study such multi-layer perceptrons, we will write down a toy model for the dynamics of protein kinase networks. The goal is to understand the essential principles, not to develop a detailed model of the system. Hence, we will use the simplest kinetics for the kinases, **first-order kinetics** (see Appendix A.7). First-order kinetics means that the rate of phosphorylation of Y by X is proportional to the concentration of active X times the concentration of its substrate, unphosphorylated Y, denoted Y_o:

$$\text{rate of phosphorylation} = v\, X\, Y_o$$

The rate of phosphorylation is governed by the rate v of kinase X, equal to the number of phosphorylations per unit time per unit of kinase.

To begin, imagine a kinase Y that is phosphorylated by two different input kinases, X_1 and X_2 (Figure 6.10). The phosphorylated form of Y is denoted Y_p, and the unphosphorylated form is denoted Y_o. The total number of unphosphorylated and phosphorylated forms of Y is conserved[1]:

$$Y_o + Y_p = Y \tag{6.4.1}$$

The rate of change of Y_p concentration is given by the difference between its phosphorylation rate by the two input kinases and the dephosphorylation of Y_p by phosphatase at rate α:

---

[1] The total concentration of Y can change due to transcription of the Y gene. These changes are usually much slower than the timescale of the phosphorylation–dephosphorylation interactions. Hence, the total Y concentration can be considered constant over the timescale of changes in signaling protein activity. Note that often the proteins in a signaling cascade are transcriptionally regulated by signals coming through the same cascade.

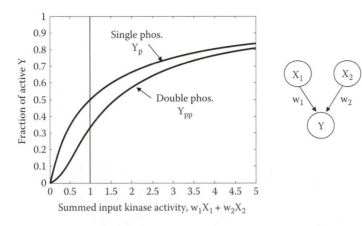

FIGURE 6.10 Fraction of active Y as a function of the weighted sum of the input kinase activities, $w_1 X_1 + w_2 X_2$, in the model with first-order kinetics. Shown is the activation curve for single and double phosphorylation of Y. The weights w_1 and w_2 are the ratios of the input kinase rates and the phosphatase rate.

$$dY_p/dt = v_1 X_1 Y_o + v_2 X_2 Y_o - \alpha Y_p \qquad (6.4.2)$$

We will study the steady-state behavior of these cascades (though dynamical functions of kinase cascades are very important; see solved exercises 6.4 and 6.5). At steady state ($dY_p/dt = 0$), these equations lead to a simple solution. The fraction of phosphorylated Y is a function of the weighted sum of the concentration of active X_1 and X_2 with weights w_1 and w_2:

$$Y_p/Y = f(w_1 X_1 + w_2 X_2) \qquad (6.4.3)$$

where the function f in Equation 6.4.3 is an increasing, saturating function (Figure 6.10):

$$f(u) = \frac{u}{1+u} \qquad (6.4.4)$$

The weight of each input corresponds to the rate of the kinase divided by the rate of the phosphatase:

$$w_1 = v_1/\alpha \qquad \text{and} \qquad w_2 = v_2/\alpha \qquad (6.4.5)$$

In other words, the concentration of phosphorylated Y is an increasing function of the weighted sum of the two input kinase activities.[1]

When Y is a kinase that needs to be phosphorylated on two sites to be active, the input function f is even steeper (Figure 6.10; solved exercise 6.2):

$$f(u) = \frac{u^2}{1+u+u^2} \qquad (6.4.6)$$

These S-shaped input functions lead to significant activation of Y only when the weighted sum of the two inputs is greater than a threshold value, which is approximately 1:

[1] Note that real signaling proteins can display more sophisticated input functions by virtue of interactions between their various protein domains (Prehoda and Lim, 2002; Ptashne and Gann, 2002; Dueber et al., 2004).

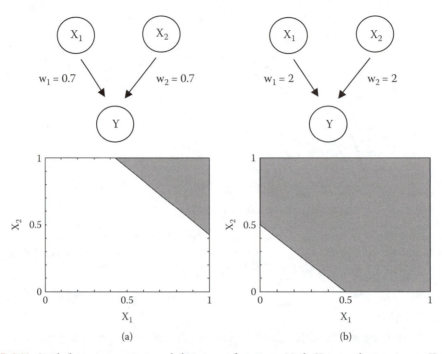

FIGURE 6.11 Single-layer perceptrons and their input functions. Node Y sums the two inputs X_1 and X_2 according to the weights w_1 and w_2. Y is activated in the shaded region of the X_1-X_2 plane. In this region, $w_1 X_1 + w_2 X_2 > 1$. Protein kinase Y is phosphorylated and hence active in the shaded region. In this figure, as well as in Figure 6.12 and Figure 6.13, the range of activities of X_1 and X_2 is between 0 and 1. An activity of 1 means that all of the input kinase molecules are active. This defines the square region in the X_1-X_2 plane portrayed in the figures.

$$w_1 X_1 + w_2 X_2 > 1 \qquad (6.4.7)$$

Hence, one can define a **threshold of activation**, $w_1 X_1 + w_2 X_2 = 1$. This threshold can be represented graphically by a straight line that divides the X_1–X_2 plane into a region of low Y_p and a region of high Y_p (Figure 6.11).

The slope and position of the threshold line that marks the boundary of the activation region depend on the weights w_1 and w_2. When the weights are relatively small, the region of high Y_p occupies a corner of the plane requiring high activity of *both* X_1 and X_2 (Figure 6.11a). This is akin to an AND gate over the two inputs. If the weights are larger, *either* X_1 or X_2 can activate Y, resulting in a larger region of high activity, similar to an OR gate (Figure 6.11b).

The precise position of the activation threshold can be tuned by changing the weights w_1 and w_2. Such changes can be made in the living cell by regulatory mechanisms that affect the rates of the input kinases X_1 and X_2 or the phosphatase rates. Changes in the weights can also be made on an evolutionary timescale by mutations that affect, for example, the chemical affinity of each of the input kinases to the phosphorylation site on the surface of protein Y.

In short, kinase Y is regulated by a threshold-like function of the weighted sum of the inputs. This generates an activation threshold that can be represented as a straight line in the plane of the two input activities. More generally, similar functions describe a layer of input kinases X_1, X_2, …, X_m, which regulates a layer of outputs Y_1, Y_2, …, Y_n. The activity of Y_j is a threshold-like function of a weighted sum of inputs $f(w_{j,1} X_1 + w_{j,2} X_2 + … + w_{j,m} X_m)$ Note that each Y_j has its own set of weights corresponding to its affinities to the input kinases. These single-layer perceptrons can compute a relatively simple function: they produce output if the summed weight of the inputs exceeds the threshold. This threshold of activity can be represented as a hyperplane in the m-dimensional space of input activities X_1, X_2, …, X_m. We will continue to use just two inputs for clarity, though the conclusions are valid for percep-trons of any size.

6.4.2   Multi-Layer Perceptrons Can Perform Detailed Computations

In our toy model, a single-layered perceptron divides the X_1–X_2 plane into two regions, separated by a straight line. This allows computations of functions similar to AND gates and OR gates. We will now see that adding additional perceptron layers can allow more intricate computations.

Consider the two-layered perceptrons in Figure 6.12. Each of the kinases in the middle layer, Y_1 and Y_2, has its own set of weights for the two inputs, X_1 and X_2. Therefore, each has its own input function in which a straight line bisects the plane into a region of low activity and a region of high activity.

The two kinases Y_1 and Y_2 can phosphorylate the output kinase Z. These kinases only phosphorylate Z when they themselves are phosphorylated (recall that protein kinases are only active when they themselves are phosphorylated). Thus, the phosphorylation of Z is a weighted sum of its two inputs, with the same function f as before (Equation 6.4.4 or 6.4.6) and with weights w_{z1} and w_{z2}:

$$Z_p/Z = f(w_{z1} Y_1 + w_{z2} Y_2) \qquad (6.4.8)$$

If both weights w_{z1} and w_{z2} are small, both Y_1 and Y_2 have to be phosphorylated to cross the activation threshold. This means that Z is phosphorylated only in the region of the X_1-X_2 plane, where *both* Y_1 and Y_2 are active. This region is defined by the intersection of the two activity regions of Y_1 and Y_2. This intersection results in activation of Z in a region of input space whose boundary is defined by two lines (Figure 6.12b). In contrast, we saw above that single-layer perceptrons allowed output regions defined more coarsely by a single straight line. Hence, the additional perceptron layer affords a somewhat more refined activation function.

Additional activation functions are shown in Figure 6.12a. Here, the activation region of Z is a union of the two activation regions of Y_1 and Y_2 (because the weights w_{z1} and w_{z2} are large enough so that either input kinase is sufficient to activate Z). Again, the activa-tion region is bounded by two linear segments.

An additional level of detail can be gained when the middle layer contains a specific phosphatase instead of a kinase. In this case, the phosphatase removes the phosphoryl

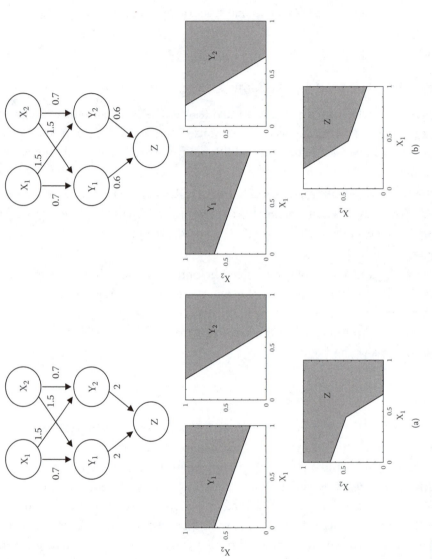

FIGURE 6.12 Two-layer perceptrons and their input functions. (a) Y_1 and Y_2 are active (phosphorylated) in the shaded regions, bounded by straight lines. The weights of Z are such that either Y_1 or Y_2 active (phosphorylated above threshold) is enough to activate Z. The activation region of Z is hence the union of the activation regions of its two inputs. (b) A configuration in which the weights for Z activation by Y_1 and Y_2 are such that *both* Y_1 and Y_2 need to be active. Hence, the activation region of Z is the intersection of the activation regions of Y_1 and Y_2.

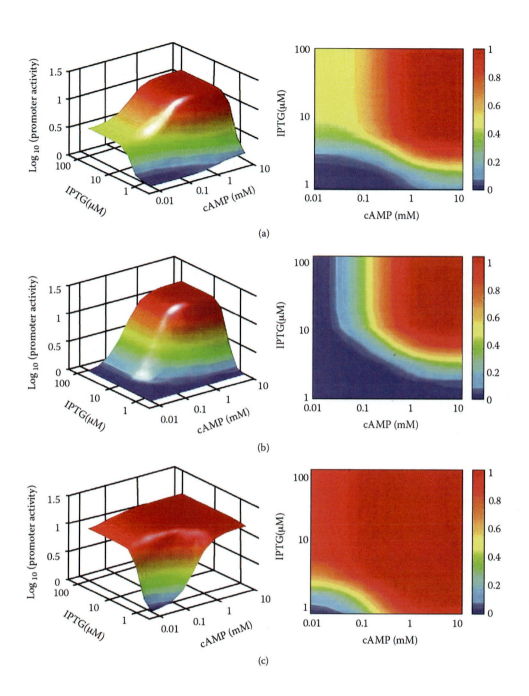

FIGURE 2.5 Two-dimensional input functions. (a) Input function measured in the lac promoter of *E. coli*, as a function of two inducers cAMP and IPTG. (b) An AND-like input function, which shows high promoter activity only if both inputs are present. (c) An OR-like input function that shows high promoter activity if either input is present. (From Setty et al., 2003.)

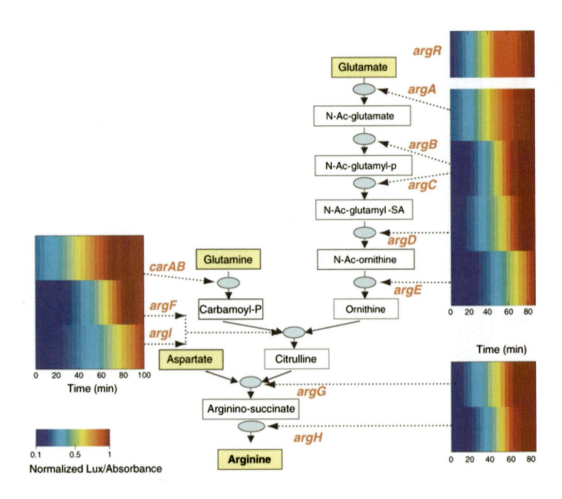

FIGURE 5.4 Temporal order in the arginine biosynthesis system. The promoters are activated in a defined order with delays of minutes between promoters. Color bars show expression from the promoters of the different operons in the system, measured by means of a luminescent reporter gene. The position of each gene product in the pathways that produce arginine is shown. (From Zaslaver et al., 2004.)

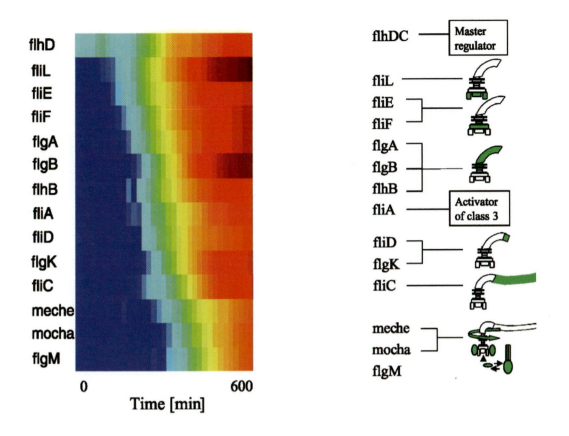

FIGURE 5.9 Temporal order in the flagella system of *E. coli*. Colored bars are the normalized expression of each promoter, where blue is low and red is high expression. Activity of each promoter was measured by means of a green fluorescent (GFP) reporter. The temporal order matches the assembly order of the flagella, in which proteins are added going from the intracellular to the extracellular sides. (From Kalir et al., 2001.)

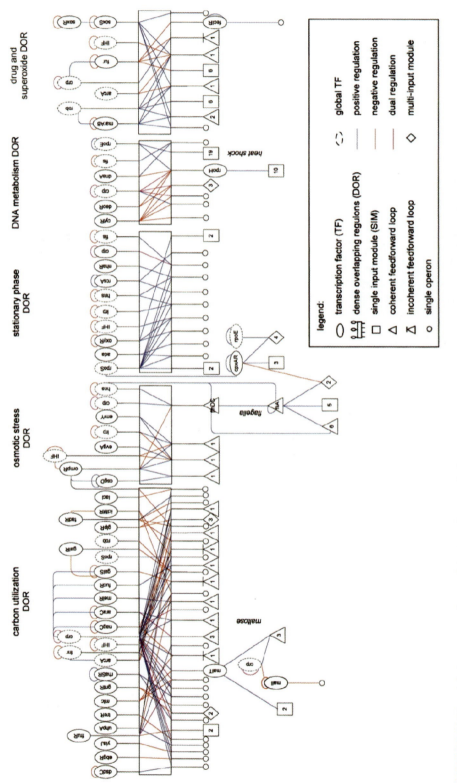

FIGURE 5.14 The global structure of part of the *E. coli* transcription network. Ellipses represent transcription factors that read the signals from the environment. Circles are output genes and operons. Rectangles are DORs. Triangles are outputs of single- or multi-output FFLs. Squares are outputs of SIMs. Blue and red lines correspond to activation and repression interactions. (From Shen-Orr et al., 2002.)

modification and effectively has a negative weight.[1] A negative weight can lead to more intricately shaped activation regions for Z. An example is shown in Figure 6.13a, in which Z is activated when the inputs X_1 and X_2 are such that node Y_1 is activated but node Y_2 is not. This design leads to a wedge-like region of Z activation. Other weights can lead to an activation of Z when either X_1 and X_2 are active, *but not both* (similar to an exclusive-or or XOR gate; Figure 6.13b). These types of computations are not possible with a single-layer perceptron (Hertz et al., 1991).

Adding additional layers can produce even more detailed functions, in which the output activation region is formed by the intersections of many different regions defined by the different weights of the perceptron.

In summary, multi-layer perceptrons can perform more detailed computations than single-layer perceptrons. Our toy model considered each kinase as a very simple unit that sums inputs and activates downstream targets when the sum exceeds a threshold. Multi-layer perceptrons allow even such simple units to perform arbitrarily complex computations, based on the power of combinatorial layered information processing (Bray, 1995).

Multi-layer perceptrons similar to the ones we have discussed have been studied in the context of artificial intelligence. They were found to display several properties that may be useful for signal transduction in cells (Bray, 1995). I will now describe three of these properties very briefly; more details can be found in texts such as Hertz et al., (1991). Multi-layer perceptrons can show discrimination, generalization, and graceful degradation. **Discrimination** is the ability to accurately recognize certain stimuli patterns. A multi-layer perceptron can be designed with weights and connections so that it can tell the difference between a set of very similar stimuli patterns. **Generalization** is the ability to "fill in the gaps" in partial stimuli patterns. A multi-layer perceptron can be designed with weights and connections so that it can respond differently to a set of different stimuli patterns. If presented with a stimulus that only partially resembles one of the original stimuli, the circuit will act as if it saw the entire input pattern. Finally, **graceful degradation** refers to the fact that damage to elements of the perceptron or its connections does not bring the network to a crashing halt. Instead, the performance of the network deteriorates, at a level proportional to the amount of damage. Indeed, mutations in a signal transduction component sometimes have only small effects on cell responses, particularly in complex organisms.[2] These three phenomena, generalization, discrimination, and

---

[1] In the toy model, a phosphatase X_2 and a kinase X_1 lead to the following equation: $dY_p/dt = v_1 X_1 (Y - Y_p) - v_2 X_2 Y_p - \alpha Y_p$. The solution is an input function: $Y_p/Y = v_1 X_1/(v_1 X_1 + v_2 X_2 + \alpha) = w_1 X_1/(w_1 X_1 + w_2 X_2 + 1)$. Thus, the higher w_2, the smaller Y_p. A negative weight is merely used as an approximation to capture the essentials of the slightly more complicated input function for phosphatases. An analogous situation occurs if a kinase phosphorylates a target kinase on a site that reduces the activity of the target kinase, instead of increasing the activity.

[2] Complex organisms encounter mutations within their lifetimes. It seems that mechanisms have evolved to allow somatic cells to function despite mutations (or at least to commit programmed cell death if they malfunction, and thus protect the organism). One such mechanism may be distributed computation by multi-layer perceptrons, which allows for graceful degradation of performance upon loss of components. Another common mechanism, sometimes called "redundancy," is based on feedback loops, which sense the loss of a component and upregulate a different component that can partially complement the function of the lost component (Kafri et al., 2005). Note that in contrast to multicelled organisms, bacteria have no somatic cells that need to survive despite mutations. Indeed, bacteria are fragile to mutations, in the sense that most mutations that inactivate a gene lead to a loss of a specific function (e.g., inability to grow on lactose upon mutation of the *lacZ* gene) or to cell death. Bacteria appear to have very few backup systems or redundant pathways.

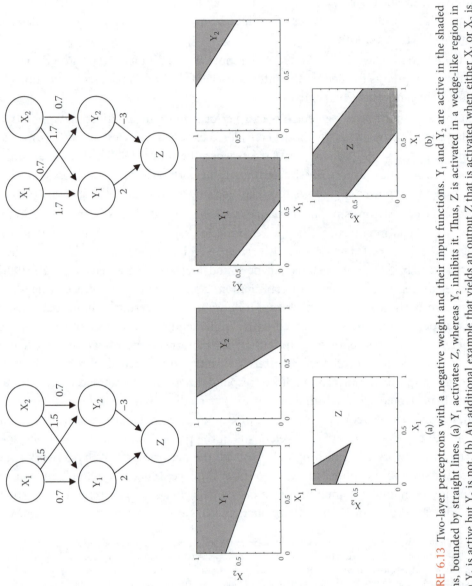

FIGURE 6.13 Two-layer perceptrons with a negative weight and their input functions. Y_1 and Y_2 are active in the shaded regions, bounded by straight lines. (a) Y_1 activates Z, whereas Y_2 inhibits it. Thus, Z is activated in a wedge-like region in which Y_1 is active but Y_2 is not. (b) An additional example that yields an output Z that is activated when either X_1 or X_2 is highly active, but not both (similar to an exclusive-OR gate).

graceful degradation, might characterize the functioning of signal transduction networks in cells.

In summary, multi-layer perceptrons allow even relatively simple units to perform detailed computations in response to multiple inputs. The deeper one goes into the layers of the perceptron, the more intricate the computation can become.

6.5 COMPOSITE NETWORK MOTIFS: NEGATIVE FEEDBACK AND OSCILLATOR MOTIFS

We have so far discussed protein signaling networks and transcription networks separately. In the cell, both networks operate in an integrated fashion. For example, the output of signal transduction pathways is often a transcription factor.

Because transcription and protein interaction networks function together, they can be described as a joint network with two colors of edges: one color represents transcription interactions, and a second color represents protein–protein interactions (Yeger-Lotem et al., 2004). In this section, we will briefly mention some of the network motifs that occur in such two-color networks. Network motifs can also be found in networks that integrate more than two levels of interactions (Zhang et al., 2005; Ptacek et al., 2005).

An example of a two-color motif is the following three-node pattern: a transcription factor X transcriptionally regulates two genes Y and Z whose protein products directly interact, for example Y phosphorylating Z. (Figure 6.14a). This reflects the fact that transcription factors often coregulate proteins that function together.

A very common composite motif is a feedback loop made of two proteins that interact with each other using two colors of edges (Figure 6.14b).[1] In this motif, protein X is a transcription factor that activates the transcription of gene Y. The protein product Y interacts with X on the protein level (not transcriptionally), often in a negative fashion. This negative regulation can take several forms. In some cases, Y enhances the rate of degradation of protein X. In other cases, Y binds X and inhibits its activity as a transcription factor by preventing its access to the DNA. This type of feedback occurs in most known gene system from bacteria to humans (Lahav et al., 2004).

This two-protein negative feedback motif is a hybrid of two types of interactions. One interaction is transcriptional, in which X activates Y on a slow timescale. The other interaction occurs on the protein level, in which Y inhibits X on a rapid timescale. It should be noted that purely transcriptional negative feedback loops are relatively rare (developmental transcription networks usually display positive transcriptional feedback loops, as discussed above). In other words, it is rare for Y to repress X on the transcription level. What could be the reason that composite negative feedbacks are much more common than purely transcriptional ones?

---

[1] A composite feedback loop, called feedback inhibition, has long been known in metabolic networks (Fell, 2003). Here, the product P of a metabolic pathway inhibits the first enzyme in the pathway E_1. The fast arm is diffusion of P and its binding to E_1, on a subsecond timescale. The slow arm is the flux changes in the pathway, such as the time needed for an enzyme to produce its product at a level equal to the K_m of the next enzyme. These changes are typically on the timescale of minutes (Zaslaver et al., 2004). Feedback inhibition has been demonstrated to enhance the stability of metabolic pathways (Savageau, 1976).

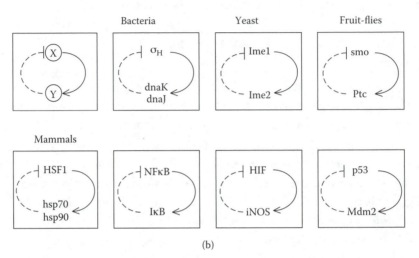

FIGURE 6.14 Composite network motifs made of transcription interactions (full arrows) and protein–protein interactions (dashed arrows). (a) Transcription factor X regulates two genes, Y and Z, whose protein products interact. (b) A negative feedback loop with one transcription arm and one protein arm is a common network motif across organisms.

To understand composite feedback, we can turn to engineering control theory. Feedback in which a slow component is regulated by a fast one is commonly used in engineering. A principal use of this type of feedback is to stabilize a system. For example, a heater that takes 15 min to heat a room is controlled in a negative feedback loop by a much faster thermostat (Figure 6.15). The thermostat compares the desired temperature to the actual temperature and adjusts the power accordingly: if the temperature is too high, the power of the heater is reduced so that the room cools down. After some time, the temperature stabilizes around the desired temperature (Figure 6.16).

One reason for using two timescales (fast thermostat on a slow heater) in this feedback loop is enhanced stability. The rapid response time of the thermostat ensures that the control of the heater is based on the current temperature. Had the thermostat been made of a vat of mercury that takes 15 min to respond to temperature changes, the heater would receive feedback based on the relatively distant past, and temperature would oscillate. In analogy, a negative feedback loop made of two slow transcription interactions is more prone to instability than a feedback loop where a fast interaction controls a slow one. For an excellent introduction to the mathematical treatment of stability, see Strogatz (2001b).

Whereas stability around a fixed state (also called homeostasis) is desirable in many biological systems, other biological systems display a strikingly different behavior — oscillatory dynamics (Winfree, 2001; Goldbeter, 2002; Murray, 2004). The cell cycle, in which cells periodically duplicate their genomes and divide, is an important oscillator (Tyson

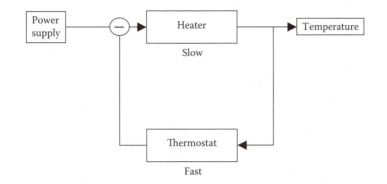

FIGURE 6.15 Negative feedback in engineering uses fast control on slow devices. A heater is controlled by a thermostat that compares the desired temperature to the actual temperature. If the temperature is too high, the power to the heater is reduced, and power is increased if temperature is too low. The thermostat works on a much faster timescale than the heater. Engineers usually tune the feedback parameters to obtain rapid and stable temperature control (and avoid prolonged oscillations in room temperature).

et al., 2002). Another well-known example is the circadian clock, a remarkably accurate biochemical circuit that produces oscillations on the scale of one day, and which can be entrained to follow daily variations in signals such as temperature and light. Other oscillators occur in regulatory systems, such as transcription factors whose concentration or activity oscillates in response to specific signals (Hoffman et al., 2002; Monk, 2003; Lahav et al., 2004; Nelson et al., 2004). Oscillators are also found in beating heart cells, spiking neurons, and developmental processes that generate repeating modular tissues (Pourquie and Goldbeter, 2003).

Biological oscillations have a typical character: their timing is usually significantly more precise than their amplitude (Lahav et al., 2004; Mihalcescu et al., 2004). Different pulses in the oscillation occur at rather accurate time intervals, but with varying amplitudes. The source of the variation in the amplitude appears to be slowly varying internal noise in protein production rates that is inherent in biochemical circuitry (Appendix D).

Many biological oscillators appear to be implemented by a two-color network motif (usually embedded in numerous other interactions). This motif is a composite negative feedback loop, in which the transcription factor X also displays positive autoregulation (Figure 6.17a) (Pomerening et al., 2003; Tyson et al., 2003). This circuit can produce oscillations with robust timing despite fluctuations in the biochemical parameters of the components (Barkai and Leibler, 2000; Vilar et al., 2002; Atkinson et al., 2003). The mechanism of robust oscillations is based on a hysteresis in the dynamics of X caused by the autoregulatory loop. This oscillator belongs to a family of models known as relaxation oscillators.

Another possible design for oscillatory circuits is a feedback loop made of multiple regulators hooked up in a cycle to form a negative feedback loop. For example, three repressors hooked up in a cycle are called a repressilator (Elowitz and Leibler, 2000) (Figure 6.17b, exercise 4.9). This circuit belongs to the family of delay oscillators, a family that often tends to show noisier oscillations with less precise timing. Extensive theoretical work on biological oscillators can be found in the further reading section at the end of the chapter.

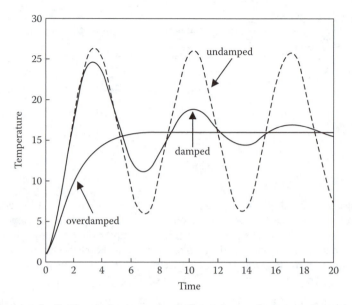

FIGURE 6.16 Negative feedback can show overdamped monotonic dynamics, damped oscillations with peaks of decreasing amplitude, or undamped oscillations. Generally, the stronger the interactions between the two nodes in relation to the damping forces on each node (such as degradation rates), the higher the tendency for oscillations.

6.6   NETWORK MOTIFS IN THE NEURONAL NETWORK OF C. ELEGANS

Many fields of science deal with networks of interactions, including sociology (Holland and Leinhardt, 1975; Wasserman and Faust, 1994), neurobiology, engineering, and ecology. Network motifs can be sought in networks from these fields by comparing them to randomized networks. One finds that:

1. Most real-world networks contain a small set of network motifs.

2. The motifs in different types of network are generally different.

Examples can be seen in Figure 6.18. To compare the motifs in networks of different sizes, one can define subgraph profiles. These profiles display the relative statistical significance of each type of subgraph relative to randomized networks (Figure 6.19).

Thus, networks from different fields usually have different network motifs. This seems reasonable because each type of network performs different functions. However, there are a few intriguing cases in which unrelated networks share similar network motifs (and similar anti-motifs, patterns that are rarer than at random). This occurs, for example, when studying the neuronal network of the nematode C. elegans, a tiny worm composed of about 1000 cells (Figure 6.20).

The synaptic connections between all of the 300 neurons in this organism were mapped by White, Brenner, and colleagues (White et al., 1986). The wiring does not seem to vary significantly from individual to individual. In this directed network the nodes are neurons, and a directed edge X → Y means that neuron X has a synaptic connection with

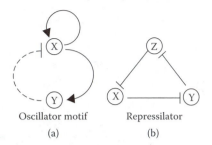

Oscillator motif   Repressilator
(a)   (b)

FIGURE 6.17 (a) A network motif found in many biological oscillatory systems, composed of a composite negative feedback loop and a positive autoactivation loop. In this motif, X activates Y on a slow timescale, and also activates itself. Y inhibits X on a rapid timescale. (b) A repressilator made of three cyclically linked inhibitors. The repressilator usually shows noisy oscillations in the presence of fluctuations in the production rates of the proteins.

neuron Y. Thus, this network can be searched for network motifs made of neurons. Strikingly, the neuronal network of *C. elegans* was found to share many of the motifs found in transcription and signal transduction networks.

The network motifs in the *C. elegans* neuronal network are similar to those found in biochemical interaction networks, despite the fact that these networks operate on very different spatial and temporal scales. The scale of the neuronal network is that of cells, and the response times are milliseconds. The scale of transcription networks is that of nanometer-size biomolecules within a cell, and the timescale is minutes to hours. Yet many of the motifs in these networks are similar. For example, the most significant three-node motif in the neuronal network is the feed-forward loop. An example of neuronal feed-forward loops is shown in Figure 6.21.

Why is the FFL a motif in both this neuronal network and transcription networks? One point of view is that this is a coincidence, and different histories gave rise to similar motifs.[1] We favor a different view, that the similarity in network motifs reflects the fact that both networks evolved toward a similar goal: they perform information processing on noisy signals using noisy components. Both networks need to convey information between sensory components that receive the signals and motor components that generate the responses. Neurons process information between sensory neurons and motor neurons. Transcription networks process information between transcription factors that receive signals from the external world and structural genes that act on the inner or outer environment of the cell. This similarity in the function of the two types of networks raises the possibility that evolution may have converged on similar circuits in both networks to perform important signal processing functions, such as sign-sensitive filtering. This

[1] White et al., (1986) raised the posibility that triangular patterns such as the FFL might arise due to the spatial arrangement of the neurons. Neurons that are neighbors tend to connect more often than distant neurons. Such neighborhood effects can produce triangle-shaped patterns, because if X is close to Y and Y is close to Z, X also tends to be close to Z. However, while this effect would produce feed-forward loops, it also would produce three-node *feedback* loops, in which X, Y, and Z are connected in a cycle (Figure 4.1), and many other patterns (Itzkovitz and Alon, 2005). Such feedback loops, however, are not found very commonly in this neuronal network: they are in fact anti-motifs. Thus, the origin of FFLs and other motifs in the neuronal network is not solely due to neighborhood effects.

Network	Nodes	Edges	N_{real}	$N_{rand} \pm$ SD	Z-score	N_{real}	$N_{rand} \pm$ SD	Z-score	N_{real}	$N_{rand} \pm$ SD	Z-score
Transcription			Feed-forward loop (X→Y→Z)			Bi-fan (X,Y→Z,W)					
E. coli	424	519	40	7 ± 3	10	203	47 ± 12	13			
S. cerevisiae	685	1,052	70	11 ± 4	14	1812	300 ± 40	41			
Neurons			Feed-forward loop			Bi-fan			Diamond		
C.elegans	252	509	125	90 ± 10	3.7	127	55 ± 13	5.3	227	35 ± 10	20
Food webs			Cascade (X→Y→Z)			Diamond					
Little rock	92	984	3219	3120 ± 50	2.1	7295	2220 ± 210	25			
Ythan	83	391	1182	1020 ± 20	7.2	1357	230 ± 50	23			
St. Martin	42	205	469	450 ± 20	NS	382	130 ± 20	12			
Chesapeake	31	67	80	82 ± 4	NS	26	5 ± 2	8			
Coachella	29	243	279	235 ± 12	3.6	181	80 ± 20	5			
Skipwith	25	189	184	150 ± 7	5.5	397	80 ± 25	13			
B. Brook	25	104	181	130 ± 7	7.4	267	30 ± 7	32			
Electronic circuits (forward logic chips)			Feed-forward loop			Bi-fan			Diamond		
s15850	10,383	14,240	424	2 ± 2	285	1040	1 ± 1	1200	480	2 ± 1	335
s38584	20,717	34,204	413	10 ± 3	120	1739	6 ± 2	800	711	9 ± 2	320
s38417	23,843	33,661	612	3 ± 2	400	2404	1 ± 1	2550	531	2 ± 2	340
s9234	5,844	8,197	211	2 ± 1	140	754	1 ± 1	1050	209	1 ± 1	200
s13207	8,651	11,831	403	2 ± 1	225	4445	1 ± 1	4950	264	2 ± 1	200
Electronic circuits (digital fractional multipliers)			Three-node feedback loop			Bi-fan			Four-node feedback loop		
s208	122	189	10	1 ± 1	9	4	1 ± 1	3.8	5	1 ± 1	5
s420	252	399	20	1 ± 1	18	10	1 ± 1	10	11	1 ± 1	11
s838	512	819	40	1 ± 1	38	22	1 ± 1	20	23	1 ± 1	25
World wide web			Feedback with two mutual dyads			Clique			Regulating feedback		
nd.edu	325,729	1.46e6	1.1e5	2e3 ± 1e2	800	6.8e6	5e4 ± 4e2	15,000	1.2e6	1e4 ± 2e2	5000

FIGURE 6.18 Network motifs found in biological and technological networks. The numbers of nodes and edges for each network are shown. For each motif, the numbers of appearances in the real network (N_{real}) and in the randomized networks ($N_{rand} \pm \sigma$, all values rounded) are shown. As a qualitative measure of statistical significance, the Z-score = $(N_{real} - N_{rand})/\sigma$ is shown. NS, not significant. Shown are motifs that occur at least U = 4 times with completely different sets of nodes. The networks include synaptic interactions between neurons in *C. elegans*, including neurons connected by at least five synapses; trophic interactions in ecological food webs, representing pelagic and benthic species (Little Rock Lake), birds, fishes, invertebrates (Ythan Estuary), primarily larger fishes (Chesapeake Bay), lizards (St. Martin Island), primarily invertebrates (Skipwith Pond), pelagic lake species (Bridge Brook Lake), and diverse desert taxa (Coachella Valley); electronic sequential logic circuits parsed from the ISCAS89 benchmark set, where nodes represent logic gates and flip-flops; and World Wide Web hyperlinks between Web pages in a single domain (only three-node motifs are shown). (From Milo et al., 2002.)

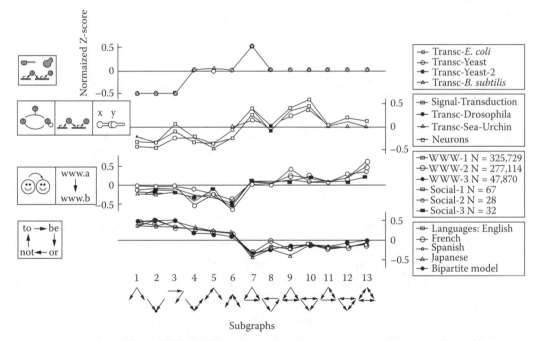

FIGURE 6.19 Significance profiles of three-node subgraphs found in different networks. The profiles display the relative statistical significance of each subgraph. The significance profile allows comparison of networks of very different sizes ranging from tens of nodes (e.g., social networks where children in a school class select friends) to millions of nodes (World Wide Web networks). The y-axis is the normalized Z-score = $(N_{real} - N_{rand})/\sigma$ for each of the 13 three-node subgraphs, where σ is the standard deviation in the random networks. Language networks have words as nodes and edges between words that tend to follow each other in texts. The neural network is from *C. elegans*, counting synaptic connections with more than five synapses. For more details, see Milo et al., 2004.

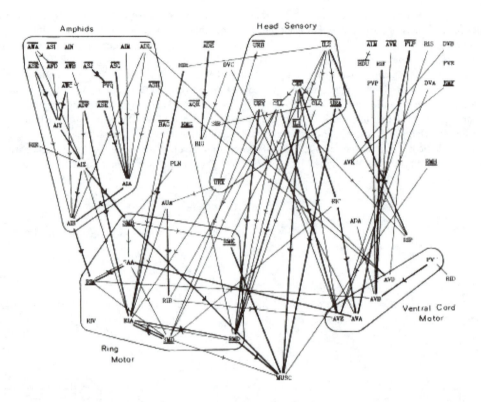

FIGURE 6.20 Map of synaptic connections between *C. elegans* neurons. Shown are connections between neurons in the worm head. (From Durbin, R., Ph.D. thesis, www.wormbase.org.)

hypothesis can in principle be tested by experiments on the dynamical functions of neural circuits, analogous to the experiments described in previous chapters on network motifs in transcription networks.

6.6.1   The Multi-Input FFL in Neuronal Networks

Having stressed the similarities, we note that the set of network motifs found in neuronal networks is not identical to those found in transcription networks. For example, although the FFL is a motif in both of these networks, the FFLs are joined together in neuronal networks in different ways than in transcription networks. This can be seen by considering the topological generalizations of the FFL pattern (Section 5.4, Figure 5.6).

The most common FFL generalization in transcription networks is the multi-output FFL, as we saw in Chapter 5. In contrast, a distinct generalization, the **multi-input FFL**, is the most common generalization in the *C. elegans* neuronal network. An example with two inputs is shown in Figure 6.21. Thus, while the FFL is a motif in both transcription networks and the neuronal network of *C. elegans*, it generalizes to larger patterns in different ways in the two networks.

Can we analyze the function of neuron motifs such as the multi-input FFL? To do this, we need to discuss the equations that govern neuron dynamics in *C. elegans*. It turns out that the canonical, simplified model for the activity of such neurons has equations that

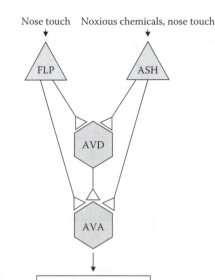

Nose touch   Noxious chemicals, nose touch

FLP

ASH

AVD

AVA

Backward movement

FIGURE 6.21 Feed-forward loops in *C. elegans* avoidance reflex circuit. When sensory neurons in the head of the worm sense a touch or a noxious odor, the circuit stimulates motor neurons that elicit backward motion.Triangles represent sensory neurons and hexagons represent interneurons. Lines with triangles represent synaptic connections between neurons. (From Hope, 1999.)

are almost identical to those we have used to model transcription networks and signal transduction network dynamics, although the molecular mechanisms are very different.

Neurons act primarily by transmitting electrical signals to other neurons via synaptic connections. Each neuron has a time-dependent transmembrane voltage difference that we will consider as the neurons' activity. *C. elegans* neurons do not appear to have sharp voltage spikes as vertebrate neurons do. Instead, they appear to have graded voltages, $X(t)$, $Y(t)$, and $Z(t)$. The classic model for the dynamics of a neuron is based on summation of synaptic inputs from its input neurons (this model is called integrate-and-fire in the context of spiking neurons). Consider neuron Y in a two-input FFL that receives synaptic inputs from two neurons X_1 and X_2 (Figure 6.22a). In the simplest integrate-and-fire model, the change in voltage of Y is activated by a step function over the weighted sum of the voltages of the two input neurons:

$$dY/dt = \beta\ \theta(w_1\ X_1 + w_2\ X_2 > K_y) - \alpha\ Y \tag{6.6.1}$$

where α is the relaxation rate related to the leakage of current through the neuron cell membrane,[1] and the weights w_1 and w_2 correspond to the strengths of the synaptic connections from input neurons X_1 and X_2 to Y.

As we have seen for signaling cascades, these weighted sums can generate either AND or OR gates (see Figure 6.11). For example, if both weights are large, so that each weight times the maximal input activity exceeds the activation threshold ($w_1\ X_{1,max} > K_y$ and $w_2\ X_{2,max} > K_y$), the result is an OR gate because either input can cause Y to be activated. An

---

[1] As in electrical capacitor–resistor systems, with resistance R and capacitance C, $\alpha \sim 1/RC$.

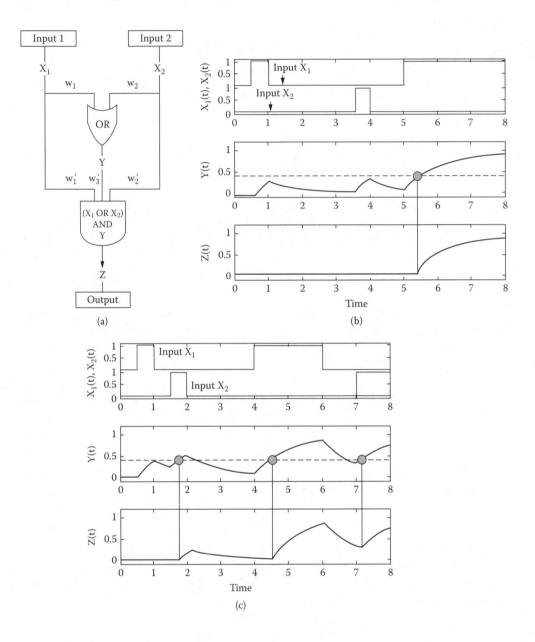

FIGURE 6.22 Dynamics in a model of a *C. elegans* neuronal multi-input FFL following pulses of input stimuli. (a) A two-input FFL. The input functions and activation thresholds are shown. (b) Dynamics of the two-input FFL in response to well-separated input pulses that stimulate X_1 and X_2, followed by a persistent X_1 stimulus. The separated pulses do not activate Z. (c) Dynamics with a brief X_1 stimulus followed closely by a short X_2 stimulus. The dashed horizontal line corresponds to the activation threshold for Y. The brief pulses are able to activate Z. Relaxation rate is $\alpha = 1$, and $w_1 = w_2 = w_1' = w_2' = 0.5$, $w_3' = 0.4$. (From Kashtan et al., 2004.)

AND gate occurs if neither weight is large enough to activate the neuron with only one input, so that both inputs are needed for activation.

With these equations, we can consider the two-input FFL network motif (Figure 6.22a). In this circuit, neuron Y compares the weighted sum of the inputs from X_1 and X_2 to a threshold. The neuron Z is also controlled by X_1 and X_2, and in addition receives inputs from neuron Y:

$$dZ/dt = \beta' \, \theta(w'_1 X_1 + w'_2 X_2 + w'_3 Y > K_z) - \alpha' \, Z \qquad (6.6.2)$$

Experiments on the nose-touch system of Figure 6.21 suggest that Z can be activated by the following logic input function: $(X_1 \text{ OR } X_2) \text{ AND } Y$ (Chalfie et al., 1985). In other words, either of the two inputs X_1 or X_2 can activate Z, provided that Y is also active (Figure 6.22a). In this case, the response to a pulse of input signal from either input can be easily understood based on the function of the simple, three-node coherent FFL. Each of the two FFLs acts as a persistence detector with respect to its stimulus. Hence, the output neuron Z can be activated by either input, but only if the input is persistent enough. This persistence detection occurs because the voltage of Y must accumulate and cross its activation threshold to activate Z. A transient activation does not give Y sufficient time to accumulate. Therefore, transient inputs do not cause activation (Figure 6.22b).

In addition to its function as a persistence detector, the multi-input FFL can perform **coincidence detection** of brief input signals: a short pulse of X_1 activation, which by itself is not sufficient to activate Z, can still do so if there is a short input of X_2 in close proximity, as demonstrated in Figure 6.22c. Thus, a transient input can cause activation if it is followed closely enough by a second transient input from the other input neuron. This coincidence detection function can ensure that a short pulse can activate the system provided that it has support from an additional input at about the same time. This feature is made possible because Y acts as a transient memory, storing information for a timescale of $1/\alpha$ (on the order of 10 msec).

The timescales of the dynamics are determined by the response times of the neuron voltages, $T_{1/2} = \log(2)/\alpha$. Whereas in transcription circuits $T_{1/2}$ often has timescales of hours, in the neuronal circuits of *C. elegans*, $T_{1/2}$ has timescales of tens of milliseconds. Despite the vast difference in temporal and spatial scales, the same simplified mathematical reasoning can be applied to understand, at least approximately, the dynamics of network motifs in both networks.

6.6.2 Multi-Layer Perceptrons in the *C. elegans* Neuronal Network

When examining patterns with four or more nodes, one finds that the most abundant network motifs in the synaptic wiring of *C. elegans* are multi-layer perceptrons (Figure 6.23). These motifs are similar to those we have seen in signal transduction networks (Section 6.4). The main structural difference is that *C. elegans* multi-layer perceptrons have a higher abundance of mutual connections (feedback loops) between pairs of nodes in the same layer.

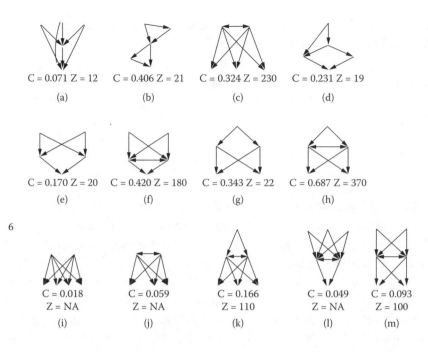

C = 0.071 Z = 12 (a) C = 0.406 Z = 21 (b) C = 0.324 Z = 230 (c) C = 0.231 Z = 19 (d)

C = 0.170 Z = 20 (e) C = 0.420 Z = 180 (f) C = 0.343 Z = 22 (g) C = 0.687 Z = 370 (h)

C = 0.018 Z = NA (i) C = 0.059 Z = NA (j) C = 0.166 Z = 110 (k) C = 0.049 Z = NA (l) C = 0.093 Z = 100 (m)

FIGURE 6.23 Five- and six-node network motifs in the neuronal network of *C. elegans*. Note the multi-layer perceptron patterns d–h and k–m, many of which have mutual connections within the same layer. The parameter C is the concentration of the subgraph multiplied by 10^3, where concentration is defined as the number of appearances of the subgraph divided by the total number of appearances of all subgraphs of the same size in the network. Z is the number of standard deviations that the concentration exceeds randomized networks with the same degree sequence (Z-score). These motifs were detected by an efficient sampling algorithm suitable for large subgraphs and large networks. (From Kashtan et al., 2004.)

It seems plausible that the function of these multi-layer structures is information processing of the type that we described in Section 6.4. Analysis of the precise computations performed by neuronal perceptron circuits will depend on accurate measurements of the weights on the edges (synaptic strengths). These weights could be modulated on several different timescales by cellular processes within the neuron, and by signaling molecules transmitted between neurons. Whether signal transduction networks and neuronal networks use similar principles in their computations is an interesting question.

The motifs in neuronal networks can, of course, perform many additional functions, depending on the input functions, thresholds, and relaxation timescales. After all, each neuron is a sophisticated cell able to perform computations and to adapt over time. The present discussion considered only the simplest scenario for these network motifs.

Finally, analysis of the structure of neuronal networks of higher organisms is still a task for the future. It is becoming apparent that these networks have a modular organization. Preliminary studies indicate the existence of network motifs both on the level of individual neurons (Song et al., 2005) and on the level of connections between modules or brain functional areas (Sporns and Kotter, 2004; Sakata et al., 2005).

6.7   SUMMARY

We have seen in this chapter that each type of biological network is built of a distinctive set of network motifs. Each of the network motifs can carry out defined dynamical functions. Developmental transcription networks display many of the motifs found in sensory transcription networks. They also have additional network motifs that correspond to the irreversible decisions and slower dynamics of developmental processes. In particular, two-node positive feedback loops, regulated by a third transcription factor, can provide lock-on to a cell fate or provide toggle switches between two different fates. Long cascades can orchestrate developmental programs that take place over multiple cell generations.

Signal transduction networks show faster dynamics and may utilize multi-layer perceptrons to perform computations on numerous input stimuli. Multi-layer perceptrons can allow even relatively simple units to perform detailed computations. Multi-layer perceptrons can perform more intricate computations than single-layer perceptrons. The dynamical features of these networks are affected by feedback loops and additional interactions. We are only beginning to understand the computational functions of signal transduction networks.

Integrated networks made of different types of interactions can show composite network motifs. A common motif is a negative feedback loop made of a slow transcription interaction and a faster protein-level interaction. This feedback loop can generate robust oscillations when coupled to a second positive autoregulation loop.

We have also examined the network of synaptic connections between neurons in the worm. Again, we saw that the patterns found in the neuronal network are only a tiny fraction of all of the possible patterns of the same size. Hence, this neuronal network has a structural simplicity reminiscent of that found in the biomolecular networks we have studied. Moreover, many of the neuron network motifs are the same as those found in transcription and signaling networks. This includes FFLs and multi-layer perceptrons. This similarity raises the possibility that these motifs perform analogous information processing functions in these different networks.

The neuronal network also has motifs not found in the other networks we have studied. One example is the multi-input FFL structure, which can perform coincidence detection on different input stimuli.

In all of these cases, networks are a convenient approximation to the complex set of biological interactions. The network representation masks a great deal of the detailed mechanisms at each node and edge. Because of this simplified level of description, the network representation helps highlight the similarity in the circuit patterns in different parts of the network and between different networks. The dynamics of the networks at this level of resolution lend themselves to analysis with rather simple models. We care only that X activates or inhibits Y, not precisely how it does it on the biochemical level. This abstraction helps us to define network motifs as specific functional building blocks of each type of network. These building blocks often appear to be joined together in ways that allow understanding of the network dynamics in terms of the dynamics of each individual motif. Hence, both on the level of local connectivity patterns and on the level of combinations of patterns into larger circuits, biological networks appear to display a degree of simplicity.

FURTHER READING

Network Motifs in Diverse Networks

Milo, R., Itzkovitz, S., Kashtan, N., Levitt, R., Shen-Orr, S., Ayzenshtat, I., Sheffer, M., and Alon, U. (2004). Superfamilies of designed and evolved networks. *Science*, 303: 1538–1542.

Milo, R., Shen-Orr, S., Itzkovitz, S., Kashtan, N., Chklovskii, D., and Alon, U. (2002). Network motifs: simple building blocks of complex networks. *Science*, 298: 824–827.

Developmental Networks

Bolouri, H. and Davidson, E.H. (2002). Modeling transcriptional regulatory networks. *Bioessays*, 24: 1118–1129.

Lawrence, P.A. (1995). *The Making of a Fly: The Genetics of Animal Design*. Blackwell Science, Ltd., The Alden Press.

Levine, M. and Davidson, E.M. (2005). Gene regulatory networks for development. *Proc. Natl. Acad. Sci*, 4936–4942.

Stathopoulos, A. and Levine, M. (2005). Genomic regulatory networks and animal development. *Dev. Cell.* 9: 479–462.

Signaling Networks as Computational Devices

Bhalla, U.S., and Iyengar, R. (1999). Emergent properties of networks of biological signalling pathways, *Science*, 283: 381–387.

Bray, D. (1995). Protein molecules as computational elements in living cells. *Nature*, 376: 307–312.

Wiley, H.S., Shvartsman, S.Y., and Laffenburger, D.A. (2003). Computational modeling of the EGF-receptor system: a paradigm for systems biology. *Trends Cell Biol.* 13: 43–50.

Perceptrons and Their Computational Power

Hertz, J., Krogh, A., and Palmer, R.G. (1991). *Introduction to the Theory of Neural Computation*. Perseus Books.

Biological Oscillators

Barkai, N. and Leibler, S. (2000). Circadian clocks limited by noise. *Nature*, 403: 267–268.

Tyson, J.J., Chen, K.C., and Novak, B. (2003). Sniffers, buzzers, toggles and blinkers: dynamics of regulatory and signaling pathways in the cell. *Curr. Opin. Cell Biol.*, 15: 221–231.

Winfree, A.T. (2001). *The Geometry of Biological Time*. Springer.

Neuronal Network of *C. elegans*

Bargmann, C.I. (1998). Neurobiology of the *Caenorhabitis elegans* genome. *Science*, 282: 2028–2033.

Durbin, R.M. (1987). Studies on the development and organization of the nervous system of *caenorhabitis elegans*. Ph.D. thesis, www.wormbase.org.

Hope, A.I., Ed. (1999). *C. elegans: A Practical Approach*, 1st ed. Oxford University Press.

EXERCISES

6.1. *Memory in the regulated-feedback network motif.* Transcription factor X activates transcription factors Y_1 and Y_2. Y_1 and Y_2 mutually activate each other. The input function at the Y_1 and Y_2 promoters is an OR gate (Y_2 is activated when either X or Y_1 bind the promoter). At time t = 0, X begins to be produced from an initial concentration of X = 0. Initially, $Y_1 = Y_2 = 0$. All production rates are $\beta = 1$ and degradation rates are $\alpha = 1$. All of the activation thresholds are K = 0.5. At time t = 3, production of X stops.

a. Plot the dynamics of X, Y_1, and Y_2. What happens to Y_1 and Y_2 after X decays away?

b. Consider the same problem, but now Y_1 and Y_2 repress each other and X activates Y_1 and represses Y_2. At time t = 0, X begins to be produced, and the initial levels are X = 0, $Y_1 = 0$, and $Y_2 = 1$. At time t = 3, X production stops. Plot the dynamics of the system. What happens after X decays away?

6.2. *Kinases with double phosphorylation.* Kinase Y is phosphorylated by two input kinases X_1 and X_2, which work with first-order kinetics with rates v_1 and v_2. Y needs to be phosphorylated on two sites to be active. The rates of phosphorylation and dephosphorylation of the two phosphorylation sites on Y are the same. Find the input function, the fraction of doubly phosphorylated Y, as a function of the activity of X_1 and X_2.

Solution:

The kinase Y exists in three states, with zero, one, and two phosphorylations, denoted Y_o, Y_1, and Y_2. The total amount of Y is conserved:

$$Y_o + Y_1 + Y_2 = Y \tag{P6.1}$$

The rate of change of Y_1 is given by an equation that balances the rate of the input kinases and the action of the phosphatases, taking into account the flux from Y_o to Y_1 and from Y_1 to Y_2, as well as dephosphorylation of Y_2 to Y_1:

$$dY_1/dt = v_1\,X_1\,Y_o + v_2\,X_2\,Y_o - v_1\,X_1\,Y_1 - v_2\,X_2\,Y_1 - \alpha\,Y_1 + \alpha\,Y_2 \tag{P6.2}$$

And the dynamic equation of Y_2 is

$$dY_2/dt = v_1\,X_1\,Y_1 + v_2\,X_2\,Y_1 - \alpha\,Y_2 \tag{P6.3}$$

At steady state, $dY_2/dt = 0$ and Equation P6.3 yields

$$(v_1X_1 + v_2X_2)\,Y_1 = \alpha\,Y_2 \tag{P6.4}$$

using the weights $w_1 = v_1/\alpha$ and $w_2 = v_2/\alpha$, we find:

$$Y_1 = Y_2/(w_1 X_1 + w_2 X_2) \qquad (P6.5)$$

Summing equations P6.3 and P6.2 yields $d(Y_1 + Y_2)/dt = Y_o (v_1 X_1 + v_2 X_2) - \alpha Y_1$, so that at steady state

$$Y_o = Y_1/(w_1 X_1 + w_2 X_2) = Y_2/(w_1 X_1 + w_2 X_2)^2 \qquad (P6.6)$$

Using equation P6.1, we find

$$Y = Y_o + Y_1 + Y_2 = (1 + 1/u + 1/u^2) Y_2 \qquad (P6.7)$$

where

$$u = w_1 X_1 + w_2 X_2 \qquad (P6.8)$$

Thus, the desired input function is:

$$Y_2/Y = u^2/(u^2 + u + 1) \qquad (P6.9)$$

Note that for n phosphorylations, the input function is $Y_n/Y = u^n/(1 + u + \ldots + u^n)$

6.3. Design a multi-layer perceptron with two input nodes, one output node, and as many intermediate nodes as needed, whose output has a region of activation in the shape of a triangle in the middle of the X_1-X_2 plane.

6.4. *Dynamics of a protein kinase cascade.* Protein kinases X_1, X_2, ..., X_n act in a signaling cascade, such that X_1 phosphorylates X_2, which, when phosphorylated, acts to phosphorylate X_3, etc.

   a. Assume sharp activation function. What is the response time of the cascade, the time from activation of X_1 to a 50% rise in the activity of X_n?

   b. What is the effect of the kinase rates on the response time? Of the phosphatase rates? Which have a larger effect on the response time (Heinrich et al., 2002)?

Solution:

   a. The rate of change of active (phosphorylated) X_i is given by the difference between the sharp phosphorylation rate by kinase X_{i-1}, with rate v_i, and the dephosphorylation process by the phosphatases that work on X_i at rate α_i:

$$dX_i/dt = v_i \, \theta(X_{i-1} > K_{i-1}) - \alpha_i X_i \qquad (P6.10)$$

where θ is the step function that equals one if the logic expression $X_{i-1} > K_{i-1}$ is true, and zero otherwise.

Thus, X_i begins to increase at the time that X_{i-1} crosses its threshold K_{i-1}. At this point, X_i begins to increase with the familiar exponential convergence to steady state (e.g., Equation 2.4.6):

$$X_i = (v_i/\alpha_i) \, [1 - e^{-\alpha_i (t - t_i)}] \tag{P6.11}$$

When the concentration of the kinase X_i (in its phosphorylated form) crosses the activation threshold, it begins to activate the next kinase in the cascade. Thus, the onset of phosphorylation of X_{i+1}, denoted t_{i+1}, can be found by solving

$$K_i = (v_i/\alpha_i) \, [1 - e^{-\alpha_i (t - t_i)}] \tag{P6.12}$$

yielding

$$t_{i+1} = t_i + \alpha_i^{-1} \log[1/(1 - \alpha_i \, K_i/v_i)] \tag{P6.13}$$

We thus find that

$$t_n = \Sigma_i \, 1/\alpha_i \log[1/(1 - \alpha_i \, K_i/v_i)] \tag{P6.14}$$

b. According to equation P6.14, the phosphatase rates α_i have a large effect on the response times. If these rates are very different for each kinase in the cascade, the response time is dominated by the slowest rate, because it has the largest $1/\alpha_i$. In contrast to the strong dependence of phosphatase rates, the response time is only weakly affected by the kinase velocities v_i, because they appear inside the logarithm in equation P6.14.

6.5. *Dynamics of a linear protein kinase cascade (Heinrich et al., 2002).* In the previous problem, we analyzed the dynamics of a cascade with sharp input functions. Now we consider the case of zero-order kinetics. This applies when the activated upstream kinase is found in much smaller concentrations than its unphosphorylated target. In zero-order kinetics, the rate of phosphorylation depends only on the upstream kinase concentration and not on the concentration of its substrate. In this case, we need to analyze a *linear set* of equations:

$$dX_i/dt = v_{i-1} \, X_{i-1} - \alpha_i \, X_i \tag{P6.15}$$

The **signal amplitude** is defined by

$$A_i = \int_0^\infty X_i(t) \, dt \tag{P6.16}$$

and the **signal duration** by

$$\tau_i = \int_0^\infty t\, X_i(t)\, dt\, /\, A_i \tag{P6.17}$$

In many signaling systems the duration of the signaling process is important, in the sense that brief signals can sometimes activate different responses than prolonged signals.

a. The cascade is stimulated by a pulse of X_1 activity with amplitude A_1, that is,

$$\int_0^\infty X_1(t)\, dt = A_1$$

What is the amplitude of the final stage in the cascade, A_n?

b. What is the signal duration of X_n?

c. How do the kinase and phosphatase rates affect the amplitude and duration of the signal? Compare to exercise 6.4.

Solution:

a. To find the amplitude, let us take an integral over time of both sides of Equation P6.15.

$$\int_0^\infty dt\, dX_i\, /\, dt = \int_0^\infty v_{i-1} X_{i-1} dt - \int_0^\infty \alpha_i X_i dt \tag{P6.18}$$

Note that the integral on the left-hand side is equal to $X_i(\infty) - X_i(0)$. Now, because the signal begins at $t = 0$ and decays at long times, we have $X_i(0) = X_i(\infty) = 0$. The integrals on the right-hand side give rise to amplitudes as defined in Equation P6.17:

$$0 = v_{i-1} A_{i-1} - \alpha_i A_i \tag{P6.19}$$

Thus,

$$A_i = (v_{i-1}/\alpha_i)\, A_{i-1} \tag{P6.20}$$

Therefore, by induction, we find that the amplitude is the product of the kinase rates divided by the product of the phosphatase rates:

$$A_n = (v_{n-1}/\alpha_n)\, A_{n-1} = (v_{n-1}\, v_{n-2}/\alpha_n\, \alpha_{n-1})\, A_{n-2} = \ldots \tag{P6.21}$$

$$= (v_{n-1}\, v_{n-2}\, \ldots\, v_2\, v_1/\alpha_n\, \alpha_{n-1}\, \ldots\, \alpha_2)\, A_1$$

b. To find the signal duration, we take an integral over time of the dynamic equation (Equation P6.15) multiplied by t to find

$$\int_0^\infty dt \ t dX_i / dt = \int_0^\infty dt \ v_{i-1} X_{i-1} - \int_0^\infty dt \ \alpha_i t X_i$$

(P6.22)

The left-hand-side integral can be solved using integration by parts to yield

$$\int_0^\infty dt \ t dX_i/dt = -A_i$$

(P6.23)

The right-hand side of P6.22 is proportional to the durations of X_{i-1} and X_i, (equation P6.17) so that we find

$$-A_i = v_{i-1} \ \tau_{i-1} \ A_{i-1} - \alpha_i \ \tau_i \ A_i$$

(P6.24)

Hence, we have, dividing both sides by A_i and using equation P6.20 to eliminate A_{i-1},

$$1 = \alpha_i \ \tau_i - \alpha_i \ \tau_{i-1}$$

(P6.25)

which can be rearranged to yield

$$\tau_i - \tau_{i-1} = 1/\alpha_i$$

Hence, the signal duration of the final step in the cascade is just the sum over the reciprocal phosphatase rates

$$\tau_n = \Sigma_i \ \alpha_i^{-1}$$

c. We have just found that phosphatase rates α_i affect both amplitude and duration in zero-order kinetics cascades. The larger the phosphatase rates, the smaller the amplitude and the shorter the duration. In contrast, the kinase rates do not affect duration at all, and affect the signal amplitude proportionally. This is similar to problem 6.4, where we saw that phosphatase rates affect timing much more strongly than kinase velocities. In both models, the sum over $1/\alpha_i$ determines the timing. This principle is identical to that which we saw in transcription networks, whose response times are governed inversely by the degradation/dilution rates: These rates are the eigenvalues of the dynamic equations. The strong effect of phosphatases on signal duration and the weak effect of kinases were demonstrated experimentally (see experiments cited in Hornberg et al., 2005).

6.6. *Coincidence detection.* Consider the two-input FFL motif of Figure 6.22. The two inputs receive brief activation pulses at a slight delay. The pulse of S_{x1} has duration d. At time t_0 after the start of the pulse, a pulse of S_{x2} begins and lasts for duration d.

a. What is the minimal S_{x1} input pulse duration d that can activate Z without need for the second pulse of S_{x2}?

b. Plot the region in which Z shows a response on a plane whose axes are pulse duration d and interpulse spacing t_0.

6.7. Consider the diamond generalization (Figure 6.7) that has two inputs X_1 and X_2 and a single output Z. This two-layer perceptron pattern has 6 edges. Assume that all neurons are 'integrate-and-fire,' and each has a threshold K = 1. Assume that neurons have voltage 0, unless the weighted inputs exceed K, in which case they assume voltage 1. Weights on the edges can be positive or negative real numbers.

a. Design weights such that this circuit computes the XOR (exclusive-or) function, where Z = 1 if either X_1 = 1 or X_2 = 1, but Z = 0 if both X_1 = 1 and X_2 = 1. This function is denoted Z = X_1 XOR X_2.

b. Design weights such that this circuit computes the 'equals' function, in which Z = 1 only if X_1 and X_2 are the same (both 0 or both 1) and Z = 0 otherwise (that is, Z = X_1 EQ X_2).

CHAPTER 7

Robustness of Protein Circuits: The Example of Bacterial Chemotaxis

7.1 THE ROBUSTNESS PRINCIPLE

The computations performed by a biological circuit depend on the biochemical parameters of its components, such as the concentration of the proteins that make up the circuit. In living cells, these parameters often vary significantly from cell to cell due to stochastic effects, even if the cells are genetically identical. For example, the expression level of a protein in genetically identical cells in identical environments can often vary by tens of percents from cell to cell (see Appendix D). Although the genetic program specifies, say, 1000 copies of a given protein per cell in a given condition, one cell may have 800 and its neighbor 1200. How can biological systems function despite these variations?

In this chapter, we will introduce an important design principle of biological circuitry: *biological circuits have* **robust** *designs such that their essential function is nearly independent of biochemical parameters that tend to vary from cell to cell.*

We will call this principle **robustness** for short, though one must always state what *property* is robust and with respect to which *parameters*. Properties that are not robust are called *fine-tuned*: these properties change significantly when biochemical parameters are varied.

Robustness to parameter variations is never absolute: it is a relative measure. Some mechanisms can, however, be much more robust than others.

Robustness was suggested to be an important design principle by M. Savageau in theoretical analysis of gene circuits (Savageau, 1971, 1976). H. Kacser and colleagues experimentally demonstrated the robustness of metabolic fluxes with respect to variations of enzyme levels in yeast (Kacser and Burns, 1973). Robustness was also studied in a different context: the patterning of tissues as an egg develops into an animal. Waddington

Bacterial chemotaxis

Attractant   Repellent

FIGURE 7.1 Bacterial chemotaxis. Bacteria swim toward a pipette with attractants and swim away from repellents.

studied the sensitivity of developmental patterning to various perturbations (Waddington, 1959). In these studies, robustness was called canalization and was considered at the level of the phenotype (e.g., the shape of the organism) but not at the level of biochemical mechanism (which was largely unknown at the time). Recent work has demonstrated how properly designed biochemical circuitry can give rise to robust and precise patterning. This subject will be discussed in the next chapter.

Here we will demonstrate the design principle of robustness by using a well-characterized protein signaling network, the protein circuit that controls bacterial chemotaxis. We will begin by describing the biology of bacterial chemotaxis. It is a relatively simple prototype for signal transduction circuitry in other cell types. Then we will describe models and experiments that demonstrate how the computation performed by this protein circuit is made robust to changes in biochemical parameters. We will see that the principle of robustness can help us to rule out a large family of plausible mechanisms and to home in on the correct design.

7.2   BACTERIAL CHEMOTAXIS, OR HOW BACTERIA THINK

7.2.1   Chemotaxis Behavior

When a pipette containing nutrients is placed in a plate of swimming *Escherichia coli* bacteria, the bacteria are attracted to the mouth of the pipette and form a cloud (Figure 7.1). When a pipette with noxious chemicals is placed in the dish, the bacteria swim away from the pipette. This process, in which bacteria sense and move along gradients of specific chemicals, is called **bacterial chemotaxis**.

Chemicals that attract bacteria are called **attractants**. Chemicals that drive the bacteria away are called **repellents**. *E. coli* can sense a variety of attractants, such as sugars and the amino acids serine and aspartate, and repellents, such as metal ions and the amino acid leucine. Most bacterial species show chemotaxis, and some can sense and move toward stimuli such as light (phototaxis) and even magnetic fields (magnetotaxis).

Bacterial chemotaxis achieves remarkable performance considering the physical limitations faced by the bacteria. Bacteria can detect concentration gradients as small as a change of one molecule per cell volume per micron and function in background concentrations spanning over five orders of magnitude. All this is done while being buffeted by Brownian noise, such that if the cell tries to swim straight for 10 sec, its orientation is randomized by 90° on average.

How does *E. coli* manage to move up gradients of attractants despite these physical challenges? It is evidently too small to sense the gradient along the length of its own

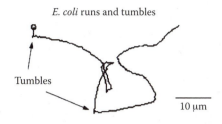

FIGURE 7.2 Trail of a swimming bacteria that shows runs and tumbles during 5 sec of motion in a uniform fluid environment. Runs are periods of roughly straight motion, and tumbles are brief events in which orientation is randomized. During chemotaxis, bacteria reduce the tumbling frequency when climbing gradients of attractants.

body.[1] The answer was discovered by Howard Berg in the early 1970s: *E. coli* uses **temporal gradients** to guide its motion. It uses a biased-random-walk strategy to sample space and convert spatial gradients to temporal ones. In liquid environments, *E. coli* swims in a pattern that resembles a random walk. The motion is composed of **runs**, in which the cell keeps a rather constant direction, and **tumbles**, in which the bacterium stops and randomly changes direction (Figure 7.2). The runs last about 1 sec on average and the tumbles about 0.1 sec.

To sense gradients, *E. coli* compares the current attractant concentration to the concentration in the past. When *E. coli* moves up a gradient of attractant, it detects a net positive change in attractant concentration. As a result, it reduces the probability of a tumble (it reduces its **tumbling frequency**) and tends to continue going up the gradient. The reverse is true for repellents: if it detects that the concentration of repellent increases with time, the cell increases its tumbling frequency, and thus tends to change direction and avoid swimming toward repellents. Thus, chemotaxis senses the temporal derivative of the concentration of attractants and repellents.

The runs and tumbles are generated by different states of the motors that rotate the bacterial flagella. Each cell has several flagella motors (Figure 7.3; see also Section 5.5) that can rotate either clockwise (CW) or counterclockwise (CCW). When the motors turn CCW, the flagella rotate together in a bundle and push the cell forward. When one of the motors turns CW, its flagellum breaks from the bundle and causes the cell to tumble about and randomize its orientation. When the motor turns CCW, the bundle is reformed and the cell swims in a new direction (Figure 7.4).

7.2.2 Response and Exact Adaptation

The basic features of the chemotaxis response can be described by a simple experiment. In this experiment, bacteria are observed under a microscope swimming in a liquid with

---

[1] Noise prohibits a detection system based on differences between two antennae at the two cell ends. To see this, note that *E. coli*, whose length is about 1 micron, can sense gradients as small as 1 molecule per micron in a background of 1000 molecules per cell volume. The Poisson fluctuations of the background signal, $\sqrt{1000} \sim 30$, mask this tiny gradient, unless integrated over prohibitively long times. Larger eukaryotic cells, whose size is on the order of 10 μm and whose responses are on the order of minutes, appear to sense spatial gradients directly.

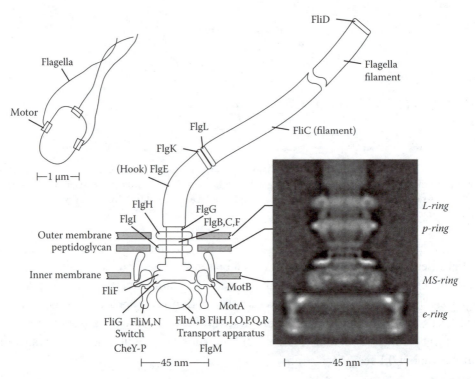

FIGURE 7.3 The bacteria flagella motor. Right panel: an electron microscope recontructed image of the flagella motor. (From Berg, 2003.)

no gradients. The cells display runs and tumbles, with an average **steady-state tumbling frequency** f, on the order of f ~ 1 sec^{-1}.

We now add an attractant such as aspartate to the liquid, uniformly in space. The attractant concentration thus increases at once from zero to l, but no spatial gradients are formed. The cells sense an increase in attractant levels, no matter which direction they are swimming. They think that things are getting better and suppress tumbles: the tumbling frequency of the cells plummets within about 0.1 sec (Figure 7.5).

After a while, however, the cells realize they have been fooled. The tumbling frequency of the cells begins to increase, even though attractant is still present (Figure 7.5). This process, called **adaptation**, is common to many biological sensory systems. For example, when we move from light to dark, our eyes at first cannot see well, but they soon adapt to sense small changes in contrast. Adaptation in bacterial chemotaxis takes several seconds to several minutes, depending on the size of the attractant step.[1]

Bacterial chemotaxis shows **exact adaptation**: the tumbling frequency in the presence of attractant returns to the same level as before attractant was added. In other words, *the steady-state tumbling frequency is independent of attractant levels.*

---

[1] Each individual cell has a fluctuating tumbling frequency signal, so that the tumbling frequency varies from cell to cell and also varies along time for any given cell (Ishihara et al., 1983; Korobkova et al., 2004). The behavior of each cell shows the response and adaptation characteristics within this noise.

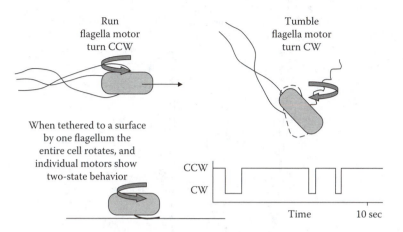

FIGURE 7.4 Bacterial runs and tumbles are related to the rotation direction of the flagella motors. When all motors spin counterclockwise (CCW), the flagella turn in a bundle and the cell is propelled forward. When one or more motors turn clockwise (CW), the cell tumbles and randomizes its orientation. The switching dynamics of a single motor from CCW to CW and back can be seen by tethering a cell to a surface by one flagellum hook, so that the motor spins the entire cell body (at frequencies of only a few Hertz due to the large viscous drag of the body).

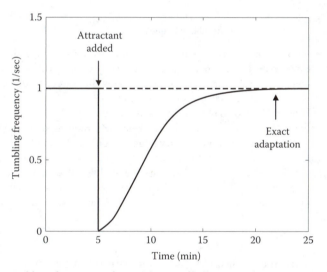

FIGURE 7.5 Average tumbling frequency of a population of cells exposed at time t = 5 to a step addition of saturating attractant (such as aspartate). After t = 5, attractant is uniformly present at constant concentration. Adaptation means that the effect of the stimulus is gradually forgotten despite its continued presence. Exact adaptation is a perfect return to prestimulus levels, that is, a steady-state tumbling frequency that does not depend on the level of attractant.

If more attractant is now added, the cells again show a decrease in tumbling frequency, followed by exact adaptation. Changes in attractant concentration can be sensed as long as attractant levels do not saturate the receptors that detect the attractant.

Exact adaptation poises the sensory system at an activity level where it can respond to multiple steps of the same attractant, as well as to changes in the concentration of other attractants and repellents that can occur at the same time. It prevents the system

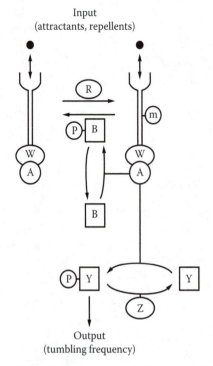

Input
(attractants, repellents)

Output
(tumbling frequency)

FIGURE 7.6 The chemotaxis signal transduction network. Information about the chemical environment is transduced into the cells by receptors, such as the aspartate receptor Tar, which span the membrane. The chemoreceptors form complexes inside the cells with the kinases CheA (A) and the adapter protein CheW (W). CheA phosphorylates itself and then transfers phosphoryl (P) groups to CheY (Y), a diffusible messenger protein. The phosphorylated form of CheY interacts with the flagellar motors to induce tumbles. The rate of CheY dephosphorylation is greatly enhanced by CheZ (Z). Binding of attractants to the receptors decreases the rate of CheY phosphorylation and tumbling is reduced. Adaptation is provided by changes in the level of methylation of the chemoreceptors: methylation increases the rate of CheY phosphorylation. A pair of enzymes, CheR (R) and CheB (B), add and remove methyl (m) groups. To adapt to an attractant, methylation of the receptors must rise to overcome the suppression of receptor activity caused by the attractant binding. CheA enhances the demethylating activity of CheB by phosphorylating CheB. (From Alon et al., 1999.)

from straying away from a favorable steady-state tumbling frequency that is required to efficiently scan space by random walk.

7.3   THE CHEMOTAXIS PROTEIN CIRCUIT OF *E. COLI*

We now look inside the *E. coli* cell and describe the protein circuit that performs the response and adaptation computations. The input to this circuit is the attractant concentration, and its output is the probability that motors turn CW, which determines the cells' tumbling frequency (Figure 7.6). The chemotaxis circuit was worked out using genetics, physiology, and biochemistry, starting with J. Adler in the late 1960s, followed by several labs, including those of D. Koshland, S. Parkinson, M. Simon, J. Stock, and others. The broad biochemical mechanisms of this circuit are shared with signaling pathways in all types of cells.

Attractant and repellent molecules are sensed by specialized detector proteins called **receptors**. Each receptor protein passes through the cell's inner membrane, and has one part outside of the cell membrane and one part inside the cell. It can thus pass information from the outside to the inside of the cell. The attractant and repellent molecules bound by a receptor are called its **ligands**.

E. coli has five types of receptors, each of which can sense several ligands. There are a total of several thousand receptor proteins in each cell. They are localized in a cluster on the inner membrane, such that ligand binding to one receptor appears to somehow affect the state of neighboring receptors. Thus, a single ligand binding event is amplified, because it can affect more than one receptor (Bray, 2002), increasing the sensitivity of this molecular detection device (Segall et al., 1986; Jasuja et al., 1999; Sourjik and Berg, 2004).

Inside the cell, each receptor is bound to a protein kinase called CheA.[1] We will consider the receptor and the kinase as a single entity, called X. X transits rapidly between two states, active (denoted X*) and inactive, on a timescale of microseconds. When X is active, X*, it causes a modification to a response regulator protein, CheY, which diffuses in the cell. This modification is the addition of a phosphoryl group (PO_4) to CheY to form phospho-CheY (denoted CheY-P). This type of modification, called **phosphorylation**, is used by most types of cells to pass bits of information among signaling proteins, as we saw in Chapter 6. CheY-P can bind the flagella motor and increase the probability that it switches from CCW to CW rotation. Thus, the higher the concentration of CheY-P, the higher the tumbling frequency (Cluzel et al., 2000).

The phosphorylation of CheY-P is removed by a specialized enzyme called CheZ. At steady-state, the opposing actions of X* and CheZ lead to a steady-state CheY-P level and a steady-state tumbling frequency.

Thus, the main pathway in the circuit is phosphorylation of CheY by X*, leading to tumbles. We now turn to the mechanism by which attractant and repellent ligands can affect the tumbling frequency.

7.3.1 Attractants Lower the Activity of X

When a ligand binds receptor X, it changes the probability[2] that X will assume its active state X*. The concentration of X in its active state is called the **activity of X**. Binding of an attractant *lowers* the activity of X. Therefore, attractants reduce the rate at which X phosphorylates CheY, and levels of CheY-P drop. As a result, the probability of CW motor rotation drops. In this way, the attractant stimulus results in reduced tumbling frequency, so that the cells keep on swimming in the right direction.

Repellents have the reverse effect: they increase the activity of X, resulting in increased tumbling frequency, so that the cell swims away from the repellent. These responses occur

---

[1] The chemotaxis genes and proteins are named with the three-letter prefix *che*, signifying that mutants in these genes are not able to perform chemotaxis.

[2] Note the strong separation of timescales in this system. Ligands remain bound to the receptor for about 1 msec. The conformation transitions between X and X* are thought to be on a microsecond timescale. Therefore, many such transitions occur within a single-ligand binding event. The activity of X is obtained by averaging over many transitions (Asakura and Honda, 1984; Mello et al., 2004; Keymer et al., 2006). Phosphorylation–dephosphorylation reactions equilibrate on the 0.1-sec timescale, and methylations occur on the many-minute timescale.

within less than 0.1 sec. The response time is mainly limited by the time it takes CheY-P to diffuse over the length of the cells, from the patch of receptors at the cell pole where CheY is phosphorylated to the motors that are distributed all around the cell.

The pathway from X to CheY to the motor explains the initial response in Figure 7.5, in which attractant leads to reduction in tumbling. What causes adaptation?

7.3.2 Adaptation Is Due to Slow Modification of X That Increases Its Activity

The chemotaxis circuit has a second pathway devoted to adaptation. As we saw, when attractant ligand binds X, the activity of X is reduced. However, each receptor has several biochemical "buttons" that can be pressed to increase its activity and compensate for the effect of the attractant. These buttons are **methylation** modifications, in which a methyl group (CH_3) is added to four or five locations on the receptor. Each receptor can thus have between zero and five methyl modifications. The more methyl groups that are added, the higher the activity of the receptor.

Methylation of the receptors is catalyzed by an enzyme called CheR and is removed by an enzyme called CheB. Methyl groups are continually added and removed by these two antagonistic enzymes, regardless of whether the bacterium senses any ligands. This seemingly wasteful cycle has an important function: it allows cells to adapt.

Adaptation is carried out by a negative feedback loop through CheB. Active X acts to phosphorylate CheB, making it more active. Thus, reduced X activity means that CheB is less active, causing a reduction in the rate at which methyl groups are removed by CheB. Methyl groups are still added, though, by CheR at an unchanged rate. Therefore, the concentration of methylated receptor, X_m, increases. Since X_m is more active than X, the tumbling frequency increases. Thus, the receptors X first become less active due to attractant binding, and then methylation level gradually increases, restoring X activity.

Methylation reactions are much slower than the reactions in the main pathway from X to CheY to the motor (the former are on the timescale of seconds to minutes, and the latter on a subsecond timescale). The protein CheR is present at low amounts in the cell, about 100 copies, and appears to act at saturation (zero-order kinetics). The slow rate of the methylation reactions explains why the recovery phase of the tumbling frequency during adaptation is much slower than the initial response.

The feedback circuit is designed so that exact adaptation is achieved. That is, the increased methylation of X *precisely balances* the reduction in activity caused by the attractant. How is this precise balance achieved? Understanding exact adaptation is the goal of the models that we will next describe.

7.4 TWO MODELS CAN EXPLAIN EXACT ADAPTATION: ROBUST AND FINE-TUNED

One can develop mathematical models to describe the known biochemical reactions in the chemotaxis circuit. We will now describe two different models based on this biochemistry. These are toy models, which neglect many details, and whose goal is to understand the essential features of the system. Both models reproduce the basic response of the chemotaxis system and display exact adaptation. In one model, exact adaptation is fine-tuned and depends on a precise balance of different biochemical parameters. In the second model, exact adaptation is robust and occurs for a wide range of parameters.

7.4.1 Fine-Tuned Model

Our first model is the most direct description of the biochemical interactions described above. In other words, it is a natural first model. Indeed, this model is a simplified form of a theoretical model of chemotaxis first proposed by Albert Goldbeter, Lee Segel, and colleagues (Knox et al., 1986). This study formed an important basis for later theoretical work on the chemotaxis system.

In the model (Figure 7.7), the receptor complex X can become methylated X_m under the action of CheR, and demethylated by CheB. For simplicity, we ignore the precise number of methyl groups per receptor and group together all methylated receptors into one variable X_m. Only the methylated receptors are active, with activity a_0 per methylated receptor, whereas the unmethylated receptors are inactive.

To describe the dynamics of receptor methylation, one needs to model the actions of the methylating enzyme CheR and the demethylating enzyme CheB. The enzyme CheR works at saturation, (that is, at a rate that is independent of the concentration of its substrate), with rate V_R. In contrast, CheB works with Michaelis–Menten kinetics (readers not familiar with Michaelis–Menten kinetics will find an explanation in Appendix A.7). Hence, the rate of change of X_m is the difference of the methylation and demethylation rates:

$$dX_m/dt = V_R R - V_B B X_m/(K + X_m) \tag{7.4.1}$$

The parameters R and B denote the concentrations of CheR and CheB. At steady state, $dX_m/dt = 0$, the dynamics reach a steady-state level of methylated receptor:

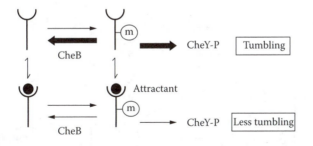

FIGURE 7.7 Fine-tuned mechanism for exact adaptation. Receptors are methylated by CheR and demethylated by CheB. Methylated receptors (marked with an m) catalyze the phosphorylation of CheY, leading to tumbles. When attractant binds, the activity of each methylated receptor is reduced and tumbling is reduced. In addition, the activity of CheB is reduced due to the negative feedback loop in the system. Thus, the concentration of methylated receptors gradually increases, until the tumbling frequency returns to the prestimulus state. Exact adaptation depends on tuning between the reduction in CheB activity and the reduction in activity per methylated receptor upon attractant binding, so that the activity returns to the prestimulus level.

$$X_m = K V_R R/(V_B B - V_R R) \qquad (7.4.2)$$

Recall that the unmethylated receptor has zero activity, whereas X_m has activity a_0 per receptor, resulting in a total steady-state activity of

$$A_0 = a_0 X_m \qquad \textit{steady-state activity with no attractant} \qquad (7.4.3)$$

The activity of the receptors, A_0, describes the rate at which CheY is phosphorylated to create CheY-P. The phosphorylated messenger CheY-P, in turn, binds the motor to generate tumbles. The activity A_0 therefore determines the steady-state tumbling frequency, $f = f(A_0)$.

Now imagine that saturating attractant is added to the cells, so that all of the receptors bind attractant ligand. The attractant causes receptors to assume their inactive conformation. As a result, the activity per methylated receptor drops to $a_1 \ll a_0$. Therefore, the total activity at short times after attractant is added drops to a low value:

$$A_1 = a_1 X_m \qquad (7.4.4)$$

Thus, the total activity is reduced after addition of attractant, $A_1 \ll A_0$ (Figure 7.8). This accounts for the sharp initial drop in tumbling frequency in Figure 7.5. Gradually, however, the methylation feedback loop kicks in. In this loop, because the receptors are less active, the rate of CheB action is decreased, from V_B to V_B'; that is, the demethylation rate is reduced. As a result, receptor methylation X_m begins to increase due to continual methylation by CheR. Receptor methylation at steady state reaches a balance between methylation and demethylation, just as in Equation 7.4.2, but with the demethylation rate set to its new value, V_B':

$$X_m' = K V_R R/(V_B' B - V_R R) \qquad (7.4.5)$$

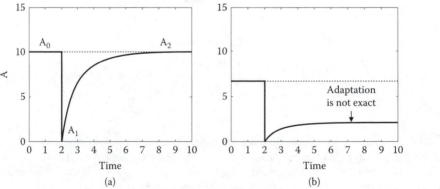

FIGURE 7.8 Activity dynamics in the fine-tuned model in response to a step addition of saturating attractant at time $t = 2$ (dimensionless units throughout). (a) Fine-tuned model shows exact adaptation with a tuned parameter set. (b) Dynamics when CheR level is lowered by 20% with respect to the fine-tuned parameter set of (a).

resulting in a new steady-state activity:

$$A_2 = a_1 X_m' \qquad \textit{steady-state activity with attractant} \qquad (7.4.6)$$

Exact adaptation means that the steady-state activity before attractant addition, A_0, is equal to the steady-state activity in the presence of ligand, A_2:

$$A_0 = A_2 \qquad \textit{exact adaptation} \qquad (7.4.7)$$

To attain exact adaptation, the increase in receptor methylation must *precisely* balance the decrease in receptor activity caused by the ligand. This results in a relation that must be fulfilled by the parameters of the system[1], based on equating Equations 7.4.2 and 7.4.5:

$$a_0 K V_R R/(V_B B - V_R R) = a_1 K V_R R/(V_B' B - V_R R) \qquad (7.4.8)$$

Let us play with numbers to get a feel for how exact adaptation works in this model. Let us use a 10-fold reduction in receptor activity due to ligand binding: activity per receptor before ligand binding is $a_0 = 10$, and after ligand binding, $a_1 = 1$. Let us use $K = 1$, $V_R R = 1$, and $V_B B = 2$ (units are not important for the present discussion). These values lead to an activity in the absence of attractant of

$$A_0 = a_0 K V_R R/(V_B B - V_R R) = 10 \cdot 1/(2 - 1) = 10 \qquad (7.4.9)$$

After attractant addition, activity per receptor drops 10-fold to $a_1 = 1$. In order to reach exact adaptation, Equation 7.4.8 constrains $V_B' B$ to a specific value, namely, $V_B' B = 1.1$, so that the activity adapts to the prestimulus level:

$$A_2 = a_1 K V_R R/(V_B' B - V_R R) = 1 \cdot 1/(1.1 - 1) = 10 \qquad (7.4.10)$$

Exact adaptation in this model depends on a strict relation between the biochemical parameters. What happens if the parameters change? For example, suppose the concentration of protein CheR is reduced by a factor of 20%, so that $V_R R$ goes from 1 to 0.8. In this case,

$$A_0 = 10 \cdot 0.8/(2 - 0.8) = 6.66 \qquad (7.4.11)$$

and

$$A_2 = 1 \cdot 0.8/(1.1 - 0.8) = 2.33 \qquad (7.4.12)$$

---

[1] The reader might worry about increasing CheR, so that the denominator in Equations 7.4.2 and 7.4.5 becomes negative, leading to a negative activity. However, increasing CheR causes unmethylated X levels to drop to the point where the approximation that CheR works at saturation is no longer valid. The action of CheR should then be described by a Michaelis–Menten equation, $V_R R X/(K_R + X)$, instead of zero-order kinetics, ensuring that activity remains positive. Exact adaptation remains fine-tuned when using Michaelis–Menten activity for CheR in the model.

We see that exact adaptation is lost, since A_2 is no longer equal to A_0. In this example, a modest 20% change in the level of a protein (CheR) caused almost a threefold difference in the steady-state activities with and without ligand (Figure 7.8b). Exact adaptation is a fine-tuned property in this model.

7.4.2   The Barkai-Leibler Robust Mechanism for Exact Adaptation

A mechanism that allows exact adaptation for a wide range of biochemical parameters was suggested by Naama Barkai and Stanislas Leibler (1997). The full model includes several methylation sites and other details, and reproduces many observations on the dynamical behavior of the chemotaxis system (a two-methylation-site version is solved in exercise 7.1). Here we will analyze a simplified version of the Barkai–Leibler model, aiming to understand how a biochemical circuit can robustly adapt.

Our toy model (Figure 7.9) has a single methylation state, so that receptors can be either unmethylated, X, or methylated, X_m. The unmethylated receptor X is inactive, whereas X_m transits between an inactive state and an active state, X_m^*. The activity is proportional to the number of receptors in the methylated, active state:

$$A = X_m^* \qquad (7.4.13)$$

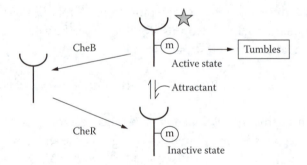

FIGURE 7.9 The Barkai–Leibler mechanism for exact adaptation. Unmethylated receptors are methylated by CheR at a constant rate. Demethylation is due to CheB, which acts only on the active methylated receptors. Methylated receptors (marked with m) transit rapidly between active and inactive states (the former marked with a star). Attractant binding increases the probability to become inactive, whereas repellents increase the probability to become active. The active receptors catalyze the phosphorylation of CheY, leading to tumbles. When attractant is added, many active receptors rapidly become inactive, and hence the tumbling frequency decreases. The reduced number of active receptors means that CheB has less substrate to work on, and thus the demethylation rate drops. Since CheR continues to work at a constant rate, the total number of methylated receptors increases. This increase only stops when demethylation rate exactly balances the methylation rate, that is, when the number of active receptors returns to its prestimulus value. Thus, exact adaptation occurs because the concentration of active receptors adjusts itself so that the demethylation rate is equal to the constant methylation rate.

The model is based on two key features. First, CheR must work at saturation. Second, CheB can only demethylate the *active* receptors, X_m^* (Figure 7.5). To repeat, CheB does not work on the inactive methylated receptors. This leads to the following equation for the total concentration of methylated receptors (both active and inactive):

$$\frac{d\left(X_m + X_m^*\right)}{dt} = V_R R - \frac{V_B B X_m^*}{K + X_m^*} \tag{7.4.14}$$

where R and B denote the concentrations of CheR and CheB. Note that the Michaelis–Menten term for CheB contains the concentration of its substrate, active receptors X_m^*. The steady-state of this dynamic equation occurs when $d(X_m + X_m^*)/dt = 0$. At steady-state, the value of X_m^* reaches a point where demethylation exactly balances the constant flux of methylation:

$$V_R R = \frac{V_B B X_m^*}{K + X_m^*} \tag{7.4.15}$$

which can be solved for the steady-state activity $A = X_m^*$:

$$A_2 = X_m^* = \frac{K V_R R}{V_B B - V_R R} \tag{7.4.16}$$

When attractant is added, it binds the receptors and decreases the probability of the active state. Therefore, the number of active receptors $A = X_m^*$ rapidly decreases. This causes the abrupt initial drop in tumbling frequency that is observed in the experiments (Figure 7.10).

After this sharp initial response, adaptation occurs due to the fact that CheB only works on the active receptors. The rate of demethylation by CheB is reduced because of the decrease in X_m^* caused by the attractant. CheR, on the other hand, continues to methylate receptors at a constant rate. Therefore, the total number of methylated receptors gradually increases:

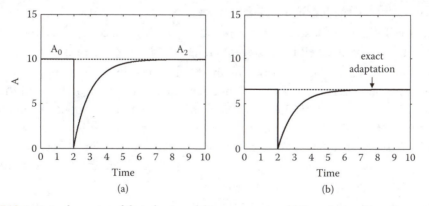

(a)   (b)

FIGURE 7.10 Activity dynamics of the robust model in response to addition of saturating attractant at time t = 2. (a) Model parameters K = 10, $V_R R = 1$, and $V_B B = 2$. (b) Same parameters with R reduced by a factor of 2. Exact adaptation is preserved. Note that the value of the steady-state tumbling frequency is fine-tuned and depends on the model parameters.

$$\frac{d\left(X_m + X_m^*\right)}{dt} = V_R R - \frac{V_B B X_m^*}{K + X_m^*} \tag{7.4.17}$$

As the total number of methylated receptors increases, so does the number of active methylated receptors X_m^*, which are a fraction of the total methylated receptors. As before, steady-state is reached when X_m^* reaches a level that balances the effects of CheR and CheB, resulting in a steady-state activity level:

$$A_2 = X_m^* = \frac{K V_R R}{V_B B - V_R R} \tag{7.4.18}$$

This activity is equal to the pre-attractant activity. Thus, *the steady-state activity does not depend on ligand levels*:

$$A_2 = A_0 \tag{7.4.19}$$

How does this mechanism work? The crucial elements are a fixed flux of methylation due to CheR, set against a counterflux of demethylation by CheB that directly depends on the activity $A = X_m^*$. At steady state, the number of active receptors always adjusts itself so that demethylation balances the reference flux of methylation. In other words, the active receptors X_m^* return to the fixed point of Equation 7.4.17, no matter what the attractant level. The activity $A = X_m^*$ reaches a steady-state value that does *not* depend on the ligand stimulus. Exact adaptation is achieved. Figure 7.10 shows the dynamics of this model for two sets of parameters, in which CheR levels are varied by a factor of 2. It is seen that the steady-state activity changes, but adaptation remains exact.

Exact adaptation occurs for a wide range of variations in any of the parameters of the model, namely, K, V_R, V_B, R, and B. In contrast, the value of the steady-state activity to which the cells adapt depends on these parameters. In other words, steady-state activity is a *fine-tuned* feature of this model (Figure 7.10). Exact adaptation, in which the steady state does not depend on ligand levels, is a *robust* feature of the model and does not depend on the precise values of the biochemical parameters.

There are limits to robustness: for example, when $V_R R$ exceeds $V_B B$, the saturation assumption for enzyme CheR is no longer valid, and robustness breaks down.

Robustness of exact adaptation in this model depends on the assumption that CheB works only on active receptors, and does not demethylate receptors that are in their inactive state. This is a specific biochemical detail that is essential for robust adaptation. The assumption that CheB works only on active receptors is not unrealistic, because enzymes can be exquisitely specific in discriminating between molecular states. Relaxing this assumption by allowing a small relative rate ε for CheB action on inactive receptors entails a loss of exact adaptation by a factor on the order of ε.

7.4.3   Robust Adaptation and Integral Feedback

The feedback in the robust mechanism is special: the demethylation rate is related directly to the activity, rather than to some other entity, such as the level of CheB-P. The negative feedback loop therefore acts directly on the variable to be controlled.

The direct feedback in the robust mechanism of exact adaptation is related to the engineering control principle of **integral feedback** (Yi et al., 2000). In integral feedback, a device is controlled by a signal that integrates over time the error between the output and the desired output. This type of feedback is guaranteed to guide the device to the desired output level, regardless of variations in the system parameters, because otherwise the integral of the error grows without bound. Moreover, in many cases integral feedback can be shown to be the *only* robust solution to this problem. The integrator in bacterial chemotaxis that effectively sums the error in activity (the activity minus the steady-state activity) is the methylation level of the receptors. The properties of integral feedback in chemotaxis are examined in exercises 7.2 and 7.3.

7.4.4 Experiments Show That Exact Adaptation Is Robust, Whereas Steady-State Activity and Adaptation Times Are Fine-Tuned

An experimental test of robustness employed genetically engineered *E. coli* strains, which allowed controlled changes in the concentration of each of the chemotaxis proteins (Alon et al., 1999). This control was achieved by first deleting the gene for one chemotaxis protein (for example, CheR) from the chromosome, and then introducing into the cell a copy of the gene under control of an inducible promoter (the *lac* promoter). Thus, expression of the protein was controlled by means of an externally added chemical inducer (IPTG). The more inducer that was added, the higher the CheR concentration in the cells. In this way, CheR levels were varied from about 0.5 to 50 times their wild-type levels. The population response of these cells to a saturating step of attractant was monitored using video microscopy on swimming cells. The experiment was carried out with changes in the expression levels of different chemotaxis proteins.

It was found that the steady-state tumbling frequency and the adaptation time varied with the levels of the proteins that make up the chemotaxis network (Figure 7.11). For example, steady-state tumbling frequency increased with increasing CheR levels, whereas adaptation time decreased. Despite these variations, exact adaptation remained robust to within experimental error. These results support the robust model for exact adaptation.

7.5   INDIVIDUALITY AND ROBUSTNESS IN BACTERIAL CHEMOTAXIS

Spudich and Koshland (1976) observed that genetically identical cells appear to have an individual character as they perform chemotaxis. Some cells are "nervous" and tumble more frequently than others, whereas other cells are "relaxed" and swim with fewer tumbles than the norm. These individual characteristics of each cell last for tens of minutes. The adaptation time to an attractant stimulus also varies from cell to cell. Interestingly, these two features are correlated: the steady-state tumbling frequency f in a given cell is inversely correlated with its adaptation time, τ, that is, $f \sim 1/\tau$.

The robust model for bacterial chemotaxis can supply an explanation for the varying chemotaxis personalities of *E. coli* cells. This is based on the cell–cell variation in chemotaxis protein levels, and particularly in the least abundant protein in the system, CheR. Variations in CheR affect the tumbling frequency f and the adaptation time τ in opposite

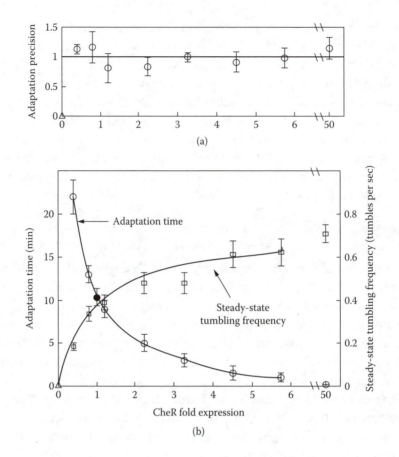

FIGURE 7.11 Experimental test of robustness in bacterial chemotaxis. The protein CheR was expressed at different levels by means of controlled expression with an inducible promoter, and the average tumbling frequency of a cell population was measured using video microscopy. Fold expression is the ratio of CheR protein level to that in the wild-type bacterium. Adaptation precision is the ratio of tumbling frequency before and after saturating attractant (1 mM aspartate). Adaptation time is the time to return to 50% of the steady-state tumbling frequency after saturating attractant addition. Adaptation time and steady-state tumbling frequency varied with CheR, whereas adaptation remained exact. Wild-type tumbling frequency in this experiment is about 0.4/sec (black dot in b). (From Alon et al., 1999.)

directions (Figure 7.11). The Barkai–Leibler model with multiple methylation sites suggests that f ~ CheR and τ ~ 1/CheR. Thus, the model predicts that f ~ 1/τ, explaining the observed correlation in these two features (see solved exercise 7.1).[1]

Despite the cell–cell variability in tumbling frequency, the vast majority of the cells in a population perform chemotaxis and climb gradients of attractants. On the other hand, mutant cells that have wild-type tumbling frequency but cannot adapt precisely (such as certain mutants in both CheR and CheB) are severely defective in chemotaxis abil-

---

[1] Detailed stochastic simulations of this protein circuit were performed by D. Bray and colleagues (Shimizu et al., 2003).

ity. Evidently, tumbling frequency need not be precisely tuned for successful chemotaxis, whereas exact adaptation is important for most ligands.[1]

In summary, it appears that the bacterial chemotaxis circuit has a design such that a key feature (exact adaptation) is robust with respect to variations in protein levels. Other features, such as steady-state activity and adaptation times, are fine-tuned. These latter features show variations within a population due to intrinsic cell–cell variations in protein levels. Because of the robust design, the intrinsic variability in the cell's protein levels does not abolish exact adaptation.

As a theorist, one can usually write many different models to describe a given biological system, especially if some of the biochemical interactions are not fully characterized. Of these models, only very few will typically be robust with respect to variations in the components. Thus, the robustness principle can help narrow down the range of models that work on paper to the few that can work in the cell. Robust design is an important factor in determining the specific types of circuits that appear in cells. In the next chapter, we will study how robustness constraints can shape the circuits that guide pattern formation in embryonic development.

FURTHER READING

Alon, U., Surette, M.G., Barkai, N., and Leibler, S. (1999). Robustness in bacterial chemotaxis. *Nature*, 397: 168–171.

Barkai, N. and Leibler, S. (1997). Robustness in simple biochemical networks. *Nature*, 387: 913–917.

Berg, H.C. (2003). *E. coli in Motion*. Springer.

Berg, H.C. and Brown, D.A. (1972). Chemotaxis in *Escherichia coli* analyzed by three-dimensional tracking. *Nature*, 239: 500–504.

Berg, H.C. and Purcell, E.M. (1977). Physics of chemoreception. *Biophys. J.*, 20: 193–219.

Kitano, H., (2002). Biological robustness. *Nat Rev. Genet.*, 5: 826–837.

Knox, B.E., Devreotes, P.N., Goldbeter, A., and Segel, L.A. (1986). A molecular mechanism for sensory adaptation based on ligand-induced receptor modification. *Proc. Natl. Acad. Sci. U.S.A.*, 83: 2345–2349.

Kollmann, M., Lovdok, L., Bartholome, K., Timmer, J., and Sourjik, V., (2005). Design principles of a bacterial signalling network. *Nature*, 438: 504–507.

Spudich, J.L. and Koshland, D.E., Jr. (1976). Non-genetic individuality: chance in the single cell. *Nature*, 262: 467–471.

Yi, T.M., Huang, Y., Simon, M.I., and Doyle, J. (2000). Robust perfect adaptation in bacterial chemotaxis through integral feedback control. *Proc. Natl. Acad. Sci. U.S.A.*, 97: 4649–4653.

---

[1] In some conditions, the attractant serine at high concentrations reduces the steady-state tumbling frequency (Berg and Brown, 1972). At these concentrations, serine inhibits chemotaxis to other attractants.

EXERCISES

7.1. *Robust model with two methylation sites.* The receptor X can be methylated on two positions, and can thus have zero, one, or two methyl groups, denoted X_0, X_1, and X_2. The enzyme R works at saturation (zero-order kinetics) to methylate X_0 and X_1. The demethylating enzyme B works only on the active receptor conformation, removing methyl groups with equal rate from X_1^* and X_2^*. For simplicity, assume that B works with first-order kinetics. The reactions are:

methylation $X_0 \rightarrow X_1$ at rate $R V_R X_0/(X_1 + X_0)$,

the last factor occurs because R is distributed between its substrates X_0 and X_1

methylation $X_1 \rightarrow X_2$ at rate $R V_R X_1/(X_1 + X_0)$

$X_1 \rightleftharpoons X_1^*$ rapid transitions at a rate that depends on the ligand level

$X_2 \rightleftharpoons X_2^*$ rapid transitions at a rate that depends on the ligand level

de-methylation $X_1^* \rightarrow X_0$ at rate $B V_B X_1^*$

de-methylation $X_2^* \rightarrow X_1$ at rate $B V_B X_2^*$

a. What is the steady-state activity $A = X_1^* + X_2^*$? Does it depend on the concentration of ligand l? Is there exact adaptation?

b. Estimate the adaptation time, the time needed for 50% adaptation after addition of saturating attractant. Note that to adapt to saturating attractant, virtually all of the receptors need to be doubly methylated.

c. Spudich and Koshland (1976) found that different cells in a population have different steady-state activities and different adaptation times. Moreover, these two features were found to be correlated: the higher the activity A, the shorter the adaptation time τ in a given cell, with $A \sim 1/\tau$. Explain this finding using the model, based on cell–cell variations in the concentration of R (Barkai and Leibler, 1997).

Solution:

a. The rates of change of the doubly methylated receptor concentration and the nonmethylated receptor concentration are:

$$d(X_2 + X_2^*)/dt = R V_R X_1/(X_1 + X_0) - B V_B X_2^* \tag{P7.1}$$

$$dX_o/dt = -R V_R X_o/(X_o + X_1) + B V_B X_1^* \tag{P7.2}$$

Subtracting these two equations yields

$$d(X_2 + X_2^*)/dt - dX_0/dt = R V_R - B V_B (X_1^* + X_2^*) = R V_R - B V_B A \tag{P7.3}$$

The activity $A = X_1^* + X_2^*$ is therefore (setting d/dt terms to zero):

$$A_{st} = R V_R/B V_B \tag{P7.4}$$

This activity does not depend on the ligand concentration. Therefore, this mechanism displays exact adaptation.

b. In the case of saturating ligand, all receptors in all of their forms bind attractant ligand. The attractant reduces the activity of all methylated receptors, and thus at initial times X_1^* is small. In addition, when adaptation is completed, X_1^* is small because the majority of receptors need to be doubly methylated in order to balance the strong inhibitory effect of the saturating attractant. Thus, it seems that X_1^* is relatively small throughout most of the dynamics. Since X_1^* is small, the demethylation flux from X_1^* to X_o is small. Hence, to a good approximation, X_o dynamics reflect only a reduction due to the action of CheR, because the term with B in Equation P7.2 is negligible:

$$dX_o/dt \approx -R V_R X_o/(X_o + X_1) \tag{P7.5}$$

so that X_o drops with time. At initial times (before attractant addition), let us denote by q the fraction of X_o among the possible substrates of CheR, $q = X_o/(X_o + X_1)$. Thus, the initial slope of the drop in X_o is $- q R V_R$. The adaptation time to saturating ligand (time to recover to 50% activity) is the time needed to build enough methylated receptors to restore activity, at the expense of most of the unmethylated ones. Thus, it is approximately the time for X_o to decline to 50% of its initial value. This adaptation time is equal to the number of methylation reactions needed (that is, methylations equal to 50% of X_o) divided by the rate at which they occur, namely (ignoring the changes in q over this time):

$$\tau \sim 0.5 X_o/q R V_R \tag{P7.6}$$

Thus, the adaptation time becomes shorter the more R enzymes exist in the cell. This makes sense because the more R enzymes there are, the faster methylation occurs and the faster the adaptation.

Note that the single methylation model discussed in the text has a different adaptation time, governed by B and not R. This is because we cannot ignore the flux from X_1^* to X_o, which is necessary to produce exact adaptation in the single methylation model. But B governs the adaptation time only if we restrict ourselves to a single methylation site, as we did for clarity in the text. In reality there are multiple methylation sites. The adaptation time is generally governed by R in models with more than one methylation site (Barkai and Leibler, 1997). In experiments, the adaptation time is found to decrease with R (Figure 7.11), in agreement with the multi-site models.

c. We saw above that the adaptation time varies as $\tau \sim 1/R$ (Equation P7.6) and the steady-state activity varies as $A_{st} \sim R$ (Equation P7.4). Thus, if R is the protein with the largest variation between genetically identical cells, one would expect that $A_{st} \sim 1/\tau$, as observed. The protein R is the least abundant chemotaxis signaling protein in *E. coli*, with on the order of 100 copies per cell, whereas there are on the order of several thousand copies of CheB, CheY, CheZ, and CheA per cell. CheR may therefore be the most prone to stochastic variations.

7.2. *Integral feedback.* A heater heats a room. The room temperature T increases in proportion to the power of the heater, P, to other sources of heat, S, and decreases due to thermal diffusion to the outside at a rate proportional to T:

$$dT/dt = a P + S - b T \tag{P7.7}$$

An integral feedback device is placed in order to keep the room temperature at a desired point T_o. In this feedback loop, the power to the heater is proportional to the integral over time of the error in temperature, $T - T_o$:

$$P = P_o - K \int (T - T_o) \, dt \tag{P7.8}$$

This feedback loop thus reduces the power to the heater if the room temperature is too high, $T > T_o$, and increases the power when the room temperature is too low. Taking the time derivative of the power, we find

$$dP/dt = -K (T - T_o) \tag{P7.9}$$

a. Show that the steady-state temperature is T_o and that this steady-state does not depend on any of the system parameters, including the room's thermal coupling to the heater, a, the additional heat sources, S, the room's thermal coupling with the outside, b, or the strength of the feedback, K. In other words, integral feedback shows robust exact adaptation of the room temperature.

b. Demonstrate that integral feedback is the *only* solution that shows robust exact adaptation of the room temperature, out of all possible linear control systems. That is, assume a general linear form for the controller:

$$dP/dt = c_1 T + c_2 P + c_3 \qquad (P7.10)$$

and show that integral feedback as a structural feature of the system is necessary and sufficient for robust exact adaptation.

7.3. *Integral feedback in chemotaxis.* Demonstrate that a simple linear form of the robust model for chemotaxis contains integral feedback. What is the integrator in this biological system (Yi et al., 2000)?

Solution:

In a linear model, CheR works at saturation and CheB works with first-order kinetics, and only on the active receptors. The rate of change of the total number of methylated receptors ($X_{m,t} = X_m + X_m^* = X_m + A$) is given by the difference between the methylation and de-methylation rates:

$$dX_{m,t}/dt = V_R R - V_B B A \qquad (P7.11)$$

This can be rewritten in terms of the difference between the activity A and its steady-state value A_{st}:

$$dX_{m,t}/dt = -V_B B (A - A_{st}) \qquad (P7.12)$$

where the steady-state activity is

$$A_{st} = V_R R/V_B B \qquad (P7.13)$$

The total number of methylated receptors, $X_{m,t}$, thus acts as the integrator in the system that integrates the error in activity over time (in analogy to Equation P7.8):

$$X_{m,t} \sim -V_B B \int (A - A_{st}) \, dt \qquad (P7.14)$$

The activity A is analogous to the room temperature in problem 7.2. To complete the analogy with problem 7.2, let us write a detailed equation for the rate of change of activity, $A = X_m^*$. The number of methylated active receptors X_m^* increases due to transitions from X_m to X_m^* at a ligand-dependent rate, k(l). The number X_m^* decreases due to the demethylating action of CheB and due to transitions to the inactive state X_m at a ligand-dependent rate k'(l). The dynamics of $X_m^* = A$ are therefore given by the sum over the rates of all of these transitions with appropriate signs:

$$dA/dt = k(l) X_m - k'(l) A - V_B B A \qquad (P7.15)$$

We want to rearrange this equation so that the first term is proportional to $X_{m,t} = X_m + A$ (analogous to the heater power, P, in exercise 7.2; Equation P7.7). For this purpose, we add and subtract $k(l)A$ to find

$$dA/dt = k(l) X_{m,t} - (k'(l) + k(l)) A - V_B B A \qquad (P7.16)$$

Thus, we end up with an integral feedback system, analogous to Equations P7.7 and P7.9, in which

$$dA/dt = a X_{m,t} - b A \qquad (P7.17)$$

$$dX_{m,t}/dt = -K (A - A_{st}) \qquad (P7.18)$$

where $a = k(l)$, $b = k'(l) + k(l) + V_B B$, and $K = V_B B$.

To restate the analogy, think of A as the temperature and $X_{m,t}$ as the power to the heater in problem 7.2. As shown in problem 7.2, the steady-state activity A_{st} does not depend on any of the parameters a, b, or K, and in particular on the ligand level that enters only through $k(l)$ and $k'(l)$ in the parameters a and b. Thus, A_{st} does not depend on the level of attractant (or repellent), and exact adaptation is achieved.

7.4. *Zero-order ultrasensitivity (Goldbeter and Koshland, 1981)*: In this exercise, we will see how two antagonistic enzymes can generate a sharp switch. A protein X can be in a modified X_1 or unmodified X_0 state. Modification is carried out by enzyme E_1, and de-modification by enzyme E_2. The rate V_2 of E_2 is constant, whereas the rate V_1 of E_1 is governed by an external signal. Consider V_1 as the input and X_1 as the output of this system.

(a) Assume that E_1 and E_2 work with first-order kinetics. What is the output X_1 as a function of input V_1.

(b) What is the sensitivity of this circuit, defined as the relative change in X_1 per relative change in V_1.

$$S(X_1, V_1) = \frac{V_1}{X_1} \frac{dX_1}{dV_1}$$

(c) Assume now that E_1 and E_2 work with zero-order kinetics. What is X_1 as a function of V_1? Note that $X_0 + X_1$ cannot exceed the total concentration X_{tot}.

(d) What is the sensitivity of the zero-order circuit? Explain why this is called "zero-order ultra-sensitivity".

(e) Compare the switching time (time to 50% change in X_1 upon a change in V_1) between the cases of (a) and (c) above.

7.5.   *Robust model with a single methylation site:* Consider the model of Section 7.4.2. The methylated receptor transits rapidly between the inactive form X_m and the active form X_m^*. Transitions from X_m to X_m^* occur at a rate $k(l)$, and transitions back occur at a rate $k'(l)$. Note that $k(l)$ and $k'(l)$ depend on ligand level l.

(a)   What is the average activity, averaged over many transition events between X_m and X_m^*?

(b)   These transitions occur much faster than changes in the methylation level of X. How can this be useful in analyzing the model?

(c)   Solve for the dynamics of $A(t) = X_m^*$ following a step addition of attractant ligand. What is the response time? Plot the dynamics schematically.

(d)   Same as (c), for a step addition of repellent ligand.

CHAPTER 8

Robust Patterning in Development

8.1 INTRODUCTION

Development is the remarkable process in which a single cell, an egg, becomes a multicellular organism. During development, the egg divides many times to form the cells of the embryo. All of these cells have the same genome. If they all expressed the same proteins, the adult would be a shapeless mass of identical cells. During development, therefore, the progeny of the egg cell must assume different fates in a spatially organized manner to become the various tissues of the organism. The difference between cells in different tissues lies in which proteins they express. In this chapter, we will consider how these spatial patterns can be formed precisely.

To form a spatial pattern requires positional information. This information is carried by gradients of signaling molecules (usually proteins) called **morphogens**. How are morphogen gradients formed? In the simplest case, the morphogen is produced at a certain source position and diffuses into the region that is to be patterned, called the **field**. A concentration profile is formed, in which the concentration of the morphogen is high near the source and decays with distance from the source. The cells in the field are initially all identical and can sense the morphogen by means of receptors on the cell surface. Morphogen binds the receptors, which in turn activate signaling pathways in the cell that lead to expression of a set of genes. Which genes are expressed depends on the concentration of morphogen. The fate of a cell therefore depends on the morphogen concentration at the cell's position.

The prototypical model for morphogen patterning is called the French flag model (Figure 8.1) (Wolpert, 1969; Wolpert et al., 2002). The morphogen concentration $M(x)$ decays with distance from its source at $x = 0$. Cells that sense an M concentration greater than a threshold value T_1 assume fate A. Cells that sense an M lower than T_1 but higher than a second threshold, T_2, assume fate B. Fate C is assumed by cells that sense low morphogen levels, $M < T_2$. The result is a three-region pattern (Figure 8.1). Real morphogens often lead to patterns with more than three different fates.

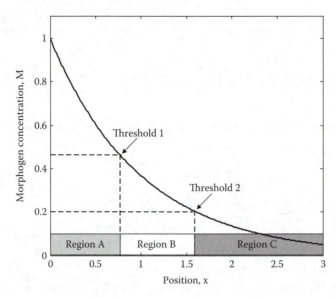

FIGURE 8.1 Morphogen gradient and the French flag model. Morphogen M is produced at x = 0 and diffuses into a field of cells. The morphogen is degraded as it diffuses, resulting in a steady-state concentration profile that decays with distance from the source at x = 0. Cells in the field assume fate A if M concentration is greater than threshold 1, fate B if M is between thresholds 1 and 2, and fate C if M is lower than threshold 2.

Figure 8.1 depicts a one-dimensional tissue, but real tissues are three-dimensional. Patterning in three dimensions is often broken down into one-dimensional problems in which each axis of the tissue is patterned by a specific morphogen.

Complex spatial patterns are not formed all at once. Rather, patterning is a sequential process. Once an initial coarse pattern is formed, cells in each region can secrete new morphogens to generate finer subpatterns. Some patterns require the intersection of two or more morphogen gradients. In this way, an intricate spatial arrangement of tissues is formed. The sequential regulation of genes during these patterning processes is carried out by the developmental transcription networks that we have discussed in Chapter 6. Additional processes (which we will not discuss), including cell movement, contact, and adhesion, further shape tissues in complex organisms.

Patterning by morphogen gradients is achieved by diffusing molecules sensed by biochemical circuitry, raising the question of the sensitivity of the patterns to variations in biochemical parameters. A range of experiments has shown that patterning in development is very robust with respect to a broad variety of genetic and environmental perturbations (Waddington, 1959; von Dassow et al., 2000; Wilkins, 2001; Eldar et al., 2004). The most variable biochemical parameter in many systems is, as we have mentioned previously, the production rates of proteins. Experiments show that changing the rate of morphogen production often leads to very little change in the sizes and positions of the regions formed. For example, a classic experimental approach shows that in many systems the patterning is virtually unchanged upon a twofold reduction in morphogen production, generated by mutating the morphogen gene on one of the two sister chromosomes.

In this chapter, we will consider mechanisms that can generate precise long-range patterns that are robust to such perturbations, following the work of Naama Barkai and her colleagues (Eldar et al., 2002, 2003, 2004). We will see that the most generic patterning mechanisms are not robust. Requiring robustness leads to special and rather elegant biochemical mechanisms.

8.2   EXPONENTIAL MORPHOGEN PROFILES ARE NOT ROBUST

Let us begin with the simplest mechanism, in which morphogen is produced at a source located at $x = 0$ and diffuses into a field of identical cells. The morphogen is degraded at rate α. We will see that the combination of diffusion and degradation leads to an exponentially decaying spatial morphogen profile.

The concentration of morphogen M in our model is governed by a one-dimensional diffusion–degradation equation. In this equation, the diffusion term, $D \, \partial^2 M / \partial x^2$, seeks to smooth out spatial variations in morphogen concentrations. The larger the diffusion constant D, the stronger the smoothing effect. The degradation of morphogen is described by a linear term $-\alpha M$, resulting in an equation that relates the rate of change of M to its diffusion and degradation:

$$\partial M / \partial t = D \, \partial^2 M / \partial x^2 - \alpha M \tag{8.2.1}$$

To solve this diffusion–degradation equation in a given region, we need to consider the values of M at the boundaries of the region. The boundary conditions are a steady concentration of morphogen at its source at $x = 0$, $M(x = 0) = M_o$, and zero boundary conditions far into the field, $M(\infty) = 0$, because far into the field all morphogen molecules have been degraded.

At steady-state ($\partial M / \partial t = 0$), Equation 8.2.1 becomes a linear ordinary differential equation:

$$D \, d^2 M / d x^2 - \alpha M = 0$$

And the solution is an exponential decay that results from a balance of the diffusion and degradation processes:

$$M(x) = M_o \, e^{-x/\lambda} \tag{8.2.2}$$

Thus, the morphogen level is highest at the source at $x = 0$, and decays with distance into the field. The decay is characterized by a decay length λ:

$$\lambda = \sqrt{D / \alpha} \tag{8.2.3}$$

The decay length λ is the typical distance that a morphogen molecule travels into the field before it is degraded. The larger the diffusion constant D and the smaller the degradation rate α, the larger is this distance. The decay is dramatic: at distances of 3λ and 10λ from the source, the morphogen concentration drops to about 5% and $5 \cdot 10^{-5}$ of its initial

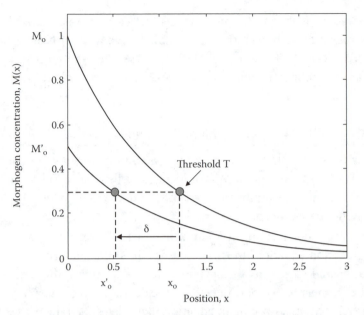

FIGURE 8.2 Changes in steady-state morphogen profile and the resulting pattern boundary upon a twofold reduction in morphogen concentration at $x = 0$, denoted M_o. The pattern boundary, defined by the position where $M(x)$ equals the threshold T, shifts to the left by δ when M_o is reduced to M_o'.

value. Roughly speaking, λ is the typical size of the regions that can be patterned with such a gradient.

The fate of each of the cells in the field is determined by the concentration of M at the cell's position: the cell fate changes when M crosses threshold T. Therefore, a boundary between two regions occurs when M is equal to T. The position of this boundary, x_o, is given by $M(x_o) = T$, or, using Equation 8.2.2,

$$x_o = \lambda \log (M_o/T) \tag{8.2.4}$$

What happens if the production rate of the morphogen source is perturbed, so that the concentration of morphogen at the source M_o is replaced by M_o'? Equation 8.2.4 suggests that the position of the boundary shifts to $x_o' = \lambda \log (M_o'/T)$. The difference between the original and the shifted boundary is (Figure 8.2)

$$\delta = x_o' - x_o = \lambda \log (M_o'/M_o) \tag{8.2.5}$$

Thus, a twofold reduction in M_o leads to a shift of the position of the boundary to the left by about $-\lambda \log(1/2) \sim 0.7 \lambda$, a large shift that is on the order of the size of the entire pattern. Region A in Figure 8.1 would be almost completely lost.

Hence, this type of mechanism does not seem to explain the robustness observed in developmental patterning. To increase robustness, we must seek a mechanism that decreases the shift δ that occurs upon changes in parameters such as the rate of morphogen production.

8.3 INCREASED ROBUSTNESS BY SELF-ENHANCED MORPHOGEN DEGRADATION

The simple diffusion and degradation process described above generates an exponential morphogen gradient that is not robust to the morphogen level at its source M_o.

To generate a more robust mechanism, let us try a more general diffusion–degradation process with a *nonlinear* degradation rate F(M):

$$\partial M / \partial t = D\, \partial^2 M / \partial x^2 - F(M) \tag{8.3.1}$$

The boundary conditions are as before, a constant source concentration, M(x = 0) = M_o, and decay to zero far into the field, M(∞) = 0. This diffusion process has a general property that will soon be seen to be important for robustness: the shift δ in the morphogen profile upon a change in M_o is uniform in space — it does not depend on position x. That is, *all regions are shifted by the same distance upon a change in M_o.*

This uniform shift certainly occurs in the exponential morphogen profile of the previous section. The shift in boundary position δ described by Equation 8.2.5 does not depend on x. Thus, if several regions are patterned by this morphogen, as in Figure 8.1, all boundaries will be shifted by the same distance δ if morphogen production is perturbed.

More generally, spatially uniform shifts result with any degradation function F(M) in Equation 8.3.1. This property is due to the fact that the cells in the field are initially identical (unpatterned), and that the field is large (zero morphogen at infinity). This means that Equation 8.3.1 governing the morphogen has translational symmetry: the diffusion–degradation equations are invariant to a coordinate change x → x + δ. Such shifts only produce changes in the boundary value at x = 0, that is, in M_o, as illustrated in Figure 8.3. The spatial shift that corresponds to a reduction of M_o to M_o' is given by the position δ at which the original profile equals M_o', M(δ) = M_o'. The solution of Equation 8.3.1 with boundary condition M_o' is identical to the solution with M_o shifted to the left by δ.

Our goal is to increase robustness, that is, to make the shift δ as small as possible upon a change in M_o to M_o'. To make the shift as small as possible, one must make the decay rate near x = 0 as large as possible, so that M_o' is reached with only a tiny shift. This could be done with an exponential profile only by decreasing the decay length λ. However, decreased λ comes at an unacceptable cost: the range of the morphogen, and hence the size of the patterns it can generate, is greatly reduced.

Thus, we seek a profile with both long range and high robustness. Such a profile should have two features:

1. Rapid decay near x = 0 to provide robustness to variations in M_o

2. Slow decay at large x to provide long range to M

A simple solution would be to make M degrade faster near the source x = 0 and slower far from the source. However, we cannot make the degradation of M explicitly depend on position x (that is, we cannot set α = α(x) in Equation 8.2.1), because the cells in the field are initially identical. A spatial dependence of the parameters would require positional

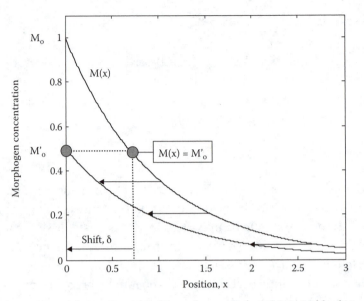

FIGURE 8.3 A change in morphogen concentration at the source from M_o to M_o' leads to a spatially uniform shift in the morphogen profile. All arrows are of equal length. The size of the shift is equal to the position at which $M(x) = M_o'$.

information that is not available without prepatterning the field. Our only recourse is nonlinear, self-enhanced degradation: a feedback mechanism that makes the *degradation rate of M increase with the concentration of M.*

A simple model for self-enhanced degradation employs a degradation rate that increases polynomially with M, for example,

$$\partial M / \partial t = D\, \partial^2 M / \partial x^2 - \alpha\, M^2 \tag{8.3.2}$$

This equation describes a nonlinear degradation rate that is large when M concentration is high, and small when M concentration is low.[1]

At steady state ($\partial M / \partial t = 0$), the morphogen profile that solves Equation 8.3.2 is not exponential, but rather a power law:

$$M = A\,(x + \varepsilon)^{-2} \qquad \varepsilon = (\alpha\, M_o / 6\, D)^{-1/2} \qquad A = 6\, D/\alpha \tag{8.3.3}$$

This power-law profile of morphogen has a very long range compared to exponential profiles. To obtain robust, long-range patterns, it is sufficient to make M_o very large, so that the parameter ε in Equation 8.3.3 is much smaller than the pattern size (note that $\varepsilon \sim 1/\sqrt{M_0}$). In this limit, the morphogen profile in the field does not depend on M_o at all:

---

[1] A nonlinear degradation $F(M) \sim M^2$ can be achieved by several mechanisms. For example, if M molecules dimerize weakly and reversibly, and only dimers are degraded, one has that the concentration of dimers (and hence the degradation of M) is proportional to the square of the monomer concentration $[M_2] \sim M^2$. Note that the parameter α in Equation 8.3.2 is in units of 1/(time · concentration).

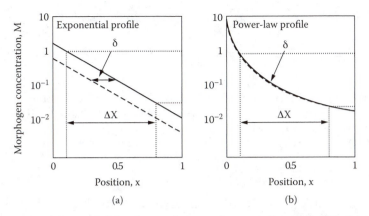

FIGURE 8.4 Comparison of exponential and power-law morphogen profiles. (a) A diffusible morphogen that is subject to linear degradation reaches an exponential profile at steady state (solid line). A perturbed profile (dashed line) was obtained by reducing the morphogen at the boundary, M_o, by a factor e. The resulting shift in cell fate boundary (δ) is comparable to the distance ΔX between two boundaries in the unperturbed profile, defined by the points in which the profile crosses thresholds given by the horizontal dotted lines. Note the logarithmic scale. (b) When the morphogen undergoes nonlinear self-enhanced degradation, a power-law morphogen profile is established at steady state. In this case, δ is significantly smaller than ΔX. The symbols are the same as in (a), and quadratic degradation was used (Equation 8.3.2). (From Eldar et al., 2003.)

$$M \sim A/x^2 \qquad (8.3.4)$$

so that there are negligible shifts even upon large perturbations in M_o. Patterning is very robust to variations in M_o, as long as M_o does not become too small (Figure 8.4).

The power-law profile is not robust to changes in the parameter $A \sim D/\alpha$, the ratio of the diffusion and degradation rates. However, parameters such as diffusion constants and specific degradation rates usually vary much less than production rates of proteins such as the morphogen.

In summary, self-enhanced degradation allows a steady-state morphogen profile with a nonuniform decay rate. The profile decays rapidly near the source, providing robustness to changes in morphogen production. It decays slowly far from the source, allowing long-ranged patterning.

8.4 NETWORK MOTIFS THAT PROVIDE DEGRADATION FEEDBACK FOR ROBUST PATTERNING

We saw that robust long-range patterning can be achieved using feedback in which the morphogen enhances its own degradation rate. Morphogens throughout the developmental processes of many species participate in certain network motifs that can provide this self-enhanced degradation. The robustness gained by self-enhanced degradation might explain why these regulatory patterns are so common.

The morphogen M is usually sensed by a receptor R on the surface of the cells in the field. When M binds R, it activates a signal transduction pathway that leads to changes in gene expression. Two types of feedback loops are found throughout diverse developmental processes (Figure 8.5).

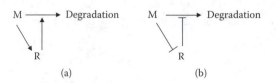

FIGURE 8.5 Two network motifs that provide self-enhanced degradation of morphogen M. (a) M binds receptor R and activates signaling pathways that increase R expression. M bound to R is taken up by the cells (endocytosis) and M is degraded. (b) M activates signaling pathways that repress R expression. The receptor R binds and inhibits an extracellular protein (a protease) that degrades M, and thus R effectively inhibits M degradation. In both (a) and (b), M enhances its own degradation rate.

The first motif is a feedback loop in which the receptor R enhances the degradation of M. An example is the morphogen M = Hedgehog and its receptor R = Patched, which participate in patterning the fruit fly and many other organisms. Morphogen binding to R triggers signaling that leads to an increase in the expression of R. Degradation of M is caused by uptake of the morphogen bound to the receptor and its breakdown within the cell (endocytosis). Thus, M enhances R production and R enhances the rate of M endocytosis and degradation (Figure 8.5a), forming a self-enhancing degradation loop.

The second type of feedback occurs when R inhibits M degradation (Figure 8.5b). A well-studied example in fruit flies is the morphogen M = Wingless and its receptor R = Frizzled. Binding of M to R triggers signaling that represses the expression of R. R in turn inhibits the degradation of M by binding to and inhibiting a protein that degrades M (an extracellular protease) or by repressing the expression of the protease.

In both of these feedback loops, M increases its own degradation rate, promoting robust long-range patterning.

Next, we discuss a different and more subtle feedback mechanism that can lead to robust patterning. Our goal is to demonstrate how the robustness principle can help us to select the correct mechanism from among many plausible alternatives.

8.5   THE ROBUSTNESS PRINCIPLE CAN DISTINGUISH BETWEEN MECHANISMS OF FRUIT FLY PATTERNING

We end this chapter by considering a specific example of patterning in somewhat more detail (Eldar et al., 2002). We begin with describing the biochemical interactions in a small network of three proteins that participate in patterning one of the spatial axes in the early embryo of the fruit fly *Drosophila*. These biochemical interactions can, in principle, give rise to a large family of possible patterning mechanisms. Of all of these mechanisms, only a tiny fraction is robust with respect to variations in all three protein levels. Thus, the robustness principle helps to home in on a nongeneric mechanism, making biochemical predictions that turned out to be correct.

The development of the fruit fly *Drosophila* begins with a series of very rapid nuclear divisions. We consider the embryo after 2.5 h of development. At this stage, it includes about 5000 cells, which form a cylindrical layer about 500 μm across. The embryo has two axes: head–tail (called the anterior–posterior axis) and front–back (called the ventral–dorsal axis).

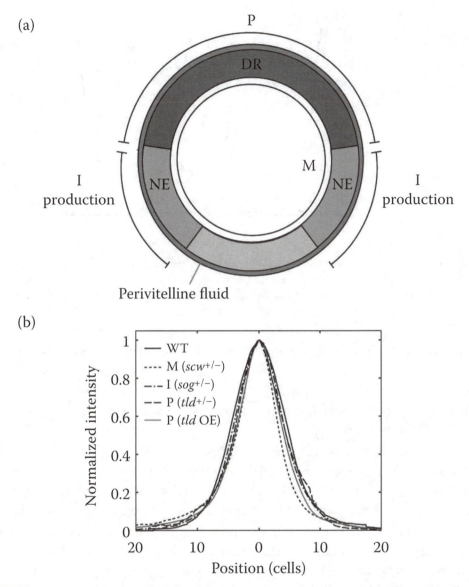

FIGURE 8.6 Cross section of the early *Drosophila* embryo, about 2 h from start of development. Cells are arranged on the periphery of a cylinder. Three cell types are found (three distinct domains of gene expression). This sets the stage for the patterning considered in this section, in which the dorsal region (DR), is to be subpatterned. Shown are the regions of expression of the genes of the patterning network: M is the morphogen (Scw, an activating BMP-class ligand); I is an inhibitor of M (Sog); and P is a protease (Tld) that cleaves I. Note that M is expressed by all cells, P is expressed only in DR, and expression of I is restricted to the regions flanking the DR (neuroectoderm, NE). (b) Robustness of signaling pathway activity profile in the DR. Pathway activity corresponds to the level of free morphogen M. Robustness was experimentally tested with respect to changes in the gene dosage of M, I, and P. Shown are measurement of signaling pathway activity for wild-type cells and mutants with half gene dosage for M (scw+/-), I (sog+/-), and P (tld+/-), as well as overexpressed P (tld OE). (From Eldar et al., 2002.)

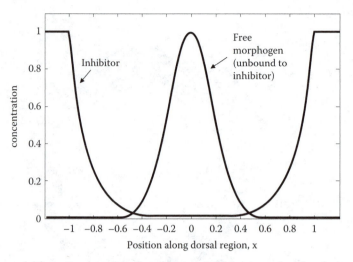

FIGURE 8.7 Simple model for patterning of the dorsal region. Inhibitor is produced at the boundaries of the region, at x = –1 and x = 1. Inhibitor is degraded, and thus its concentration decays into the dorsal region. Free morphogen, unbound to inhibitor, is thus highest at the center of the region, at x = 0, where inhibitor is lowest.

We will consider the patterning of the **dorsal region** (DR). Our story begins with a coarse pattern established by an earlier morphogen, which sets up three regions of cells along the circumference of the embryo (Figure 8.6a). The DR is about 50 cells wide. The goal of our patterning process is to subdivide this region into several subregions using a gradient of the morphogen M.

The cells in the DR have receptors that activate a signaling pathway when M is present at sufficiently high levels. Proper patterning of the DR occurs when the activity of this signaling pathway is high at the middle of the DR and low at its boundaries (Figure 8.6b), that is, when active morphogen M is found mainly near the midline of the region.

The molecular network that achieves this patterning is made of M and two additional proteins. The first is an inhibitor I that binds M to form a complex C = [MI], preventing M from signaling to the cells. The final protein in the network is a protease P that cleaves the inhibitor I. Note that P is able to cleave I when it is bound to M, liberating M from the complex. The morphogen M is not degraded in this system. The three proteins M, I, and P diffuse within a thin fluid layer outside of the cells. M is produced everywhere in the embryo, whereas I is produced only in the regions adjacent to the DR, and P is found uniformly throughout the DR.

The simplest mechanism for patterning by this system is based on a gradient of inhibitor I, set up by diffusion of I into the DR and its degradation by P (Figure 8.7). The concentration of I is highest at the two boundaries of the DR, where it is produced, and lowest at the midline of the DR. Since the inhibitor I binds and inhibits M, the activity of M (the concentration of free M) is highest at the midline of the DR, and the desired pattern is achieved. In this model, the steady-state concentration of total M (bound and free) is uniform, but its activity profile (free M) is peaked at the midline.

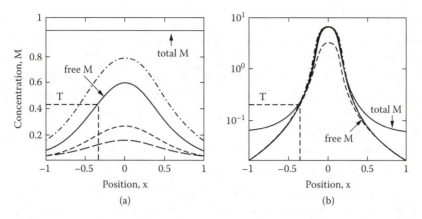

FIGURE 8.8 Patterning in nonrobust and robust mechanisms. (a) Profile of free M in a typical nonrobust network. The profile of free M (full curve) is shown for a nonperturbed network and for three perturbed networks representing half-production rates of M, I, or P (dotted, dot–dash, and dashed lines). The total concentration of M (free and bound to I, M + [MI]) is indicated by the horizontal line. The dashed line (T) indicates the threshold where robustness was measured. (b) Profile of free M in a typical robust system (note logarithmic scale on the y-axis). (From Eldar et al., 2002.)

Unfortunately, this simple mechanism is not robust to changes in the expression of M, I, or P. Changes of twofold in the production rate of any of the three proteins lead to significant changes in the morphogen profile and the resulting patterns (Figure 8.8a). In contrast, experiments show that the profile of free morphogen is highly robust to changes in the levels of any of the proteins in the system (Figure 8.6b).

To make this mechanism robust, we might propose self-enhanced degradation of M, as in the previous section. However, we cannot directly apply the nonlinear degradation mechanism of the previous section, because in this system, M is not appreciably degraded.

To understand how a robust mechanism can be formed with these molecules, let us consider the general equations that govern their behavior.

The free inhibitor I diffuses and is degraded by P at a rate α_I. Since P is known to be uniformly distributed throughout the DR, the degradation rate of I is spatially uniform and proceeds at a rate $\alpha_I P I$. Free inhibitor is further consumed when it binds free M to form a tightly bound complex, at rate k:

$$\partial I / \partial t = D_I \, \partial^2 I / \partial x^2 - k I M - \alpha_I P I \qquad (8.5.1)$$

The complex C = [IM] is formed at rate k I M and degraded by P at rate α_C:

$$\partial C / \partial t = D_C \, \partial^2 C / \partial x^2 + k I M - \alpha_C P C \qquad (8.5.2)$$

The free morphogen M diffuses, binds inhibitor I at rate k, and is liberated when the complex C is degraded:

$$\partial M / \partial t = D_M \, \partial^2 M / \partial x^2 - k I M + \alpha_C P C \qquad (8.5.3)$$

These nonlinear equations are too tough to solve analytically. Eldar and Barkai therefore studied these equations numerically (Eldar et al., 2002). The profiles of M, I, and C were found for a given set of parameters (diffusion constants, degradation rates, and k). The shift in the free morphogen profile was determined upon a twofold change in the production rate of each of the three proteins M, I, and P. This was repeated for different sets of parameters, scanning four orders of magnitude of change in each parameter. It was found that the vast majority of the parameter combinations gave nonrobust solutions (97% of the solutions were nonrobust according to the robustness threshold used).

The nonrobust solutions typically showed exponentially decaying profiles of M activity. The amount of total M (free and bound to I) was uniform in space, as shown in Figure 8.8a. However, about 0.5% of the parameter sets showed a very different behavior. The profile was highly robust to changes in any of the protein production rates. The morphogen activity profile was nonexponential and had power-law tails. In addition, the distribution of total morphogen was not spatially uniform. Morphogen protein was concentrated near the midline of the region (Figure 8.8b).

Inspection of the parameter values that provided the robust solutions showed that they all belonged to the same limiting class, in which certain parameters were much smaller than others. In particular, robustness was found when free M could not diffuse; only M within a complex C could diffuse (so that the diffusion constant of the complex is much larger than the diffusion constant of the free morphogen, $D_C \gg D_M$). Furthermore, in the robust model, free I is not degraded by the protease P. In fact, P can only degrade I within the complex C ($\alpha_C \gg \alpha_I$). The robust mechanism is well described by the following set of steady-state equations, setting time derivatives to zero. They are simpler than the full equations because they have two parameters set to zero ($D_M = 0$, $\alpha_I = 0$):

$$D_I \, \partial^2 \, I / \partial \, x^2 - k \, I \, M = 0 = \partial \, I / \partial \, t \qquad (8.5.4)$$

$$D_C \, \partial^2 \, C / \partial \, x^2 + k \, I \, M - \alpha_C \, P \, C = 0 = \partial \, C / \partial \, t \qquad (8.5.5)$$

$$- k \, I \, M + \alpha_C \, P \, C = 0 = \partial \, M / \partial \, t \qquad (8.5.6)$$

Remarkably, these nonlinear equations can be solved analytically. Summing Equations 8.5.5 and 8.5.6 shows that C obeys a simple equation:

$$D_C \, \partial^2 \, C / \partial \, x^2 = 0 \qquad (8.5.7)$$

The general solution of this equation is $C(x) = a \, x + C_o$, but due to the symmetry of the problem in which the left and right sides of the DR are equivalent, the only solution is a spatially uniform concentration of the complex:

$$C(x) = const = C_o \qquad (8.5.8)$$

Using this in Equation 8.5.6, we find that the product of free I and M is spatially uniform:

$$k\,IM = \alpha_C\,P\,C_o \qquad (8.5.9)$$

and therefore, Equation 8.5.4 can be written explicitly for M, using the relation between I and M from Equation 8.5.9, to find a simple equation for $1/M$:

$$\partial^2\,M^{-1}/\partial\,x^2 = k/D_I \qquad (8.5.10)$$

whose solution is a function peaked near $x = 0$:

$$M(x) = A/(x^2 + \varepsilon^2) \qquad A = 2\,D_I/k \qquad (8.5.11)$$

The only dependence of the morphogen profile on the total levels of M, M_{tot}, is through the parameter ε:

$$\varepsilon \sim \pi\,A/M_{tot} \qquad (8.5.12)$$

The parameter ε can be made very small by making the total amount of morphogen M_{tot} sufficiently large. In this case the morphogen profile effectively becomes a power law that is not dependent on any of the parameters of the model (except $A = 2D_I/k$),

$$M(x) \sim A/x^2 \qquad \text{far from midline, } x \gg \varepsilon \qquad (8.5.13)$$

In particular, the free M(x) profile away from the midline described by this equation does not depend on the total level of M or I. The profile also does not depend on the level of the protease P or its rate of action, since these parameters do not appear in this solution at all. In summary, the free morphogen profile is robust to the levels of all proteins in the system and can generate long-range patterns due to its power-law decay.

How does this mechanism work? The mechanism is based on shuttling of morphogen by the inhibitor. Morphogen M cannot move unless it is shuttled into the DR by complexing with the inhibitor I. Once the complex is degraded, the morphogen is deposited and cannot move until it binds a new molecule of I. Since there are more molecules of I near its source at the boundaries of the DR, morphogen is effectively pushed into the DR and accumulates where concentration of I is lowest, at the midline. Free inhibitor that wanders into the middle region finds so much M that it complexes and is therefore rapidly degraded by P. Hence, it is difficult for the inhibitor to penetrate the midline region to shuttle M away. This is a subtle but robust way to achieve an M profile that is sharply peaked at the midline and decays more slowly deep in the field. These properties are precisely the requirements for long-range robust patterning that we discussed in Section 8.3. But unlike Section 8.3, this is done without M degradation. Interestingly, both mechanisms lead to long-ranged power-law profiles.

The robust mechanism requires two important biochemical details, as mentioned above. The first is that inhibitor I is degraded only when complexed to M, and not when free. The second is that M cannot diffuse unless bound to I. Both of these properties have been demonstrated experimentally, the latter following the theoretical prediction (Eldar et al., 2002).

More generally, this chapter and the previous one aimed to point out that robustness can help to distinguish between different mechanisms, and point to unexpected designs. Only a small fraction of the designs that generate a given pattern can do so robustly. Therefore, the principle of robustness can help us to arrive at biologically plausible mechanisms. Furthermore, the robust designs seem to show a pleasing simplicity.

FURTHER READING

Berg, H.C. (1993). *Random Walks in Biology*. Princeton University Press.

Eldar, A., Dorfman, R., Weiss, D., Ashe, H., Shilo, B.Z., and Barkai, N. (2002). Robustness of the BMP morphogen gradient in *Drosophila* embryonic patterning. *Nature*, 419: 304–308.

Eldar, A., Rosin, D., Shilo, B.Z., and Barkai, N. (2003). Self-enhanced ligand degradation underlies robustness of morphogen gradients. *Dev. Cell*, 5: 635–646.

Eldar, A., Shilo, B.Z., and Barkai, N. (2004). Elucidating mechanisms underlying robustness of morphogen gradients. *Curr. Opin. Genet. Dev.*, 14: 435–439.

Additional Reading

Kirschner, M.W. and Gerhart, J.C. (2005). *The Plausibility of Life*. Yale University Press.

Lawrence, P.A. (1995). The first coordinates. In *The Making of a Fly: The Genetics of Animal Design*. Blackwell Science Ltd., Chap. 2.

Slack, J.M. (1991). *From Egg to Embryo*. Cambridge University Press, U.K., Chap. 3.

Wolpert, L. (1969). Positional information and the spatial pattern of cellular differentiation. *J. Theor. Biol.*, 25: 1–47.

EXERCISES

8.1. *Diffusion from both sides*. A morphogen is produced at both boundaries of a region of cells that ranges from $x = 0$ to $x = L$. The morphogen diffuses into the region and is degraded at rate α. What is the steady-state concentration of the morphogen as a function of position? Assume that the concentration at the boundaries is $M(0) = M(L) = M_o$. Under what conditions is the concentration of morphogen at the center of the region very small compared to M_o?

Hint: The morphogen concentration obeys the diffusion–degradation equation at steady-state:

$$D \, d^2 M/d x^2 - \alpha M = 0$$

The solutions of this equation are of the form:

$$M(x) = A \, e^{-x/\lambda} + B \, e^{x/\lambda}$$

Find λ, A, and B that satisfy the diffusion–degradation equation and the boundary conditions.

8.2. *Diffusion with degradation at boundary.* A morphogen is produced at x = 0 and enters a region of cells where it is not degraded. The morphogen is, however, strongly degraded at the other end of the region, at x = L, such that every molecule of M that reaches x = L is immediately degraded. The boundary conditions are thus M(0) = M_o and M(L) = 0.

a. What is the steady-state concentration profile of M?

b. Is patterning by this mechanism robust to changes of the concentration at the source, M(0) = M_o?

Hint: The morphogen obeys a simple equation at steady state:

$$D \; d^2 \, M/d \, x^2 = 0$$

Try solutions of the form M(x) = A x + B, and find A and B such that M(x = L) = 0 and M(x = 0) = M_o. Next, find the position where M(x) equals a threshold T, and find the changes in this position upon a change of M_o.

8.3. *Polynomial self-enhanced degradation.* Find the steady-state concentration profile of a morphogen produced at x = 0. The morphogen diffuses into a field of cells, with nonlinear self-enhanced degradation described by

$$\partial \, M/\partial \, t = D \, \partial^2 \, M/\partial \, x^2 - \alpha \, M^n$$

When is patterning with this profile robust to the level of M at the boundary, M_o?

Hint: Try a solution of the form M(x) = $a(x + b)^m$ and find the parameters a and b in terms of D, M_o, and α.

8.4. *Robust timing.* A signaling protein X inhibits pathway Y. At time t = 0, X production stops and its concentration decays due to degradation. The pathway Y is activated when X levels drop below a threshold T. The time at which Y is activated is t_Y. Our goal is to make t_Y as robust as possible to the initial level of X, X(t = 0) = X_o.

a. Compare the robustness of t_Y in two mechanisms, linear degradation and self-enhanced degradation (note that in this problem, all concentrations are spatially uniform).

$$\partial \, X/\partial \, t = - \alpha \, X$$

$$\partial \, X/\partial \, t = - \alpha \, X^n$$

Which mechanism is more robust to fluctuations in X_o? Explain.

b. Explain why a robust timing mechanism requires a rapid decay of X at times close to $t = 0$.

8.5. *Activator accumulation vs. repressor decay* (harder problem). Compare the robustness of t_Y in problem 8.4 to an alternative system, in which X is an activator that begins to be produced at $t = 0$, activating Y when it exceeds threshold T. Consider both linear or nonlinear degradation of X. Is the accumulating activator mechanism more or less robust to the production rate of X than the decaying repressor mechanism?

Answer:

An activator mechanism is generally less robust to variations in the production rate of X than the decaying repressor mechanism of problem 8.4. (Rappaport et al., 2005).

8.6. *Flux boundary condition*: Morphogen M is produced at $x = 0$ and diffuses into a large field of cells where it is degraded at rate α. Solve for the steady-state profile, using a boundary condition of constant flux J at $x = 0$, $J = D\partial M/\partial x$. Compare with the solution discussed in the text, which used a constant concentration of M at $x = 0$, M_0.

CHAPTER 9

Kinetic Proofreading

9.1   INTRODUCTION

In the preceding two chapters we have discussed how circuits can be designed to be robust with respect to fluctuations in their biochemical parameters. Here, we will examine robustness to a different, fundamental source of errors in cells. These errors result from the presence, for each molecule X, of many chemically similar molecules that can confound the specific *recognition* of X by its interaction partners. Hence, we will examine the problem of molecular recognition of a target despite the background of similar molecules. How can a biochemical recognition system pick out a specific molecule in a sea of other molecules that bind it with only slightly weaker affinity?

In this chapter, we will see that diverse molecular recognition systems in the cell seem to employ the same principle to achieve high precision. This principle is called **kinetic proofreading**. The explanation of the structure and function of kinetic proofreading was presented by John Hopfield (1974).

To describe kinetic proofreading, we will begin with recognition in information-rich processes in the cell, such as the reading of the genetic code during translation. In these processes a chain is synthesized by adding at each step one of several types of monomers. Which monomer is added at each step to the elongating chain is determined according to information encoded in a template (mRNA in the case of translation). Due to thermal noise, an incorrect monomer is sometimes added, resulting in errors. Kinetic proofreading is a general way to reduce the error rate to levels that are far lower than those achievable by simple equilibrium discrimination between the monomers.

After describing proofreading in translation, we will consider this mechanism in the context of a recognition problem in the immune system (McKeithan, 1995; Goldstein et al., 2004). We will see how the immune system can recognize proteins that come from a dangerous microbe despite the presence of very similar proteins made by the healthy body. Kinetic proofreading can use a small difference in affinity of protein ligands to make a very precise decision, protecting the body from attacking itself. Finally, we will discuss kinetic proofreading in other systems.

Kinetic proofreading is a somewhat subtle idea, and so we will use three different approaches to describe it. In the context of recognition in translation, we will use kinetic equations to derive the error rate. In the context of the immune recognition, we will use a delay time argument. But first we will tell a story about a recognition problem in a museum.

As an analogy to kinetic proofreading, consider a museum curator who wants to design a room that would select Picasso lovers from among the museum visitors. In this museum, half of the visitors are Picasso lovers and half do not care for Picasso. The curator opens a door in a busy corridor. The door leads to a room with a Picasso painting, allowing visitors to enter the room at random. Picasso lovers that happen to enter the room hover near the picture for, on average, 10 min, whereas others stay in the room for only 1 min. Because of the high affinity of Picasso lovers for the painting, the room becomes enriched with 10 times more Picasso lovers than nonlovers.

The curator wishes to do even better. At a certain moment, the curator locks the door to the room and reveals a second, one-way revolving door. The nonlovers in the room leave through the one-way door, and after several minutes, the only ones remaining are Picasso lovers, still hovering around the painting. Enrichment for Picasso lovers is much higher than 10-fold.

If the revolving door were two-way, allowing visitors to enter the room at random, only a 10-fold enrichment for Picasso lovers would again occur. Kinetic proofreading mimics the Picasso room stratagem by using nearly irreversible, nonequilibrium reactions as one-way doors.

9.2   KINETIC PROOFREADING OF THE GENETIC CODE CAN REDUCE ERROR RATES OF MOLECULAR RECOGNITION

Consider the fundamental biological process of translation. In translation, a ribosome produces a protein by linking amino acids one by one into a chain (Figure 9.1). The type of amino acid added at each step to the elongating chain is determined by the information encoded by an mRNA. Each of the twenty amino acid is encoded by a codon, a series of three letters on the mRNA. The mapping between the 64 codons and the 20 amino acids is called the genetic code (Figure 9.2).

To make the protein, the codon must be read and the corresponding amino acid must be brought into the ribosome. Each amino acid is brought into the ribosome connected to a specific tRNA molecule. That tRNA has a three-letter recognition site that is complementary, and pairs with the codon sequence for that amino acid on the mRNA (Figure 9.1). There is a tRNA for each of the codons that specify amino acids in the genetic code.

Translation therefore communicates information from mRNA codons to the amino acids in the protein sequence. The codon must recognize and bind the correct tRNA, and not bind to the wrong tRNA. Since this is a molecular process working under thermal noise, it has an error rate. The wrong tRNA can attach to the codon, resulting in a translation error where a wrong amino acid is incorporated into the translated protein. These translation errors occur at a frequency of about 10^{-4}. This means that a typical protein of

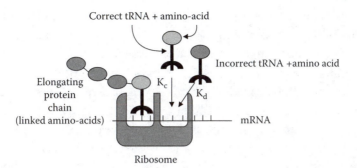

FIGURE 9.1 Translation of a protein at the ribosome. The mRNA is read by tRNAs that specifically recognize triplets of letters on the mRNA called codons. When a tRNA binds the codon, the amino acid that it carries (symbolized in the figure as an ellipse on top of the trident-like tRNA symbol) links to the elongating protein chain (chain of ellipses). The tRNA is ejected and the next codon is read. Each tRNA competes for binding with the other tRNA types in the cell. The correct tRNA binds with dissociation constant K_c, whereas the closest incorrect tRNA binds with $K_d > K_c$.

	Second letter				
First letter	U	C	A	G	Third letter
U	Phe	Ser	Tyr	Cys	U
	Phe	Ser	Tyr	Cys	C
	Leu	Ser	STOP	STOP	A
	Leu	Ser	STOP	Trp	G
C	Leu	Pro	His	Arg	U
	Leu	Pro	His	Arg	C
	Leu	Pro	Gln	Arg	A
	Leu	Pro	Gln	Arg	G
A	Ile	Thr	Asn	Ser	U
	Ile	Thr	Asn	Ser	C
	Ile	Thr	Lys	Arg	A
	Met	Thr	Lys	Arg	G
G	Val	Ala	Asp	Gly	U
	Val	Ala	Asp	Gly	C
	Val	Ala	Glu	Gly	A
	Val	Ala	Glu	Gly	G

FIGURE 9.2 The gentic code. Each 3-letter codon maps to an amino acid or a stop signal that ends translation. For example, CUU codes for the amino acid leucine (Leu). Polar amino acids are shaded, non-polar amino acids in white. This code is universal across nearly all organisms.

100 amino acids has a 1% chance to have one wrong amino acid. A much higher error rate would be disastrous, because it would result in the malfunction of an unacceptable fraction of the cell's proteins.

9.2.1 Equilibrium Binding Cannot Explain the Precision of Translation

The simplest model for this recognition process is **equilibrium binding** of tRNAs to the codons. We will now see that simple equilibrium binding cannot explain the observed error rate. This is because equilibrium binding generates error rates that are equal to the ratio of affinities of the correct and incorrect tRNAs. This would result in error rates that are about 100 times higher than the observed error rate.

To analyze equilibrium binding, consider codon C on the mRNA in the ribosome that encodes the amino acid to be added to the protein chain. We begin with the rate of binding of the correct tRNA, denoted c, to codon C. Codon C binds c with an on-rate k_c. The tRNA unbinds from the codon with off-rate k_c'. When the tRNA is bound, there is a probability v per unit time that the amino acid attached to the tRNA will be covalently linked to the growing, translated protein chain. In this case, the freed tRNA unbinds from the codon and the ribosome shifts to the next codon in the mRNA. The equilibrium process is hence

$$c + C \underset{k_c'}{\overset{k_c}{\rightleftharpoons}} [cC] \xrightarrow{v} \text{correct amino acid} \tag{9.2.1}$$

At equilibrium, the concentration of the complex [cC] is given by the balance of the two arrows marked k_c and k_c' (the rate v is much smaller than k_c and k_c' and can be neglected). Hence, at steady state, collisions of c and C that form the complex [cC] at rate k_c balance the dissociation of the complex [cC], so that $c C k_c = [cC] k_c'$. This results in a concentration of the complex [cC], which is given by the product of the concentrations of the reactants divided by the dissociation constant K_c:

$$[cC] = c \, C / K_c \tag{9.2.2}$$

where K_c is equal to the ratio of the off-rate and on-rate for the tRNA binding:[1]

$$K_c = k_c' / k_c \tag{9.2.3}$$

The smaller the dissociation constant, the higher the affinity of the reactants.

The incorporation rate of the correct amino acid is equal to the concentration of the bound complex times the rate at which the amino acid is linked to the elongating protein chain:

$$R_{correct} = v[cC] = v \, c \, C / K_c \tag{9.2.4}$$

In addition to the correct tRNA, the cells contain different tRNAs that carry the other amino acids and that compete for binding to codon C. Let us consider, for simplicity, only one of these other tRNAs, the tRNA that carries a different amino acid that has the highest affinity to codon C. It is this incorrect tRNA that has the highest probability to yield false recognition by binding the codon C, leading to incorporation of the wrong amino acid. The concentration of this incorrect tRNA is about equal to the concentration of the

---

[1] The rate v, at which the complex produces the product (an amino acid linked to the growing protein chain), is much smaller than the other rates in the process, as mentioned above. The reactants can thus bind and unbind many times before product is formed. This is the case for many enzymatic reactions (Michaelis–Menten picture, see Appendix A). When v is not negligible compared to k_c', we have $K_c = (k_c' + v)/k_c$. The error rate in kinetic proofreading is smaller the smaller the ratio v/k_c'.

correct tRNA (many of the tRNAs have approximately the same concentrations). The incorrect tRNA, denoted d, can bind the codon C in the following equilibrium process:

$$d+C \underset{k_d'}{\overset{k_d}{\rightleftharpoons}} [dC] \xrightarrow{v} \text{incorrect amino acid} \qquad (9.2.5)$$

The concentration of incorrect complex [dC] is governed by the dissociation constant equal to the ratio between the off- and on-rates of d, $K_d = k_d'/k_d$. Thus, the equilibrium concentration of bound complex is

$$[dC] = d \, C/K_d$$

The rate of incorrect linking is given by the concentration of this incorrect complex times the rate of linking the amino acids into the elongating chain. The linking process occurs at a molecular site on the ribosome that is quite distant from the recognition site, and does not distinguish between the different tRNAs d and c. Hence, the linking rate v is the same for both processes, and we obtain

$$R_{wrong} = v \, d \, C/K_d \qquad (9.2.6)$$

Since d is the incorrect tRNA, it has a larger dissociation constant for binding C than the correct tRNA, c, that is, $K_d > K_c$, and hence $R_{wrong} < R_{correct}$.

The resulting **error rate**, F_o, is the ratio of the rates of incorrect and correct amino acid incorporation. The error rate is approximately equal to the ratio of the dissociation constants, since all other concentrations (tRNA concentrations) are about the same for c and d:

$$F_o = R_{wrong}/R_{correct} = v \, d \, C \, K_c/v \, c \, C \, K_d \approx K_c/K_d \qquad (9.2.7)$$

To repeat the main conclusion, the error rate in equilibrium recognition is determined by the ratio of dissociation constants for the correct and incorrect tRNAs. As occurs for many biological binding events, the on-rates for both d and c are limited by diffusion and are about the same, $k_d = k_c$ (Appendix A). *It is the off-rate, k_d' which distinguishes the correct codon from the incorrect one*: the wrong tRNA unbinds more rapidly than the correct tRNA, $k_d' \gg k_c'$, because of the weaker chemical bonds that hold it in the bound complex. Using Equation 9.2.3, we find

$$F_o = R_{wrong}/R_{correct} = K_c/K_d \approx k_c'/k_d' \qquad (9.2.8)$$

The off-rates are akin to the dissociation rates of museum visitors from the Picasso painting in the Picasso room story above.

How does equilibrium recognition compare with the actual error rates? The affinity of codons to correct and incorrect tRNAs was experimentally measured, to find an affinity ratio of about $K_c/K_d \sim 1/100$. Hence, there is a large discrepancy between the predicted equilibrium recognition error, $F_o \sim K_c/K_d \sim 1/100$, and the actual translation error rate, $F = 1/10,000$. It therefore seems that equilibrium recognition cannot explain the high fidelity found in this system.[1]

9.2.2 Kinetic Proofreading Can Dramatically Reduce the Error Rate

We just saw that equilibrium binding can only provide discrimination that is as good as the ratio of the chemical affinity of the correct and incorrect targets. What mechanism can explain the high fidelity of the translation machinery, which is a hundred-fold higher than predicted from equilibrium recognition?

The solution lies in a reaction that occurs in the translation process, which was well known at the time that Hopfield analyzed the system, but whose function was not understood and was considered a wasteful side reaction. In this reaction, the tRNA, after binding the codon, undergoes a chemical modification. That is, c binds to C and then is converted to c^*. This reaction is virtually irreversible, because it is coupled to the hydrolysis of a GTP molecule.[2] The modified tRNA, c^*, can either fall off of the codon or donate its amino acid to the elongating protein chain:

$$c+C \underset{k_c'}{\overset{k_c}{\rightleftharpoons}} [cC] \xrightarrow{m} [c^*C] \xrightarrow{v} \text{correct amino acid} \qquad (9.2.9)$$

$$\downarrow l_c'$$

$$c+C$$

The fact that the modified tRNA can fall off seems wasteful because the correct tRNA can be lost. However, it is precisely this design that generates high fidelity. The secret is that c^* offers a second discrimination step: the wrong tRNA, once modified, can fall off of the codon, but it cannot mount back on. This irreversible reaction acts as the one-way door in the Picasso story.

To compute the error rate in this process, we need to find the concentration of the modified bound complex. The concentration of $[c^*C]$ is given by the balance of the two processes described by the arrows marked with the rates m and l_c' (since the rate v is much smaller than the other rates), leading to a balance at steady state between modification of the complex $[cC]$ and the dissociation of c^* at rate l_c', $m[cC] = l_c' [c^*C]$, yielding a steady-state solution:

$$[c^*C] = m\,[cC]/l_c' \qquad (9.2.10)$$

---

[1] Why not increase the ratio of the off-rates of the incorrect and correct tRNAs, k_d'/k_c', to improve discrimination? Such an increase may be unfeasible due to the chemical structure of codon–anticodon recognition, in which different codons can differ by only a single hydrogen bond. In addition, decreasing the off-rate of the tRNAs by increasing the number of bonds they make with codons would cause them to stick to the codon for a longer time. This would interfere with the need to rapidly bind and discard many different tRNAs in order to find the correct one, and slow down the translation process (exercise 9.3). Thus, biological recognition may face a trade-off in which high affinity means slow recognition rates.

[2] Near irreversibility is attained by coupling a reaction to a second reaction that expends free energy. For example, coupling a reaction to ATP hydrolysis can shift it away from equilibrium by factors as large as 10^8, achieved because the cell continuously expends energy to produce ATP.

The rate of correct incorporation is the linking rate v times the modified complex concentration (Equation 9.2.4):

$$R_{correct} = v\,[c*C] = v\,m\,c\,C/l_c'\,K_c \qquad (9.2.11)$$

The same applies for d. The conversion of d to d* occurs at the same rate, m, as the conversion of c to c*, since the modification process does not discriminate between tRNAs. The rate that the wrong tRNA d* falls off of the codon is, however, much faster than the rate at which c* falls off. This is because the chemical affinity of the wrong tRNA to the codon C is weaker than the affinity of the correct tRNA. The off-rate ratio of the correct and incorrect modified tRNAs is the same as the ratio for the unmodified tRNAs, since they are all recognized by the same codon C:

$$l_d'/l_c' = k_d'/k_c' \approx K_d/K_c \qquad (9.2.12)$$

Thus, d* undergoes a second discrimination step, with a significant chance that the wrong tRNA is removed. The rate of wrong amino acid linkage is the same as in Equation 9.2.11, with all parameters for c replaced with the corresponding parameters for d:

$$R_{wrong} = v\,[d*C] = v\,m\,d\,C/l_d'\,K_d \qquad (9.2.13)$$

resulting in an error rate, using Equation 9.2.12:

$$F = R_{wrong}/R_{correct} = (K_c/K_d)\,(l_c'/l_d') = (K_c/K_d)^2 = F_o^2 \qquad (9.2.14)$$

Thus, the irreversible reaction step affords a proofreading event that adds a multiplicative factor of K_c/K_d to the error rate. In effect, it allows two separate equilibrium recognition processes, the second working on the output of the first. This results in an error rate that is the square of the equilibrium recognition error rate:

$$F = F_o^2 \qquad (9.2.15)$$

It is important to note that had all reactions been reversible and at equilibrium, no improvement would be gained over the simple scheme (Equation 9.2.1). This is due to detailed balance and is discussed in exercise 9.2. The equilibrium model with detailed balance is similar to the Picasso room in which the one-way door is changed to a two-way door that allows visitors in and out at random.

Thus, the proofreading step implemented by a modification of the tRNA can reduce the error rate from the equilibrium recognition rate of about $F_o = 1/100$ to a much lower error rate, $F = F_o^2 = 1/10,000$, similar to the observed error rate.

An even higher level of fidelity can be attained by linking together n irreversible (or nearly irreversible) proofreading processes:

$$c + C \rightleftharpoons [cC] \xrightarrow[m_1]{} [c^*C] \xrightarrow[m_2]{} [c^{**}C] \longrightarrow \dots \xrightarrow[m_n]{} [c^{**\dots*}C] \xrightarrow{v} \text{product}$$

$$\downarrow l'_{c_1} \qquad \downarrow l'_{c_2} \qquad\qquad \downarrow l'_{c_n}$$

$$c + C \qquad c + C \qquad\qquad c + C$$

$$(9.2.16)$$

Each irreversible step adds a proofreading factor F_o, resulting in an overall error rate of

$$F = F_o^{n+1} \tag{9.2.17}$$

9.3 RECOGNIZING SELF AND NON-SELF BY THE IMMUNE SYSTEM

We have just seen how kinetic proofreading uses a nonequilibrium step to reduce errors in translation. We will now use a slightly different (but equivalent) way to explain kinetic proofreading, based on time delays. For this purpose, we will study a biological instance of kinetic proofreading in the immune system.

The immune system monitors the body for dangerous pathogens. When it detects pathogens, the immune system computes and mobilizes the appropriate responses. The immune system is made of a vast collection of cells that communicate and interact in myriad ways.

One of the major tools of the immune system is antibodies. Each antibody is a protein designed to bind with high affinity to a specific foreign protein made by pathogens, called the antigen.

One of the important roles of the immune system is to scan the cells of the body for antigens, for example, for proteins made by a virus that has infected the cell. The scanning task is carried out by T-cells. Each of the T-cells has receptors made of a specific antibody against a foreign protein antigen. To provide information for the T-cells, each cell in

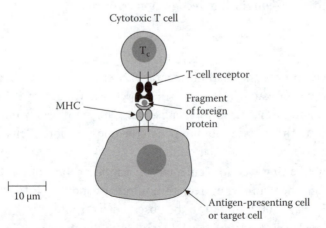

FIGURE 9.3 Recognition of foreign peptides by T-cells. Target cells present fragments of their proteins bound to MHC proteins on the cell surface. Each T-cell can recognize specific foreign peptides by means of its T-cell receptor. Recognition can result in killing of the target cell by the T-cell. Note that the receptor and MHC complex are not to scale (cells are ~ 10nm).

the body presents fragments of proteins on the cell surface. The proteins are presented in dedicated protein complexes on the cell surface called MHCs (Figure 9.3).

The goal of the T-cell is to eliminate infected cells. Each T-cell can recognize a specific antigen in the MHC because its receptor can bind that foreign peptide. If the T-cell receptor recognizes its antigen, the foreign protein fragment in the MHC on a cell, it triggers a signal transduction cascade inside the T-cell. The signaling causes the T-cell to kill the cell that presented the foreign peptide. This eliminates the infected cell and protects the body from the virus.

In the recognition process, it is essential that the T-cell does not kill cells that present proteins that are normally produced by the healthy body. If such misrecognition occurs, the immune system attacks the cells of the body, potentially leading to an autoimmune disease.

The precision of the recognition of non-self proteins by T-cells is remarkable. T-cells can recognize minute amounts of a foreign protein antigen in a background of self-proteins, even though the self-proteins have only a slightly lower affinity to the T-cell receptor than the foreign target. The error rate of recognition is less than 10^{-6}, although the affinity of the antigen is often only 10-fold higher than the affinities of the self-proteins.

9.3.1 Equilibrium Binding Cannot Explain the Low Error Rate of Immune Recognition

The receptors on a given T-cell are built to recognize a specific foreign protein, which we will call the correct ligand, c. The correct ligand binds the receptors with high affinity. In addition to c, the receptors are exposed to a variety of self-proteins, which bind the receptor with a weaker affinity. In particular, some of these self proteins are quite similar to the correct ligand and pose the highest danger for misrecognition, in which the receptors mistake a self-protein for the correct ligand. For clarity, let us treat these wrong ligands as a single entity d, with a lower affinity to the receptor. We will begin by the simplest model for recognition, in which c and d bind the receptor in an equilibrium process. As in the previous section, this yields error rates that are proportional to the ratio of affinities of the incorrect and correct targets. Since the affinities of the correct and incorrect ligands are not very different, equilibrium recognition results in an unacceptably high rate of misrecognition.

The dynamics of binding of the correct ligand c to the receptor R includes two processes. The first process is collisions of c and R at a rate k_{on} to form a complex, $[cR]$, in which the ligand is bound to the receptor. The inverse process is dissociation of the complex, in which the ligand unbinds form the receptor at rate k_{off}. The rate of change of the concentration of bound receptor is the difference between the collision and dissociation rates:

$$d[cR]/dt = k_{on} \, c \, R - k_{off} \, [cR]$$

At steady-state, $d[cR]/dt = 0$ and we find

$$[cR] = R \, c/K_c$$

where K_c is the dissociation constant of the correct ligand to the receptor, $K_c = k_{off}/k_{on}$.

When the ligand binds the receptor, it triggers a signal transduction pathway inside the T-cell, which leads to activation of the T-cell. Once ligand binds the receptor, the signaling pathway is activated with probability v per unit time. Therefore, the rate of T-cell activation in the presence of a concentration c of correct ligand is

$$A_{correct} = [cR] \, v = c \, R \, v/K_c$$

A similar set of equations describe the binding of the incorrect ligand d to the receptor. The on-rate and off-rate of the incorrect ligand are k'_{on} and k'_{off}, leading to

$$d[Rd]/dt = k'_{on} \, d \, R - k'_{off} \, [Rd]$$

The steady-state concentration of the incorrect complex, [Rd] is given by the product of the concentration of d and R divided by the dissociation constant for d:

$$[Rd] = R \, d/K_d$$

where $K_d = k'_{off}/k'_{on}$.

The affinity of the incorrect ligand is smaller than that of the correct ligand, so that $K_d > K_c$. As mentioned in the previous section, this difference in affinities is usually due to the difference in the *off-rates* of the ligands, rather than to different on-rates. The correct ligand dissociates from the receptor at a slower rate than the incorrect ligand due to its stronger chemical bonds with the receptor, $k_{off} < k'_{off}$. In other words, the correct ligand spends more time bound to the receptor than the incorrect ligand.

In the equilibrium recognition process, when the incorrect ligand binds, it can activate the signaling pathway in the T-cell with the same intrinsic probability as the correct ligand, v. In equilibrium recognition, the receptor has no way of distinguishing between the ligands other than their affinities. The resulting rate of activation due to the binding of the incorrect ligand is

$$A_{wrong} = [dR] \, v = d \, R \, v/K_d$$

Hence, the error rate of the T-cells, defined by the ratio of incorrect to correct activations, is

$$F_o = A_{wrong}/A_{correct} = K_c \, d \, R \, v/K_d \, c \, R \, v = (K_c/K_d) \, (d/c)$$

The error rate in this equilibrium recognition process is thus given by the ratio of affinities of the incorrect and correct ligands, times the ratio of their concentrations. In the immune system, the incorrect ligands often have only a 10-fold lower affinity than the correct ligand, $K_c/K_d \sim 0.1$. Furthermore, the concentration of incorrect ligand (proteins made by the healthy body) often exceeds the concentration of the correct ligand (pathogen protein). Hence, the equilibrium error rate is $F_o > 0.1$. This is far higher than the observed error rate in T-cell recognition, which can be $F = 10^{-6}$ or lower.

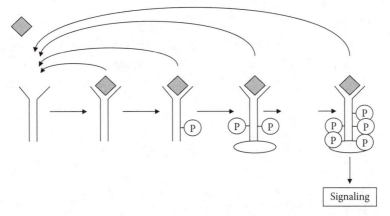

FIGURE 9.4 Kinetic proofreading model in T-cell receptors. Ligand binding initiates modifications to the receptors. When sufficient modifications have occurred, signaling pathways are triggered in the cell. At any stage, ligand can dissociate from the receptor, resulting in immediate loss of all of the modifications. The series of modifications creates a delay between ligand binding and signaling. Only ligands that remain bound throughout this delay can trigger signaling.

How can we bridge the huge gap between the high rate of equilibrium recognition errors and the observed low error rate in the real system? The next section describes a kinetic proofreading mechanism in the receptors that amplifies small differences in affinity into large differences in the recognition rates.

9.3.2 Kinetic Proofreading Increases Fidelity of T-Cell Recognition

The actual recognition process in T-cell receptors includes several additional steps, which may at first sight appear to be superfluous details. After ligand binding, the receptor undergoes a series of covalent modifications, such as phosphorylation on numerous sites (Figure 9.4). These modifications are energy-consuming and are held away from thermal equilibrium. When modified, the receptor binds several protein partners inside the cell. Activation of the signaling pathway inside the T-cell begins only after all of these modifications and binding events are complete. Kinetic proofreading relies on these extra steps to create a delay τ that allows the system to reduce its error rates. The basic idea is that only ligands that remain bound to the receptors for a long enough time have a chance to activate the T-cell (McKeithan, 1995).

To understand this, let us examine a binding event of the correct ligand. Once bound, the ligand has a probability per unit time k_{off} to dissociate from the receptor. Hence, the probability that it remains bound for a time longer than t after binding is

$$P(t) = e^{-k_{off} t}$$

Signaling in the cell only occurs at a delay τ after ligand binds the receptor, due to the series of modifications of the receptors that is needed to activate the signaling pathway. Hence, the probability per ligand binding that the T-cell is activated is equal to the probability that the ligand is bound for a time longer than τ:

$$A_{correct} = e^{-k_{off}\tau}$$

Similarly, the incorrect ligand has an off-rate k'_{off}. The off-rate of the incorrect ligand is, as mentioned above, larger than that of the correct ligand, because it binds the receptor more weakly. The probability that the incorrect ligand activates the receptor is

$$A_{wrong} = e^{-k'_{off}\tau}$$

Hence, the error rate in the delay mechanism is the ratio of these activation rates:

$$F = A_{wrong}/A_{correct} = e^{-(k'_{off} - k_{off})\tau}$$

This allows a very small error rate even for moderate differences between the off-rates, provided that the delay is long enough ($\tau > k'_{off}$). For example, if the off-rate of the correct ligand is $k_{off} = 1$ sec^{-1} and the incorrect ligand is $k'_{off} = 10$ sec^{-1}, and the delay is $\tau = 1.5$ sec, one finds

$$F \sim e^{-(10 - 1)\cdot 1.5} = e^{-13.5} \sim 10^{-6}$$

Thus, long delays can enhance fidelity. However, this comes at a cost. The longer the delay, the larger the number of binding events of the correct ligand that unbind before signaling can begin. Thus, increasing the delay can cause a loss of sensitivity. The loss of sensitivity is tolerated because of the greatly improved discrimination between the correct ligand and incorrect-but-chemically-similar ligands.

Kinetic proofreading is a general mechanism that provides specificity due to a delay step that gives the incorrect ligands a chance to dissociate before recognition is complete. In order for kinetic proofreading to work effectively, the receptors must lose their modifications when the ligand unbinds before a new ligand molecule can bind. Otherwise, the wrong ligand can bind to receptors that have some of the modifications from a previous binding event, resulting in a higher probability for misrecognition.

Experiments to test kinetic proofreading use a series of ligands with different k_{off} values (reviewed in Goldstein et al., 2004). The experiments are designed so that the fraction of the receptors bound by each ligand is the same. This is achieved by using higher concentrations of ligands with weaker binding (larger k_{off}), or by normalizing the results per binding event. Simple equilibrium recognition predicts a constant probability for triggering signaling per ligand binding event, regardless of the k_{off} of the ligand. In contrast, the experiments show that the probability of activation of the signaling pathway depends inversely on k_{off}. This means that the longer the ligand is bound to the receptor, the higher the probability that it triggers signaling. This is consistent with the kinetic proofreading picture.

Kinetic proofreading uses modification of the T-cell receptor after ligand binding to create a delay. This process is not unique to T-cell receptors. In fact, these types of modifications occur in practically every receptor in mammalian cells, including receptors that sense hormones, growth factors, and other ligands. This raises the possibility that delays and kinetic proofreading are widely employed by receptors to increase the fidelity of recognition. Kinetic proofreading can provide robustness against misrecognition of the background of diverse molecules in the organism.

9.4 KINETIC PROOFREADING MAY OCCUR IN DIVERSE RECOGNITION PROCESSES IN THE CELL

The hallmark of kinetic proofreading is the existence of a nonequilibrium reaction in the recognition process that forms an intermediate state, providing a delay after ligand binding. The system must operate away from equilibrium, so that ligands cannot circumvent the delay by rebinding directly in the modified state. New ligand binding must primarily occur in the unmodified state.

These ingredients are found in diverse recognition processes in the cell. An example is DNA binding by repair proteins (Reardon and Sancar, 2004) and recombination proteins (Tlusty et al., 2004). One such process is responsible for repairing DNA with a damaged base-pair. A recognition protein A binds the damaged strands, because it has a higher affinity to damaged DNA than to normal DNA. After binding, protein A undergoes a modification (phosphorylation). When phosphorylated, it recruits additional proteins B and C that nick the DNA on both sides of A and remove the damaged strand, allowing specialized enzymes to fill in the gap and polymerize a fresh segment in place of the damaged strand. The modification step of protein A may help prevent misrecognition of normal DNA as damaged.

An additional example occurs in the binding of amino acids to their specific tRNAs (Hopfield, 1974; Hopfield et al., 1976). A special enzyme recognizes the tRNA and its specific amino acid and covalently joins them. Covalent joining of the wrong amino acid to the tRNA would lead to the incorporation of the wrong amino acid in the translated protein. Interestingly, the error rate in the tRNA formation process is about 10^{-4}, similar to the translation error rate we examined in Section 9.2 due to misrecognition between tRNAs and their codons.[1] This low error rate is achieved by an intermediate high-energy state, in which the enzyme that connects the amino acid to the tRNA first binds both reactants, then modifies the tRNA, and only then forms the covalent bond between the two. Again, we see the hallmarks of kinetic proofreading.

Intermediate states are found also in the process of protein degradation in eukaryotic cells (Rape et al., 2006). Here, a protein is marked for degradation by means of a specific enzyme that covalently attaches to the protein a chain made of a small protein subunit called ubiquitin (Hershko and Ciechanover, 1998). A different de-ubiquitinating enzyme can remove the ubiquitin, saving the tagged protein from its destruction. Here, addition of ubiquitin subunits one by one can implement a delay, so that there is a chance for the wrong protein to be de-ubiquitinated and not destroyed. This can allow differential degradation rates for proteins that have similar affinities to their ubiquitiating enzyme.

---

[1] It is interesting to consider whether the two error rates are tuned to be similar. It may not make sense to have one error rate much larger than the other (the larger error would dominate the final errors in proteins).

In summary, kinetic proofreading is a general mechanism that allows precise recognition of a target despite the presence of a background noise of other molecules similar to the target. Kinetic proofreading can explain seemingly wasteful side reactions in biological processes that require high specificity. These side reactions contribute to the fidelity of recognition at the expense of energy and delays. Hence, kinetic proofreading is a general principle that can help us to understand an important aspect of diverse processes in a unified manner.

FURTHER READING

Goldstein, B., Faeder, J.R., and Hlavacek, W.S. (2004). Mathematical and computational models of immune-receptor signalling. *Nat. Rev. Immunol.* 4: 445–456.

Hopfield, J.J. (1974). Kinetic proofreading: a new mechanism for reducing errors in biosynthetic processes requiring high specificity. *Proc. Natl. Acad. Sci. U.S.A.* 71: 4135–4139.

McKeithan, T.W. (1995). Kinetic proofreading in T-cell receptor signal transduction. *Proc. Natl. Acad. Sci. U.S.A.* 92: 5042–5046.

Tlusty, T., Bar-Ziv, R., and Libchaber, A. (2004). High fidelity DNA sensing by protein binding fluctuations, *Phys. Rev. Lett.* 93: 258103.

EXERCISES

9.1.   *At any rate.* Determine the error rate in the proofreading process of Equation 9.2.9. What conditions (inequalities) on the rates allow for effective kinetic proofreading?

Solution:

The rate of change of [cC] is governed by the collisions of c and C with on-rate k, their dissociation with off-rate k_c', and the formation of [cC*] at rate m:

$$d[cC]/dt = k \, c \, C - (m + k_c') \, [cC] \qquad (P9.1)$$

so that at steady-state, in which d[cC]/dt = 0, we have

$$[cC] = k \, c \, C/(m + k_c') \qquad (P9.2)$$

Similarly, [cC*] is produced at rate m, dissociates at rate l_c' and produces a product at rate v:

$$d[cC^*]/dt = m[cC] - (v + l_c') \, [cC^*] \qquad (P9.3)$$

so that at steady state, using Equation P9.2, we have

$$[cC^*] = m/(v + l_c') \, [cC] = \frac{m \, k \, c \, C}{(v + l_c')(m + k_c')} \qquad (P9.4)$$

Similar considerations for the wrong ligand d can be made, noting that for d the on-rate k, the complex formation rate m, and the product formation rate v are the same as for c, but that the off-rates k_d' and l_d' are larger than the corresponding rates for c due to the weaker affinity of d to C. Thus,

$$[dC^*] = \frac{m \, k \, c \, C}{(v+1_c')(m+k_c')} \tag{P9.5}$$

The error rate is the ratio of incorrect and correct production rates $v[dC^*]/v[cC^*]$:

$$F = v[dC^*]/v[cC^*] = \frac{d(v+1_c')(m+k_c')}{c(v+1_d')(m+k_d')} \tag{P9.6}$$

When $v \ll l_c'$ and l_d', and when $m \ll k_c'$ and k_d', we have the minimal error rate in this process:

$$F = \frac{d \, l_c' \, k_c'}{c \, l_d' \, k_d'} \tag{P9.7}$$

Thus, minimal errors require that the complexes [dC] dissociate much faster than the rate of formation of [dC*], and that [dC*] dissociate much faster than the rate of product formation. This gives many opportunities for the wrong ligand to fall off of the complex, before an irreversible step takes place.

In processes where the dissociation from the state [cC] and [cC*] are based on the same molecular site (e.g., the tRNA–codon interaction), we have $l_c' = k_c'$, and the same for d, so that (assuming c~d)

$$F = \left(\frac{k_c'}{k_d'}\right)^2 = F_0^2 \tag{P9.8}$$

where F_0 is the equilibrium error rate.

9.2. *Detailed balance.* Determine the error rate in a proofreading scheme in which transitions from [cC] to [c*C] occurs at a forward rate m_c and backward rate m_c', transitions from [c*C] to c + C occur at forward rate l_c and backward rate l_c', and corresponding constants for d, and where the product formation rate v is negligible compared to the other rates. Consider the case where all reactions occur at equilibrium. Use the **detailed balance** conditions, where the flux of each reaction is exactly equal to the flux of the reverse reaction, resulting in zero net flux along any cycle (also known in biochemistry as the **thermodynamic box** conditions).

a. Show that detailed balance requires that $k_c \, m_c \, l_c' = k_c' \, m_c' \, l_c$, and the same for d.

b. Calculate the resulting error rate F. Explain.

9.3. *Optimal tRNA concentrations.* In order to translate a codon, different tRNAs randomly bind the ribosome and unbind if they do not match the codon. This means that, on average, many different tRNAs need to be sampled for each codon until the correct match is found. Still, the ribosome manages to translate several dozen codons per second (Dennis et al., 2004). We will try to consider the optimal relations between the concentrations of the different tRNAs, which allow the fastest translation process, in a toy model of the ribosome.

a. Let the concentration of tRNA number j (j goes from 1 to the number of different types of tRNAs in the cell) be c_j. The relative concentration of tRNA number j is therefore $r_j = c_j/\Sigma c_j$. Suppose that each tRNA spends an average time t_0 bound to the ribosome before it unbinds or is used for translation. What is the average time needed to find the correct tRNA for codon j? Assume that there is no delay between unbinding of a tRNA and the binding of a new tRNA, and neglect the unbinding of the correct tRNA.

b. Suppose that the average probability of codon j in the coding region of genes in the genome is p_j. What is the optimal relative concentration of each tRNA that allows the fastest translation? Use a Lagrange multiplier to make sure that $\Sigma r_j = 1$.

Solution:

a. When codon j is to be read, the ribosome must bind $tRNA_j$. The probability that a random tRNA is $tRNA_j$ is r_j. Thus, on average one must try $1/r_j$ tRNAs before the correct one binds the ribosome. Hence, the average time to find the correct tRNA for codon j is

$$T_j = t_0/r_j$$

b. The time to translate the average codon is the sum of the times T_j weighted by the codon probabilities in the genome:

$$T = \Sigma T_j\, p_j = \Sigma\, p_j\, t_0/r_j$$

To minimize the translation time, we need to minimize T. Taking the derivative of T with respect to each r_j, we look for the relative concentrations that yield a minimum and thus have zero derivative. A Lagrange multiplier L is used to make sure that $\Sigma r_j = 1$:

$$dT/dr_j = d/dr_j\, (\Sigma p_j\, t_0/r_j + L\, \Sigma r_j) = -t_0\, p_j/r_j^2 + L = 0$$

Solving for r_j, and using a value of L such that $\Sigma r_j = 1$, yields an optimal r_j that is related to the square root of the codon probability p_j:

$$r_j^{\text{opt}} = \sqrt{p_j}\, \Big/ \sum_j \sqrt{p_j}$$

Thus, the rarer the codon, the lower the relative concentration of its tRNA.

9.4. *Optimal genetic code for minimizing errors.* In this exercise we consider an additional mechanism for reducing translation errors, based on the structure of the genetic code.

(a) First consider a code based on an alphabet of two letters (0 and 1), and where codons have two letters each. Thus, there are four possible codons ([00], [01], [10], and [11]). This genetic code encodes two amino acids, A and B (and no stop codons). Each amino acid is assigned two of the four codons.

a. What are the different possible genetic codes?

b. Assume that misreading errors occur, such that a codon can be misread as a codon that differs by one letter (e.g., [00] can be misread as [01] or [10], but not as [11]). Which of the possible codes make the fewest translation errors?

c. Assume that the first letter in the codon is misread at a higher probability than the second letter (e.g., [00] is misread as [10] more often than as [01]). Which of the codes have the lowest translation errors?

d. Study the real genetic code in Figure 9.2. Compare the grouping of codons that correspond to the same amino acid. How can this ordering help reduce translation errors? Based on the structure of the genetic code, can you guess which positions in the codon are most prone to misreading errors? Can you see in the code a reflection of the fact that U and C in the third letter of the codon cannot be distinguished by the translation machinery (a phenomenon called "third-base wobble")?

e. In the real genetic code, chemically similar amino acids tend to be encoded by similar codons (Figure 9.2). Discuss how this might reduce the impact of translation errors on the fitness of the organism.

CHAPTER 10

Optimal Gene Circuit Design

10.1   INTRODUCTION

In Chapters 1 through 6 we saw that evolution converges again and again to the same network motifs in transcription networks. This suggests network motifs are selected because they confer an advantage to the cells, as compared to other circuit designs. Can one develop a theory that explains which circuit design is selected under a given environment?

In this chapter, we will consider simple applications of a theory of natural selection of gene circuits. We will discuss the forces that can drive evolutionary selection in bacteria. The circuit that is selected, according to this theory, offers an optimal balance between the costs and benefits in a given environment.

Are cellular circuits optimal? It is well known that most mutations and other changes to the cells' networks cause a decrease in the performance of the cells. To understand evolutionary optimization, one needs to define a *fitness function* that is to be maximized. One difficulty in optimization theories is that we may not know the fitness function in the real world. For example, we currently do not know the fitness functions of cells in complex organisms. Such cells live within a society of other cells, the different tissues of the body, in which they play diverse roles. Fitness functions might not even be well defined in some cases; disciplines such as psychology and economics deal with processes that do not appear to optimize a fitness function, but only "satisfice" (Simon, 1996) in the sense of fulfilling several conflicting and incomparable constraints. This might apply to cells under some conditions.

Our view is that optimality is an idealized assumption that is a good starting point for generating testable hypotheses on gene circuits. This chapter will therefore treat the simplest systems in which one can form a phenomenological description of the fundamental forces at play during natural selection. For additional examples, refer to the work on optimality in metabolic networks in books by Savageau, Heinrich and Schuster, Palsson and others (see Further Reading in Chapter 1).

We will begin with simple situations in which fitness can be defined. One such situation occurs in bacteria that grow in a constant environment that is continually replenished. In this case, it is possible to define a fitness function based on the growth rate of the organism. The bacterium with the fastest growth rate eventually takes over the population, provided that its growth advantage is large enough to overcome random genetic drift effects. Hence, evolutionary selection under conditions of growth in a constant environment tends to maximize the growth rate.

As a detailed example, we will describe an experimental and theoretical study of the fitness function for the lactose (*lac*) system of *Escherichia coli*. We will ask what determines the amount of Lac proteins produced by the cells at steady-state. We will see that expressing the Lac proteins bears a cost: the cell grows slower the more proteins it expresses. On the other hand, the action of these proteins — breaking down the sugar lactose for use as an energy source — bestows a growth benefit to the cells. The fitness function, which is the difference of the cost and benefit, has a well-defined maximum. This maximum occurs at the protein level that maximizes the growth rate in a given environment. Direct evolutionary experiments show that the population is rapidly taken over by cells with mutations that tune the protein level to its optimal value. This analysis enables us to understand why evolution selects a specific expression level for the Lac proteins, and suggests that this optimization can occur rather rapidly and precisely.

After describing the cost–benefit analysis in the *lac* system, we will examine simple theories for the selection of gene regulation. Why are some genes regulated, whereas others are expressed at a constant level? We will see that gene regulation has a selective advantage in environments that vary over time. This is because the benefit of regulation, namely, the ability to respond to changes in the environment, can offset the cost of the regulatory system.

Finally, we will examine how the cost–benefit theory can be used to study the selection of the feed-forward loop network motif, described in Chapter 4, in environments that contain pulses of the input signal. We will see that it is possible to characterize the environments in which the feed-forward loop (FFL) circuit increases fitness compared to simple regulation with no FFL.

Our first question is: What sets the expression level of a protein? Why are some proteins produced at a few copies per cell, others at thousands, and yet others at tens or hundreds of thousands?

10.2   OPTIMAL EXPRESSION LEVEL OF A PROTEIN UNDER CONSTANT CONDITIONS

We begin by forming a **fitness function** f — a quantity to be optimized. In the case of bacteria growing in a favorable environment, a good choice for f is the growth rate of the cells.

Consider bacteria growing in a test tube. We start with a small number of bacteria. The number of cells grows exponentially until they get too dense. The number of cells, N, grows exponentially with time, with growth rate f:

$$N(t) = N(0)\, e^{ft} \tag{10.2.1}$$

Now, if two species with different values of f compete for growth and utilize the same resources, the one with higher f will survive and be selected and inherit the test tube. Thus, evolutionary selection in this simple case will tend to maximize f over time. This type of evolutionary selection process was elegantly described by G.F. Gause in *The Struggle for Existence* (Gause, 1934).

The fitness function can help us address our question: What determines the level of expression of a protein? To be specific, we will consider a well-studied gene system, the *lac* system of *E. coli*, which has already been mentioned in previous chapters. The *lac* system encodes proteins such as LacZ, which breaks down the sugar lactose for use as an energy and carbon source. When fully induced, *E. coli* makes about 60,000 copies of the LacZ protein per cell. Why not 50,000 or 70,000? What determines the expression level of this protein?

Optimality theory maintains that a protein expression level is selected that *maximizes the fitness function*. Therefore, our first goal is to evaluate the fitness as a function of the number of copies of the protein expressed in the cell. We will consider the simplest environment possible, in which conditions are constant and do not change with time. In the case of LacZ, this means an environment with a constant concentration of the sugar lactose. The fitness is composed of two terms: the **cost** of producing protein LacZ and the **benefit** it provides to the cells.

10.2.1 The Benefit of the LacZ Protein

Let us begin with the benefit. The benefit is defined as the relative increase in growth rate due to the action of the protein. In the case of LacZ, the benefit is proportional to the rate at which LacZ breaks down its substrate, lactose. The rate of the enzyme LacZ is well described by standard Michaelis–Menten kinetics (see Appendix A). Hence, LacZ breaks down lactose at a rate that is proportional to the number of copies of the protein, Z, times a saturating function of the concentration of lactose, L:

$$b(Z, L) = \frac{\delta Z L}{K+L} \tag{10.2.2}$$

where K is the Michaelis constant[1] and δ is the maximal growth rate advantage per LacZ protein — the growth advantage per LacZ protein at saturating lactose. Hence, the benefit grows linearly with protein level Z.

The benefit function was experimentally evaluated for the *lac* system (Figure 10.1). For this purpose, a useful experimental tool was used, the inducer IPTG. IPTG is a chemical analog of lactose, that causes expression of the Lac proteins, but is not metabolized by the cells. Thus, IPTG confers no benefit on its own. Benefit was measured by keeping the system maximally induced by means of IPTG, and by measuring growth rates in the presence of different levels of lactose. The observed benefit function was well described by Equation 10.2.2. The experiments indicate that the relative increase growth rate due to the fully vin-

---

[1] The Michaelis constant in this case is that of the transporter LacY, K = 0.4 m*M*. This is because the influx rate of lactose is limiting under most conditions. The concentrations of LacY and LacZ are proportional to each other because both genes are on the same operon.

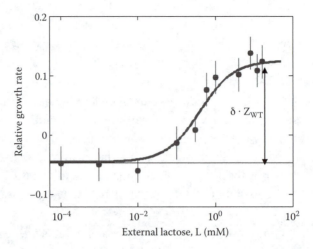

FIGURE 10.1 Benefit of Lac proteins of *E. coli* as a function of lactose concentration in the environment. Cells were grown with saturating IPTG so that LacZ is in its fully induced level Z_{WT}, and varying levels of lactose. Growth rate difference is shown relative to the growth rate of cells grown with no IPTG or lactose. $\delta Z_{WT} \sim 0.17$ is the benefit of fully induced Lac proteins at saturating lactose levels. Full line: Theoretical growth rate (Equation 10.2.2) (with $\delta = 0.17\ Z_{WT}^{-1}$ and K = 0.4 mM). (From Dekel and Alon, 2005.)

duced level of LacZ in the presence of saturating amounts of the sugar lactose is about 17% under the conditions of the experiment.

10.2.2 The Cost of the LacZ Protein

Now that we have an estimate of the benefit, let us discuss the cost. The cost function for LacZ was experimentally measured (Figure 10.2) by inducing expression of LacZ protein to different levels by means of the inducer IPTG in the absence of lactose. The inducer IPTG incurs only the costs of protein production, but gives no benefit because it cannot be utilized by the cells.[1] Expression of LacZ was found to reduce the growth rate of the cells. The cost, equal to the reduction in growth rate, is found to be a *nonlinear* function of Z: the more proteins produced, the larger the cost of each additional protein.

Why is the cost a nonlinearly increasing function of Z? The reason is that production of the protein not only requires the use of the cells' resources, but also reduces the resources available to other useful proteins. To describe this in a toy model, we can assume that the growth rate of the cell depends on an internal resource R (such as the amount of free ribosomes in the cell). The growth rate is typically a saturating function of resources such as R, following a Michaelis function:

$$f \sim \frac{R}{K_R + R} \tag{10.2.3}$$

The production of protein Z places a burden on the cells: mRNA must be produced and amino acids must be synthesized and linked to form Z. This burden can be described as

---

[1] Control experiments show that IPTG itself is not toxic to the cells. For example, IPTG does not affect the growth rate of cells in which the *lac* genes are deleted from the genome.

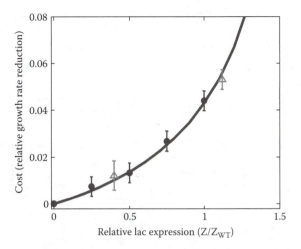

FIGURE 10.2 Cost of Lac proteins in *E. coli*. The cost is defined as relative reduction in growth of *E. coli* wild-type cells grown in defined glycerol medium with varying amounts of IPTG (an inducer that induces *lac* expression but gives no benefit to the cells) relative to cells grown with no IPTG. The x-axis is LacZ protein level relative to LacZ protein level at saturating IPTG (Z_{WT}). Also shown are the costs of strains evolved at 0.2 m*M* lactose for 530 generations (data point at $0.4 \cdot Z_{WT}$, open triangle) and 5 m*M* lactose for 400 generations (data points at $1.12 \cdot Z_{WT}$, open triangle). Full line: Theoretical cost function (Equation 10.2.4) with $\eta = 0.02 \ Z_{WT}^{-1}$. (From Dekel and Alon, 2005.)

a reduction in the internal resource R, so that each unit of protein Z reduces the resource by a small amount. The upshot is that the reduction in growth rate begins to diverge when so much Z is produced that R begins to be depleted (see mathematical derivation in solved Exercise 10.4):

$$c(Z) = \frac{\eta Z}{1 - Z/M} \tag{10.2.4}$$

This cost function tells us that when only a few copies of the protein are made, the cost is approximately linear with protein level and goes as $c(Z) \sim \eta Z$. The cost increases more steeply when Z becomes comparable to an upper limit of expression, M, when it begins to seriously interfere with other essential proteins. In real life, proteins do not come too close to the point Z = M, where the cost function diverges.

The experimental measurements of the cost function agree reasonably with Equation 10.2.4 (Figure 10.2). They show that the relative reduction in growth rate due to the fully induced *lac* system is about 4.5%. Note that this cost of a few percent makes sense, because the fully induced Lac proteins make up a few percent of the total amount of proteins in the cell.

10.2.3 The Fitness Function and the Optimal Expression Level

Having discussed the cost and benefit functions, we can now form the **fitness function**, equal to the difference between benefit and cost. The fitness function is equal to the growth rate of cells that produce Z copies of LacZ in an environment with a lactose concentration of L:

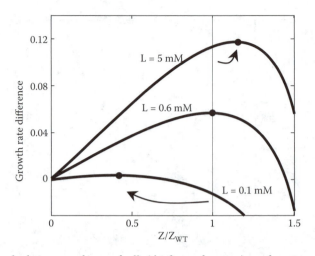

FIGURE 10.3 Predicted relative growth rate of cells (the fitness function) as a function of Lac protein expression, in different concentrations of lactose, based on the experimentally measured cost and benefit functions. The x-axis is the ratio of protein level to the fully induced wild-type protein level, Z/Z_{WT}. Shown are relative growth differences with respect to uninduced wild-type cells, for environments with lactose levels L = 0.1 mM, L = 0.6 mM, and L = 5 mM, according to Equation 10.2.5. The dot on each line is the predicted optimal expression level, which provides maximal growth (Equation 10.2.7). Cells grown in lactose levels above 0.6 mM are predicted to evolve to increased Lac protein expression (top arrow), whereas cells grown at lactose levels lower than 0.6 mM are predicted to evolve to decreased expression (lower arrow).

$$f_L(Z) = b(Z,L) - c(Z) = \frac{\delta Z L}{K+L} - \frac{\eta Z}{1-Z/M} \qquad (10.2.5)$$

This function displays a maximum, an optimal expression level of protein Z, as shown in Figure 10.3. The maximum occurs because benefit grows linearly with protein level Z, but the cost increases nonlinearly. The position of this maximum depends on L. The optimal protein level Z_{opt} can be found by taking the derivative of the fitness function with respect to Z:

$$d f_L/d Z = 0 \qquad (10.2.6)$$

Differentiating Equation 10.2.5, we find that the optimal expression level that maximizes the fitness function is

$$Z_{opt} = M\left(1 - \sqrt{\frac{\eta(K+L)}{\delta L}}\right) \qquad (10.2.7)$$

The more lactose in the environment, the higher the predicted optimal protein level. This is because the more lactose in the environment, the higher the benefit per LacZ enzyme, and the higher the selection pressure to produce more enzymes. The fully induced wild-type expression level, Z_{WT} is predicted to be optimal when L ~ 0.6 mM under these experimental conditions, as shown in Figure 10.3.

High lactose levels are thus predicted to supply a pressure for the increase of LacZ expression. Conversely, low levels of lactose show predicted optimal expression levels that are lower than the wild-type level of about $Z_{WT} = 60,000$/cell (Figure 10.3). When there is no lactose in the environment, the optimal level is $Z_{opt} = 0$, because proteins confer only costs and no benefits.

Generally, when costs exceed benefits, there is no need to produce any protein at all. This applies to environments with so little lactose that LacZ cannot provide a benefit that justifies its costs. Such a situation occurs when L is smaller than a critical level L_c given by asking when Z_{opt} in Equation 10.2.7 becomes equal to zero:

$$Z_{opt} = 0 \qquad \text{when } L < L_c = K(\delta/\eta - 1)^{-1} \qquad (10.2.8)$$

In the conditions of the experiments described above, the critical level of lactose needed for selection of the gene system is $L_c \sim 0.05$ mM. If lactose environments with L<L_c persist for many generations, the organism will tend to lose the gene encoding LacZ. The loss of unused genes is a well-known phenomenon; for example, bacteria grown in a chemostat[1] on glucose medium with no lactose lose the *lac* genes within a few days (Hartl and Dykhuizen, 1984).

10.2.4 Laboratory Evolution Experiment Shows That Cells Reach Optimal LacZ Levels in a Few Hundred Generations

To test the predictions of this cost–benefit analysis, a **laboratory evolution experiment** was carried out, by growing *E. coli* cells in tubes with a specified level of lactose. The lactose levels in the experiment were high enough to warrant full induction of LacZ expression. Every day, 1/100 of the cells from each tube were passed to a tube with fresh medium, a procedure known as **serial dilution**. The cells grew in the tube until they reached stationary phase. The next morning, 1/100 of the cells were again passed to a fresh tube, and so on. Thus, every day, the cells grew 100-fold, corresponding to $\log_2(100) = 6.6$ generations. The experiment was conducted for several months, running seven tubes in parallel, each with a different lactose level. The concentration of the LacZ protein was monitored over time. It was found that the cells heritably changed their LacZ expression level. The LacZ protein level reached the predicted optimal level within several hundred generations (Figure 10.4 and Figure 10.5).

Analysis of this evolutionary process indicated that the cells reached their optimal, adapted levels in each case by means of a single mutation that changed the LacZ protein level. For each lactose concentration, there are on the order of 100 possible mutations that can reach the desired optimal expression level. Many of these mutants arise and outgrow the wild-type cells. Finally, the mutants take over the population in the tube.

In summary, the cost and benefit functions can be directly measured to form a fitness function. This fitness function, measured in the wild-type bacterium, predicts that the protein level has an optimal value in each lactose environment. Cells rapidly evolve to this optimal value in evolutionary experiments. This gives us a sense of the speed and

---

[1] A chemostat is a device that keeps bacteria growing at a constant growth rate, by supplying a constant flow of fresh medium into a mixed aerated chamber, from which medium with cells is removed at the same rate. Cell generation time is locked onto the time for exchange of half of the medium in the chamber (Novick and Weiner, 1957; Balagadde et al., 2005; Ronen and Botstein, 2006.)

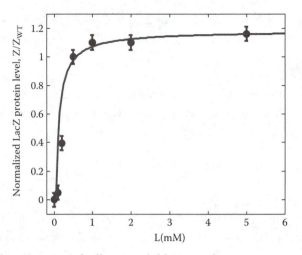

FIGURE 10.4 Adapted LacZ activity of cells in serial dilution evolution experiments as a function of the lactose concentration in the environment, L, relative to wild-type cells. Data are for 530 generations, except for the data point at 5 mM lactose, which is at generation 400. Full line: Theoretical prediction for optimal expression level (Equation 10.2.7). (From Dekel and Alon, 2005.)

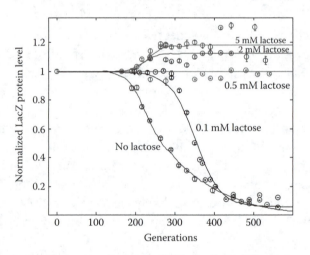

FIGURE 10.5 Experimental evolutionary adaptation of *E. coli* cells to different concentrations of lactose. LacZ protein level relative to wild-type protein level, in cells grown for 530 generations in serial dilution experiments with different lactose levels, is shown as a function of generation number. Cells were grown in 0, 0.1, 0.5, 2, and 5 mM lactose in a glycerol minimal medium supplemented with 0.15 mM IPTG. Lines are population genetics simulations of the serial dilution conditions (Crow and Kimura, 1970; Hartl and Clark, 1997). In these simulations, cells grew exponentially and underwent dilution; mutants with the optimal Z level arise with probability p per generation. The only fitting parameter in these simulations is the probability p per cell division for a mutation that yields the optimal LacZ level. (From Dekel and Alon, 2005.)

precision in which biological networks can adjust parameters such as protein expression levels.

We have just discussed the optimal expression level of a protein in constant conditions. What happens when conditions change with time, that is, in nonconstant environments? The next section will treat the principal way that cells deal with changes: gene regulation.

10.3 TO REGULATE OR NOT TO REGULATE: OPTIMAL REGULATION IN VARIABLE ENVIRONMENTS

In this section, we will ask why are some genes regulated and other genes expressed continually without regulation. When does it pay to regulate a gene?

For this purpose, consider a variable environment, one that is not constant in time. Suppose that our gene product Z provides benefit to the cells only in environmental condition C_z, which occurs only some of the time. For example, a sugar metabolism enzyme Z is beneficial only when the sugar is available in the environment.

Regulation means that protein Z is only produced in condition C_z, when it is needed, and not in other conditions. Regulation has a cost: the cost of production and maintenance of a regulatory system that can read the environment, and then calculate and implement the required changes in Z production. Thus, regulation will only be feasible if the benefits of such a system exceed its costs.

To analyze the optimal strategy, we compare three organisms with different designs for Z regulation. The environment of all three organisms displays condition C_z with probability p, and other conditions, in which Z is superfluous, with probability $1 - p$. This probability p is called the **demand** for Z. Demand will figure prominently in the next chapter.

In organism one, protein Z is not regulated and is produced at a constant rate under all conditions. This is known as constitutive expression. In the second organism, a regulatory system R is in place, so that Z is produced only under condition C_z, the condition in which its function is required. Organism three has neither the gene for Z nor the genes for its regulation system R on its genome. It cannot express protein Z at all.

The fitness function of the organisms includes two factors: cost and benefit, as in the previous section. First, producing protein Z leads to a reduction in growth rate due to the burden of synthesizing and maintaining Z. This cost is denoted c. The second factor is the benefit that the cell gets from the action of Z. The benefit is described by a growth rate advantage b.

The unregulated organism constantly produces Z, but gains its benefit only a fraction p of the time, when Z is in demand, so that it has a fitness

$$f_1 = p b - c \qquad (10.3.1)$$

The second organism regulates Z using the dedicated regulatory system R, to produce Z only under the proper conditions. This organism thus saves unneeded production and pays the cost, c, only a fraction p of the time. However, it bears the cost, r, of producing and maintaining the regulatory system R:

$$f_2 = p b - p c - r \qquad (10.3.2)$$

Finally, the third organism that lacks the system altogether will have fitness zero, where fitness zero is the baseline fitness without cost or benefit of Z:

$$f_3 = 0 \qquad (10.3.3)$$

Regulation will be selected when organism two has the highest fitness, $f_2 > f_1$ and $f_2 > f_3$. This leads to the following inequalities:

$$p < 1 - r/c \text{ and } p > r/(b - c) \qquad \textit{regulation selected} \qquad (10.3.4)$$

Similarly, the unregulated design in which Z is constitutively expressed will be selected when $f_1 > f_2$, $f_1 > f_3$, leading to the inequalities:

$$p > c/b \text{ and } p > 1 - r/c \qquad (10.3.5)$$

These inequalities (Equations 10.3.4 and 10.3.5) link a property of the environment, the demand for Z defined as the fraction of time p that condition C_z occurs, to the cost and benefit parameters of protein Z and its regulatory system. For each value of p, these equations tell us whether regulation will be selected over simpler designs.

The range of environments in which each of the three designs is optimal is shown in Figure 10.6. Regulation is selected at an intermediate range of demand, p. High demand tends to favor systems that are continually expressed. The design where Z is continually expressed is always optimal when p = 1, because if Z is always needed, regulation becomes superfluous. When p = 0, the protein is never needed and the optimal mechanism is to never express it.

The three domains in Figure 10.6 meet at a point. This point has coordinates $p^* = c/b$ and $(r/c)^* = 1 - c/b$. The larger the cost of producing the protein, c, relative to its benefit, b, the more this point, which corresponds to the apex of the triangular region in Figure 10.6, moves to the right and down. When costs exceed benefits, c > b, this region vanishes and regulation is never selected. In fact, when c > b, selection favors organisms that lack Z altogether, because its cost exceeds its benefit. The smaller the ratio of protein cost to benefit, c/b, the larger the range of environments in which regulation is selected.

There exist organisms in nature whose environment is quite constant. An example is intracellular parasites, organisms that live within cells and are supplied with nutrients and stable conditions (Moran, 2002; Wernegreen, 2002; Moran, 2003; Wilcox et al., 2003). In such constant environments, every protein has either p = 1 or p = 0. These organisms indeed lose most of their regulation systems, such as transcription factors. They hold a small set of genes continually expressed and lack many of the genes found in related, non-parasitic organisms. This agrees with the behavior shown in Figure 10.6, on the lines p = 1 and p = 0.

At the other extreme are bacteria that live in constantly changing and challenging environments such as the soil. These organisms have comparatively large genomes dense with prolific regulation systems.[1] These bacteria probably have 0 < p < 1 for most genes, so that extensive regulation is selected as shown in Figure 10.6.

---

[1] In bacteria, the number of transcription factors tends to increase with the number of genes in the genome. The number of transcription factors increases as N^a, where N is the number of genes and a ~ 2 in bacteria and a ~ 1.3 in eukaryotes (Huynen and van Nimwegen, 1998; van Nimwegen, 2003). Thus, increasing the number of genes seems to require increasingly elaborate regulation mechanisms with more transcription factors per gene.

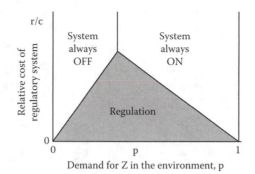

FIGURE 10.6 Selection phase diagram, showing regions where gene regulation, genes always ON, or genes always OFF, are optimal. The x-axis is the fraction of time p that the environment shows conditions in which protein Z is needed and provides benefit (the demand for Z). The ratio of protein cost to regulation system cost is r/c.

In summary, regulation makes sense only if the environment is sufficiently variable. In variable environments, the cost of the regulation system is offset by the advantage of information processing that can respond to changes in the environment.

We have thus examined the selection of the expression level of a protein and the selection of gene regulation systems. Cost–benefit analysis gives us a way to understand the forces that drive these evolutionary processes. As a final example, we now turn to the cost–benefit analysis of a gene circuit, the feed-forward loop network motif.

10.4   ENVIRONMENTAL SELECTION OF THE FEED-FORWARD LOOP NETWORK MOTIF

As we have seen throughout this book, gene regulation networks contain recurring elementary circuits termed network motifs. Evolution appears to have independently converged on these motifs in different organisms, as well as in different systems within the same organism (Conant and Wagner, 2003; Mangan et al., 2003). We will now try to understand, in a highly simplified model, under which environmental conditions a particular motif might be selected.

For this purpose, we will examine one of the most common network motifs, the coherent feed-forward loop (FFL). As we saw in Chapter 4, the coherent FFL can perform a basic dynamical function: it shows a delay following ON steps of an input signal, but not after OFF steps. The FFL is widespread in transcription networks, but not every gene is included in an FFL. In the *E. coli* transcription network, for example, about 40% of the known genes regulated by two inputs are regulated by an FFL, and 60% are regulated by a simple two-input design (both types of circuits are shown in Figure 10.7). It is therefore interesting to ask why the FFL is selected in some systems and not in others.

To answer this question, let us perform a simplified cost–benefit analysis for the selection of this gene circuit in a given dynamically fluctuating environment (Dekel et al., 2005). By *environment* we mean the time-dependent profiles of the input signals in the natural habitat of the organism. We will find conditions that the environment must

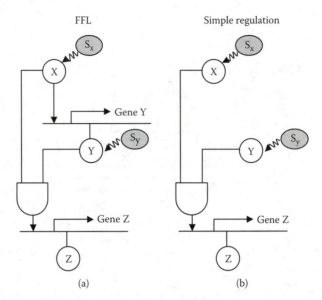

FFL   Simple regulation

(a)   (b)

FIGURE 10.7 Feed-forward loop (type-1 coherent FFL with AND input function) and a simple AND regulation circuit. (a) Feed-forward loop, where X activates Y and both jointly activate gene Z in an AND gate fashion. The inducers are S_x and S_y. In the *ara* system, for example, X = CRP, Y = AraC, Z = *araBAD*, S_x = cAMP, and S_y = arabinose. (b) A simple AND gate regulation, where X and Y activate gene Z. In the *lac* system, for example, Y = LacI is a repressor that is induced by S_y = lactose, X = CRP, and S_x = cAMP.

satisfy in order for the FFL to be selected over a simple-regulation circuit. We will see that the FFL can be selected in environments where the distribution of the input pulse duration is sufficiently broad and contains both long and short pulses. We will also determine the optimal values of the delay of the FFL circuit as a function of the environment.

We will not go through the full calculations in the main text — these calculations are given in solved exercises 10.5 to 10.9. The main results of the analysis are as follows. Suppose that the system is presented with a pulse of input S_x of duration D. The fitness function, based on the cost and benefit of protein Z, can be integrated over the pulse duration, $\varphi(D) = \int_0^D f(t)dt$. This integrated fitness shows that *short pulses of input signals have a detrimental effect on growth* (Figure 10.8): they lead to a reduction in fitness. The reason for the fitness reduction is that when the input pulses are shorter than critical pulse duration, D_c, protein Z does not have time to build up to levels in which the accumulated benefit exceeds the costs of production.

Since fitness is reduced by expression of protein Z in response to brief input pulses, a circuit that can avoid responses to brief pulses, and allow responses only to persistent pulses, can be advantageous. As we saw in Chapter 4, the coherent FFL can perform exactly this type of filtering task. In the coherent type-1 FFL, Z is expressed only at a **delay** after the signals appear. Thus, only pulses of input signals longer than the delay time of the FFL will lead to Z expression.

The delay in the FFL, which we will denote T_{ON}, results from the time it takes for transcription factor Y to accumulate and cross its activation threshold for gene Z. Recall that this delay time is related to the biochemical parameters of protein Y, namely, its

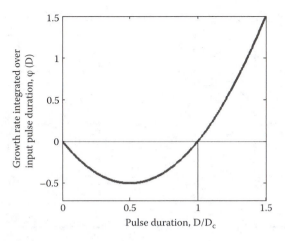

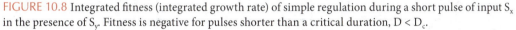

FIGURE 10.8 Integrated fitness (integrated growth rate) of simple regulation during a short pulse of input S_x in the presence of S_y. Fitness is negative for pulses shorter than a critical duration, $D < D_c$.

degradation rate, maximal level, and activation threshold for Z (Equation 4.6.5). The delay can therefore be tuned by natural selection to best fit the environment.

The delay in the FFL acts to filter out pulses that are shorter than T_{ON} (Figure 10.9). This avoids the reduction in growth for short pulses. However, the delay property of the FFL also has a disadvantage, because during long pulses, *Z is produced only at a delay and misses some of the potential benefit of the pulse* (Figure 10.9b). This means that there are some situations in which the FFL does more harm than good. To assess whether the FFL confers a net advantage to the cells, relative to simple regulation, requires analysis of the full distribution of pulses in the environment.

The environment of the cell can be characterized by the probability distribution of the duration of input pulses, P(D). Let us assume for simplicity that the pulses are far apart, so that the system starts each pulse from zero initial Z levels (and Y levels in the case of the FFL). In this case, the average fitness, averaged over many input pulses, can be found by integrating the fitness per pulse over the pulse distribution, $\Phi = \int P(D)\, \varphi(D)\, dD$. The design with higher average fitness has a selective advantage.

These considerations map the relation between the selection of these gene circuits and the environment in which they evolve. This is expressed as relations between certain integrals of the pulse distribution. Exercises 10.7 and 10.8 show that these relations can be solved exactly for certain distributions. These solutions indicate that the FFL is selected in some environments and not in others. For example, the FFL is never selected over simple regulation in environments with an exponential pulse distribution, $P(D) \sim e^{-D/D_o}$. On the other hand, the FFL can be selected in environments with a bimodal pulse distribution, which has a probability p for short pulses that reduce fitness, and a probability $1 - p$ for long, beneficial pulses. The optimal delay for an FFL in such an environment is a delay that precisely equals the duration of the short pulses. This delay filters out the non-beneficial pulses, but has a minimal negative impact on the fitness during long pulses. For this environment, one can draw a selection diagram that shows which circuit design has higher mean fitness (Figure 10.10). This selection diagram shows that the FFL is more fit

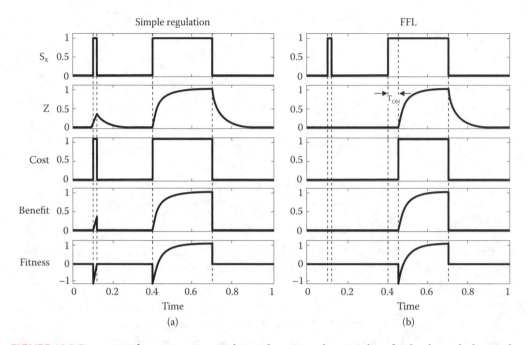

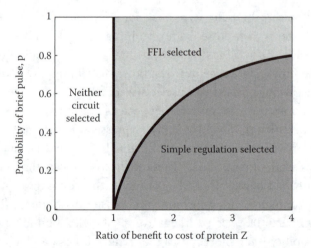

FIGURE 10.9 Dynamics of gene expression and growth rate in a short, nonbeneficial pulse and a long pulse of S_x. The signal S_y is present throughout. (a) Simple regulation shows a growth deficit (negative fitness) for short pulses. (b) FFL filters out the short pulse, but has reduced benefit during the long pulse. The figure shows (top to bottom): (1) Pulses of S_x. (2) Normalized dynamics of Z expression. Z is turned on after a delay T_{ON} ($T_{ON} = 0$ in the case of simple regulation) and approaches its steady-state level Z_m. (3) Normalized cost (reduction in growth rate) due to the production load of Z. Cost begins after the delay T_{ON}. (4) Normalized growth rate advantage (benefit) from the action of Z. (5) Net normalized growth rate (fitness).

FIGURE 10.10 Selection diagram for an environment with two types of pulses, a short pulse that occurs with probability p and a long pulse with probability $1 - p$. The x-axis is the ratio of benefit to production costs of protein Z. Three selection regions are shown, where FFL, simple regulation, or neither circuit has maximal fitness.

than simple regulation in a region where brief pulses are common and the benefit-to-cost ratio of the gene system is not too high. Simple regulation is superior when brief pulses are rare. When costs exceed benefits, neither circuit is selected. Exercise 10.10 applies this to the case of two sugar metabolism systems in *E. coli*, the lactose simply regulated system and the arabinose FFL system that was mentioned in Chapter 4.6.5.

I hope that this simplified analysis gives a taste for the possibility of studying the selection of gene circuits, and their optimal parameters, in temporally changing environments.

10.5   SUMMARY

In this chapter we discussed cost–benefit analysis as a theoretical framework for optimal circuit design. We saw that for growing bacteria, the fitness function corresponds to the cell growth rate. The cost and benefit functions can be directly measured, showing for the *lac* system a cost that increases nonlinearly with the amount of protein produced. The fitness function, equal to the difference between benefit and cost, has a well-defined optimum in each environmental condition. Optimal protein levels that maximize growth rate are reached rapidly and precisely by evolutionary selection in controlled evolutionary experiments.

We also analyzed the cost and benefit of gene regulation. We saw that gene regulation is worth maintaining only in variable environments. In constant environments, regulation would tend to be lost, as is the case in organisms living as parasites within the relatively constant conditions provided by their hosts.

Finally, we saw that cost–benefit analysis can also be carried out in a dynamically changing environment, to suggest criteria for the selection of network motifs such as the coherent FFL. According to this simplified analysis, the FFL can be selected in environments that have deleterious short pulses of induction, which need to be filtered out by the function of the FFL.

We currently have more information about the structure of transcriptional networks than about the precise environment and ecology in which they evolved. One might imagine an inverse problem — "inverse ecology" — deducing information about the environment based on the observed gene regulation networks. This is based on the idea that optimal circuits contain, in a sense, *an internal model of the environment*. For example, the optimal delay time of the FFL contains information about the distributions of input pulses. Thus, an intriguing goal is to use optimality considerations to understand the molecular details of mechanisms based on the environment in which they were selected.

We will continue with these ideas in the next chapter, in which we will use optimization concepts to deduce rules for patterns of gene regulation.

FURTHER READING

On Fitness and Evolution

Crow, J.F. and Kimura, M. (1970). *An Introduction to Population Genetics Theory*. Harper and Row.

Elena, S.F. and Lenski, R.E. (2003). Evolution Experiments with Microorganisms: The Dynamics and Genetic Bases of Adaptation. *Nat. Rev. Gent.*, 4: 457–69.

Gause, G.F. (1934). *The Struggle for Existence*. Dover Phoenix.

Optimality and Evolution in the *lac* System

Dekel, E. and Alon, U. (2005). Optimality and evolutionary tuning of the expression level of a protein. *Nature*, 436: 588–592.

Hartl, E. and Dykhuizen, D.E. (1984). The population genetics of Eschrichis coli. *Annu. Rev. Genet.* 18: 31–68.

Selection of the FFL Network Motif

Dekel, E., Mangan, S., and Alon, U. (2005). Environmental selection of the feed-forward loop circuit in gene-regulation networks. *Phys. Biol.*, 2: 81–88.

Optimality Principles in Metabolism

Heinrich, R. and Schuster, S. (1996). *The Regulation of Cellular Systems.* Kluwer Academic Publishers.

Ibarra, R.U., Edwards, J.S., and Palsson, B.O. (2002). *Escherichia coli* K-12 undergoes adaptive evolution to achieve *in silico* predicted optimal growth. *Nature*, 420: 186–189.

Melendez-Hevia, E. and Isidoro, A. (1985). The game of the pentose phosphate cycle. *J. Theor. Biol.*, 117: 251–263.

EXERCISES

10.1. *Limiting substrate.* Protein X is an enzyme that acts on a substrate to provide fitness to the organism. The substrate concentration is L. Calculate the fitness function $f(X, L)$ assuming linear cost, $c \sim -\eta X$, and a benefit that is a Michaelis–Menten term, $b(L, X) = b_0 L X/(X + K)$, appropriate for cases where the substrate, rather than the enzyme X, is limiting. Calculate the optimal enzyme level as a function of L and K.

10.2. For exercise 10.1, what is the minimal substrate level L_c required for maintenance of the gene for X by the organism? When is the gene lost? Explain.

10.3. *Optimal expression of a subunit.*

    a. Multiple units of protein X act together in a multi-unit complex. The benefit is a Hill function, $b(X) = b_0 X^n/(K^n + X^n)$, and the cost function is linear in X. What is the optimal protein level? Explain.

    b. Protein X brings benefit to the cell only when its concentration exceeds X_0, so that $b(X) = \theta(X > X_0)$, where θ is the step function. What is the optimal expression level of X?

10.4. *Cost function.*

    a. Derive the cost function in Equation 10.2.4, based on a limiting resource R, such that the growth rate is equal to $f = f_0 R/(K_R + R)$. Each unit of protein Z reduces R by a small amount ε.

    b. In bacterial cells, the resource R often increases as the growth rate decreases. For example, the fraction of free ribosomes increases as growth rate slows, because at high growth rates the ribosomes are mostly engaged in making new ribosomes. This effect can be added to the model to find similar cost functions at the low to intermediate expression levels of Z relevant to the experiments described in this chapter, but with no divergence at high Z. Assume that R = m/f, where f is the growth rate and m is a parameter. Derive the cost function in this case.

Solution for a:

a. The burden of Z production can be described as a reduction in the internal resource R, such that each unit of protein Z reduces the resource by a small amount ε, so that R goes to R − εZ. Hence, the cost, defined as the relative reduction in growth rate, is as in Equation 10.2.4:

$$c(Z) = \frac{f(Z) - f(0)}{f(0)} = \frac{f_o R / (K_R + R) - f_o (R - \varepsilon Z) / (K_R + R - \varepsilon Z)}{f_o R / (K_R + R)} = \frac{\eta Z}{1 - Z / M}$$

where the initial reduction per subunit of Z is $\eta = K_R \varepsilon / (K + R)$ and the parameter M is $M = (K_R + R) / \varepsilon$. Note that the cost can never diverge, because when Z depletes all of the resource R, that is, when $Z = R / \varepsilon$, one finds $f(Z) = 0$ and the cost is equal to c = 1.

10.5. *Short input pulses have a negative effect on growth.* Exercises 10.5 to 10.9 build a story for the selection conditions of the FFL and simple-regulation circuits. Consider a simple gene regulation mechanism with two inputs X and Y that control the expression of gene Z (that is, regulation without the third edge X → Y in the FFL). The two inputs are both needed for Z expression, so that this may be described as a simple-regulation circuit with an AND input function (Figure 10.7b). In this design, the production of Z occurs at a constant rate β in the presence of both signals S_x and S_y, and is otherwise zero. Show that pulses of the signals that are very short lead to a reduction in fitness. Only pulses that are long enough lead to a net growth advantage.

Solution:

To analyze the effects of time-dependent inputs, we will employ cost–benefit analysis that describes the effects of production of Z on the growth rate of the cells. The cost of Z production entails a reduction in growth rate c = −ηβ, where β is the rate of production of Z and η is the reduction in growth rate per Z molecule produced.[1] On the other hand, the action of the Z gene product conveys an advantage to the cells. This advantage, the benefit, is described by b(Z), the increase in growth rate due to the action of Z. The overall effect of Z on the growth rate is a sum of the cost and benefit terms:

$$f = -\eta\beta + b(Z) \qquad\qquad (P10.1)$$

---

[1] For simplicity, we neglect the nonlinear cost effects described in Section 10.2. Also, note that typically, the costs for the production of the transcription factors X and Y are negligible compared to the production cost of enzyme Z, since they are typically produced in far fewer copies per cell than enzymes (Nguyen et al., 1989; Ghaemmaghami et al., 2003). If Y costs are not negligible, the advantage of FFL over simple regulation increases, because the FFL prevents unneeded Y production.

Now consider a pulse of input signals, in which S_x is present at saturating levels for a pulse of duration D. The growth of cells with simple regulation, integrated over time D, is given by:

$$\varphi(D) = \int_0^D f(t)dt = -\eta\beta D + \int_0^D b(Z)dt \qquad (P10.2)$$

When the pulse begins, protein Z begins to be produced at rate β, and degraded or diluted out by cell growth at rate α. The dynamics of Z concentration are given by the simple dynamical equation we discussed in Chapter 2 (Equation 2.4.2):

$$\frac{dZ}{dt} = \beta - \alpha Z \qquad (P10.3)$$

resulting in the familiar exponential approach to steady-state $Z_m = \beta/\alpha$:

$$Z(t) = Z_m(1 - e^{-\alpha t}) \qquad (P10.4)$$

For long pulses ($D\alpha \gg 1$), the protein concentration Z is saturated at its steady-state value $Z = Z_m$. Protein Z has a net positive effect on cell growth:

$$\varphi(D) \approx -\eta\beta D + b(Z_m)D > 0 \qquad (P10.5)$$

provided that the benefit of Z exceeds its production costs:

$$b(Z_m) > \eta\,\beta \qquad (P10.6)$$

Short pulses, however, can have a deleterious effect on growth. To see this, consider short pulses such that $D\alpha \ll 1$. During the short pulse, the concentration of Z rises linearly with time (as we saw in Equation 2.4.7), with a slope equal to the production rate

$$Z(t) \sim \beta\,t \qquad (P10.7)$$

Since Z cannot reach high levels during the short induction pulse, we can use a series expansion of the benefit function $b(Z) \sim b' Z$, where $b' = d\,b/d\,Z$ at $Z = 0$. Using this in Equation P10.2, we find that the integrated growth rate is a quadratic function of the duration of the pulse, D (plotted in Figure 10.8):

$$\varphi(D) = \int_0^D (-\eta\beta + b'\beta t)dt = -\eta\beta D + b'\beta \frac{D^2}{2} \qquad (P10.8)$$

Importantly, the expression of Z causes a *reduction in growth* ($\varphi(D) < 0$) for pulses shorter than a critical pulse duration, D_c, found by solving $\varphi(D_c) = 0$ (Figure 10.8):

$$D_c = \frac{2\eta}{b'} \qquad (P10.9)$$

Pulses with $D = D_c$ are at the break-even point, because the cost exactly equals the benefit. Only pulses longer than D_c give a net benefit to the cells. Thus, simple regulation leads to reduction of growth in environments that have mainly short pulses, even though Z confers a net advantage for sufficiently long input pulses (Figure 10.9a).

10.6. *Conditions for selection of FFL over simple regulation.* Exercise 10.5 showed that expression of Z in response to short input pulses reduces fitness. Hence, a circuit that can avoid responses to short pulses, and allow responses only to persistence pulses, can be advantageous. As we saw in Chapter 4, the coherent FFL can perform this type of filtering task. In the coherent FFL, Z is expressed only at a delay T_{ON} after the signals appear. Thus, only pulses of input signals longer than the delay time of the FFL will lead to Z expression. However, the filtering of short pulses has a disadvantage, because during long pulses, Z is produced only at a delay and misses some of the potential benefit of the pulse (Figure 10.9b). To assess whether the FFL confers a net advantage to the cells, relative to simple regulation, requires analysis of the distribution of pulses in the environment. The environment of the cell can be characterized by the probability distribution of the duration of input pulses, P(D). Assume that the pulses are far apart, so that the system starts each pulse from zero initial Z levels (and Y levels in the case of the FFL). In this case, the overall fitness, averaged over many cell generations, can be found by integrating the fitness per pulse over the pulse distribution. Find conditions for the selection of the FFL over simple regulation.

Solution:

For simple-regulation circuits, the integrated fitness includes an integral over all possible pulses, times the fitness per pulse $\varphi(D)$:

$$\Phi_{simple} = \int_0^{\infty} P(D)\varphi(D)dD \qquad (P10.10)$$

For FFL circuits, production starts after a delay T_{ON}. Pulses shorter than T_{ON} result in no Z production and $\varphi(D < T_{ON}) = 0$. Long pulses begin to be utilized only after the delay T_{ON}, so that their duration is effectively $D - T_{ON}$ (Figure 10.9b), resulting in a contribution in the integral only from pulses longer than T_{ON}:

$$\Phi_{FFL} = \int_{T_{ON}}^{\infty} P(D)\varphi(D - T_{ON})dD \qquad (P10.11)$$

Note that the simple-regulation case, Equation P10.10, is equivalent to an FFL with $T_{ON} = 0$.

The resulting condition for selection of FFL over simple regulation is when its averaged fitness exceeds that of simple circuits and is positive:

$$\Phi_{FFL} > \Phi_{simple} \qquad \Phi_{FFL} > 0 \qquad\qquad (P10.12)$$

Simple regulation is selected when its integrated fitness exceeds that of the FFL

$$\Phi_{simple} > \Phi_{FFL} \qquad \Phi_{simple} > 0 \qquad\qquad (P10.13)$$

Neither circuit is selected otherwise ($\Phi_{FFL} < 0$ and $\Phi_{simple} < 0$). For the purpose of this comparison, the FFL is chosen to have the optimal value for T_{ON} (that maximizes Φ_{FFL}), because natural selection can tune this parameter to best adapt to the environment.

10.7. *The FFL is not selected in the case of exponential pulse distributions.* Analyze the average fitness of the FFL and simple regulation in an environment in which pulses have a constant probability per unit time to end. Such environments have an exponential pulse distribution:

$$P(D) = D_0^{-1} e^{-D/D_0} \qquad\qquad (P10.14)$$

Solution:

Using Equations P10.10 and P10.11, we find that

$$\Phi_{FFL} = \int_{T_{ON}}^{\infty} D_0^{-1} e^{-D/D_0} (D - T_{ON}) dD = e^{-T_{ON}/D_0} \int_0^{\infty} D_0^{-1} e^{-D/D_0} D \, dD = e^{-T_{ON}/D_0} \Phi_{simple} < \Phi_{simple}$$

$$(P10.15)$$

Thus, the *FFL is never selected* since $\Phi_{FFL} < \Phi_{simple}$.

An intuitive reason why FFL is not selected in environments with exponential pulse distributions is related to the fact that exponential distributions are memoryless. Knowledge that a pulse has lasted for time t does not help us to predict how long it will continue to last. The FFL, which effectively reduces the pulse duration by a delay T_{ON}, confers no advantage relative to simple regulation.

10.8. *The FFL can be selected in bimodal distributions with long and short pulses.* Consider an environment that has two kinds of pulses. A pulse can have either a short duration, $D_1 \ll D_c$, with probability p, or a long duration, $D_2 \gg 1/\alpha$, with probability 1 − p. Analyze the conditions for selection of FFL and simple regulation as a function of p and the ratio of the benefit to cost ratio of protein Z (Figure 10.10).

Solution:

The short pulses D_1 are nonbeneficial, since they are shorter than the critical pulse width at which gene Z reaches the break-even point, $D_1 < D_c$ (Figure 10.8). In contrast, the long pulses D_2 are beneficial and have a benefit of approximately (applying Equation P10.5)

$$\varphi(D_2) = -\eta\beta D_2 + b(Z_m)D_2 > 0 \qquad (P10.16)$$

In this case, it is easy to calculate the optimal delay in the FFL: the optimal delay is $T_{ON} = D_1$. That is, the optimal FFL has a delay that blocks the short pulses precisely; a longer delay would not further filter out short pulses, and would only reduce the benefit of the long pulses. The condition for selection of FFL over simple regulation, found by solving Equations P10.10 and P10.11 to find $\Phi_{simple} = (1 - p)(b(Z_m) - \eta\beta) D_2 - p\eta\beta D_1$ and $\Phi_{FFL} = (1 - p)(b(Z_m) - \eta\beta) (D_2 - D_1)$. This shows that the FFL is more fit when the probability of short pulses exceeds a factor related to the ratio of cost to benefit of Z:

$$p > 1 - \frac{\eta\beta}{b(Z_m)} \qquad (P10.17)$$

The phase diagram for selection is shown in Figure 10.10. When the ratio of benefit to cost, $b(Z_m)/\eta\beta$, is small, neither circuit is selected (costs outweigh benefits). At large relative benefits, the FFL is selected if short pulses are common enough — that is, if p is large enough (Equation P10.17). If short pulses are rare, simple regulation is selected. At a given p, the higher the benefit-to-cost ratio, the more likely the selection of simple-regulation circuits.

10.9. *Why is FFL selected in the* ara *system but not in the* lac *system of* E. coli? In this exercise, we will apply, in a qualitative way, the results of exercise 10.8 to the case of sugar systems in *E. coli*. Why is the FFL selected in some sugar systems, such as arabinose utilization (*ara* system discussed in Section 4.6.5), whereas simple regulation is selected in others, such as the lactose system (*lac* system)?

Solution:

The models are only simplified toy models, but let us proceed for demonstration purposes. Both *ara* and *lac* systems share the same transcription activator, $X = CRP$, stimulated by $S_x = cAMP$, a signaling molecule produced by the cell upon glucose starvation. Thus, both *ara* and *lac* systems have the same S_x pulse distribution. However, these systems differ in the benefit they yield per sugar molecule: the benefit-to-cost ratio, $b(Z_m)/\eta\beta$, appears to be different for the two systems. The benefit per lactose molecule, which is split into glucose + galactose, is greater than the benefit per arabinose molecule (approximately 70 ATPs per lactose molecule vs. approximately 30 ATPs per arabinose molecule). In addition to its smaller benefit, the cost of the *ara* system may be larger than the cost of the *lac* system, because there are at least seven highly expressed *ara* proteins (the metabolic enzymes AraB, AraA, and AraD, and the pumps AraE and AraFGH), compared to only three highly expressed *lac* proteins (LacZ, LacY, and LacA). Thus, the parameter $b(Z_m)/\eta\beta$ for the *ara* system may be more to the left in Figure 10.10 relative to the *lac* system, favoring selection of FFL in the former.

The delay in the FFL can be tuned by natural selection. As mentioned in Chapter 4, the delay in the *ara* system appears to be on the timescale of the deleterious short pulses in the environment.

10.10. *Cascades vs. FFLs.* Repeat the calculations of Exercises 10.6 and 10.7 for a cascade X → Y → Z. Show that cascades are never more optimal than FFLs for environments with pulses of input signals. Explain this result.

10.11. (Advanced students) *The cost and benefit of SIM.* X controls genes Z_1 and Z_2 in a single-input module (SIM). Gene products Z_1 and Z_2 assemble into a complex, such that n_1 units of protein Z_1 first assemble into subunit S_1, and then n_2 units of protein Z_2 join subunit S_1 and form the final product S_2. X begins to be produced at rate β at time t = 0. What are the optimal activation thresholds and production rates for genes Z_1 and Z_2? Use logic input functions. The production costs for Z_1 and Z_2 are η_1 and η_2, and benefit only occurs when a unit of S_2 is produced.

CHAPTER 11

Demand Rules for Gene Regulation

11.1 INTRODUCTION

The control of gene expression involves complex mechanisms that show large variation in design. In this chapter, we will discuss whether the mechanism that is used in each case is a result of random historical choice, or whether there are rules that can help us to understand the design in each case. For this purpose, we will return to transcription networks, and attempt to deduce rules for gene regulation. The specific question we will ask is: why are there positive and negative modes of regulation? That is, why are some genes regulated by a repressor, and others by an activator? What determines the mode of the regulation in each case?

It is important to first note that activators and repressors can achieve exactly the same regulatory goals. For example, a gene that is fully expressed only in the presence of a signal (Figure 11.1), can be regulated by one of two mechanisms: Either an activator binds the promoter to activate the gene, or a repressor falls off the promoter to activate the gene. These two mechanisms realize the same input-output relationship: Expression is turned on by the binding of an activator in the positive mode of control, and by the unbinding of a repressor in the negative mode of control. More generally, a gene controlled by N regulators, each of which can be either an activator or a repressor, has 2^N possible mechanisms that can generate a given input-output mapping.

Among these equivalent mechanisms, evolutionary selection chooses one for each system. Are there rules that govern this selection? One possibility is that evolution chooses randomly between equivalent designs. Hence, the selected mechanism is determined by historical precedent. Another possibility is that general principles exist, which govern the choice of mechanism in each system.

The question of rules for gene regulation was raised by M.A. Savageau in his pioneering study of transcriptional control (Savageau 1974, 1977, 1983). Savageau found that the

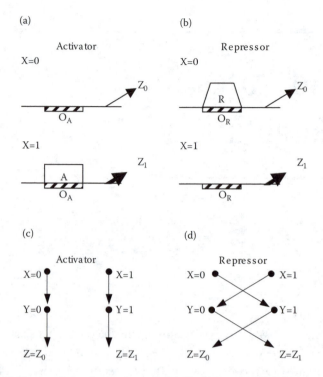

FIGURE 11.1 Transcription regulation mechanisms of gene Z. (a) Positive control (activator). In the absence of an inducing signal (input-state $X=0$), the binding site O_A of activator A is free. This causes the gene to be expressed at a low level Z_0. When the signal is present ($X=1$), O_A is bound by the activator A, which causes the gene to be fully expressed ($Z=Z_1$). (b) Negative control (repressor). When the inducer signal is absent ($X=0$), respressor binding site O_R is bound by repressor R. This causes the gene to be expressed at a low level Z_0. When the inducer is present ($X=1$), O_R is free, which causes the gene to be fully expressed ($Z=Z_1$). (c) Mapping between input-states X, binding-states Y and output-states Z in the case of positive control. $Y=0/1$ corresponds to a free/bound site, respectively. (d) Mapping between input-states, binding-states and output states in the case of negative control.

mode of control is correlated with the *demand*, defined as the fraction of time in the natural environment that the gene product is needed near the high end of its regulatory range. High-demand genes, in which the gene product is required most of the time, tend to have positive (activator) control. Low-demand genes, in which the gene product is not required most of the time, tend to have negative (repressor) control. This demand rule appears to be in agreement with over 100 gene systems (Savageau 1988) from *E. coli* and other organisms, where the mode of control is known and the demand can be evaluated.

In this chapter, we will describe demand rules for gene regulation based on intrinsic differences between the modes of control. These differences are due to the fact that biological internal-states are prone to errors, which lead to errors in the output. The errors result in a reduction of the organism's fitness. This reduction is called the **error-load**. Equivalent mechanisms, which implement the same input-output relationship, can differ in their error-loads.

After describing the Savageau demand rules, we will examine the proposal that evolution selects the mechanism that minimizes the error-load. We will see how error-load minimization can explain the connection between the mode of regulation and the demand for a gene in the organism's environment: rarely needed genes tend to be regulated by repressors, and genes commonly needed at full expression tend to be regulated by activators. This theory can be extended to the case of multiple regulators. We will see how error-load minimization can explain detailed features of the structure of the *E. coli lac* system. We will also discuss the criteria for when selection according to these rules dominates over historical precedent.

More generally, the goal of this chapter (and one of the main goals of this book) is to encourage the point of view that rules can be sought to understand the detailed structure of biological systems.

11.2 THE SAVAGEAU DEMAND RULE

M. A. Savageau noted a strong correlation between the mode of bacterial gene regulation and the probability that the gene is fully expressed in the environment. To formulate this rule, Savageau defined the **demand** for a gene system as follows:

"When a system operates close to the high end of its regulatable range most of the time in its natural environment it is said to be a **high-demand** system. When it operates at the low end of its regulatable range most of the time in it natural environment it is said to be a **low-demand** system".

Demand corresponds to the frequency at which the function carried out by the gene system is needed within the ecology of the organism. For example, a system that degrades a certain sugar for use as an energy source is in low demand if the sugar is rare in the environment. The system is in high demand if the sugar is often available. A system that synthesizes an amino-acid is in low demand if that amino-acid is commonly available in significant amounts in the environment – demand is low because de-novo synthesis of the amino-acid is not often needed. Conversely, a system that synthesizes an amino-acid that is only rarely available from the outside is in high demand.

Each of these systems can be regulated either by a repressor or by an activator- that is, either by a negative or by a positive regulation mode. The demand rule may be stated as follows. "The molecular mode of gene regulation is correlated with the demand for gene expression in the organism's natural environment. *The mode is positive when the demand is high and negative when the demand is low.*" Thus, rarely needed genes tend to be regulated by repressors, commonly needed genes by activators.

11.2.1 Evidence for the Demand Rule in *E. coli*

To test the demand rule, one needs to have knowledge of the mode of regulation and of the demand for the gene system in question. For this purpose, Savageau collected data on the natural environment of the bacterium *E. coli*. One of the principle habitats of *E. coli* is the intestinal system of its mammalian host. Studies of this environment suggest that different sugars and amino-acids have different abundances. Some sugars are taken up readily by the body, and are thus rarely available for the bacteria. Other sugars are less readily absorbed and are much more common in the bacteria's environment. This leads to the following estimated ranking of sugar abundances: D-glucose < D-galactose < glycerol < D-xylose < L-glycose < L-mannose < L-fucose < L-rhamnose < L-arabinose

TABLE 11.1 Molecular Mode of Regulation and Demand for Degradation Gene System in the Environment of *E. coli*

Degradation system (induced in presence of substance)	Mode of regulation	Regulator	Demand for expression
Arabinose	Positive	AraC	High
Fucose	Positive	FucR	High
Galactose	Negative	GalR, GalS	Low
Glycerol	Negative	GlpR	Low
Lactose	Negative	LacI	Low
Lysine	Positive	CadC	High
Maltose	Positive	MalT	High
Rhamnose	Positive	RhaS	High
Xylose	Positive	CylR	High
Proline (degradation)	Negative	putA	Low

(Savageau 1976, 1977, 1983). The sugar lactose is also a rare sugar in the environment of *E. coli*, because it is cleaved by specialized enzymes in the upper intestinal tract. Similarly, estimates for amino-acid abundances are: glycine > leucine > phenylalanine > histidine > alanine > serine > valine > aspartate > proline > threonine > cystine > isoleucine > methionine.

A comparison of the mode of control of inducible systems that degrade nutrients is shown in Table 11.1. For example, the sugar galactose is seldom present at high concentrations in the environment of *E. coli*, which corresponds to low demand for the galactose genes that degrade and utilize this sugar. According to the demand rule, the galactose system should have negative control. This is in agreement with the repressors GalR and GalS that control this system. On the other hand, arabinose is found at high concentrations, corresponding to high demand of the arabinose utilization system. Its mode of regulation is positive, with the activator AraC, in agreement with the demand rule.

Similar results are shown in Table 11.2 for a number of biosynthesis systems that produce a compound in the cell. The expression of these systems is reduced if the compound that they synthesize is available from the outside. For example, the arginine biosynthe-

TABLE 11.2 Mode of Regulation and Demand for Biosynthesis Gene Systems

Biosynthetic system (induced in the absence of substance)	Mode of regulation	Regulator	Demand for expression
Arginine	Negative	ArgR	Low
Cysteine	Positive	CysB	High
Isoleucine	Positive	IlvY	High
Leucine	Positive	Lrp, LeuO	High
Lysine	Positive	LysR	Low
Tryptophan	Negative	TrpR	Low
Tyrosine	Negative	TyrR	Low

Note: Note that Lysine biosynthesis is an exception to the demand rules (though it appears to have additional control mechanisms including a 'ribo-switch', an RNA-based lysine sensor).

sis system of *E. coli* produces the amino-acid arginine that is relatively abundant in the natural environment of the cells. The corresponding biosynthesis system is thus in low demand, and the demand rules allow one to predict a negative mode of control. Indeed, this system is regulated by a repressor ArgR. On the other hand, cysteine biosynthesis, a high demand system due to the low abundance of cysteine, is regulated in a positive mode by the activator CysB.

The rule also successfully predicts that systems with antagonistic functions, such as biosynthesis and degradation of a compound, tend to have opposite modes of regulation. In contrast, systems with aligned functions, such as transport and utilization of a compound, tend to have the same regulation mode (Savageau, 1977, 1989). These predictions follow from the demand rule because antagonistic systems tend to have opposite demands, and systems with aligned functions tend to have the same demand (both high or both low). Note that predictions of this kind do not require knowledge of the precise demand for the systems.

More complete data on gene regulation networks tends to support the demand rule, but some exceptions to the rule are also found. One possible example is the biosynthesis system of lysine, an abundant amino acid. Since lysine is relatively abundant, the demand for its *de-novo* synthesis is low. However, this system is controlled in a positive mode by the activator LysR, in contrast to the predicted negative mode. The definition of demand is often tentative, because we lack information on the ecology of the cells for many systems. On the whole, however, the Savageau demand rule seems to capture the mode of many of the known gene regulation interactions in bacteria where the demand can be reasonably estimated.

11.2.2 Mutational Explanation of the Demand Rule

The demand rule was deduced by Savageau based on the effects that mutations have on the two modes of regulation. This theory first assumes that there are no inherent functional differences between the two modes of regulation. That is, precisely the same modulation of gene expression in response to a signal can be achieved either by a repressor binding the promoter or by an activator unbinding from the promoter. This assumption suggests that one should focus on the behavior of mutants that are altered in the regulatory mechanism.

The theory next uses the fact that most mutations in highly evolved structures are detrimental, and very few mutations are beneficial. Consequently, most mutations in a regulatory mechanism lead to loss of regulation. In the case of positive mode, loss of regulation results in super-repressed low expression, because the activator is no longer functional. In the case of a negative mode, loss of regulation results in constitutive high expression because the repressor is not functional. Thus, mechanisms with different modes respond in opposite ways to mutations.

The result of these considerations is that the two modes will fare differently in a given environment (Table 11.3). The positive mode of regulation is more stable against

TABLE 11.3 Selection of Mutants for Different Modes of Control According to the Mutant-Selection Theory

Demand	Mode of regulation	
	Positive	Negative
High	Regulation selected	Regulation lost
Low	Regulation lost	Regulation selected

mutations in a high demand environment, and the negative mode is more stable in a low demand environment.

To understand this, consider a positively regulated gene in a high demand environment. The wild-type organism will induce the gene to high levels most of the time. Mutants who have lost the regulation, will not express the gene. As a result they will be at a disadvantage most of the time, and will be lost from the population.

In a low demand environment, however, expression of the gene will be shut off most of the time. The mutants, who have lost the activator, will be unable to express the gene. There will be relatively weak selection against this loss of regulation in this system, because the gene is rarely needed at high expression. Super-repressed mutants will accumulate as a result of mutations, and the functional regulatory system will be lost. Hence, a positive mode is more stable in a high demand environment than in a low demand environment.

The predictions are just the opposite when one considers a negative mode of regulation. In this case, mutants will be strongly selected against in a low demand environment because these mutants have lost the repressor and have un-needed high expression. The high expression has a fitness cost and leads to loss of the mutants from the population. In contrast, there will be a relatively lower selection pressure against mutants in a high-demand environment, because the gene is needed at high levels most of the time. Thus, according to this argument, mutants will accumulate over time, and the negative regulatory system will be lost.

In summary, the mutant selection theory suggests that negative mode is stable in low demand environments, and positive mode is stable in high demand environments.

11.2.3 The Problem with Mutant-Selection Arguments

Mutant-selection arguments are valid only if there is no intrinsic fitness advantage to one of the two modes of control. If such intrinsic differences exist, they would dominate over the differential effects of mutations. The fitter mechanism would readily take over the population. In other words, mutational effects are second-order with respect to inherent differences in the wild-type mechanisms. In the next section, we will develop a theory to understand the demand rules based on intrinsic differences between the modes of regulation. These differences between modes of regulation correspond to their resistance to errors.

11.3 RULES FOR GENE REGULATION BASED ON MINIMAL ERROR LOAD

We will now describe a framework for deducing demand rules based on inherent fitness differences between the two modes of control (Shinar et al., 2006). The idea is simple to understand. The main assumption is that in many regulatory systems, DNA sites that are bound tightly to their regulatory protein are more protected from errors than free DNA sites. This is because free sites are exposed to non-specific binding. These binding errors lead to changes in gene expression, which reduce the organism's fitness. This leads to the proposal that in order to minimize errors, such systems will evolve positive control in high-demand environments and negative control in low-demand environments: in both cases, the DNA regulatory sites are bound most of the time and thus protected from errors.

To understand the demand rules in more detail, consider a gene regulated by a transcription factor, which can be an activator or a repressor (Figure 11.1). In either case, there is one state in which the regulator binds its site tightly, and another state in which the site is free. This is the idealized picture. In reality, the system is embedded in the cell, where many additional regulators and other factors are present. When the site is tightly bound by its designated regulator, the site is protected from these factors. In contrast, when the site is free, it is exposed to non-specific binding. This non-specific binding can lead to errors in the expression level of the gene, and thus to a reduction in the fitness of the organism. The relative reduction in fitness is called the **error-load**. Thus, our main assumption is that free DNA sites are exposed to errors, whereas sites tightly bound by their regulators are protected from these errors.

There are at least two sources of errors connected with the free site. The first source of errors is cross-talk with the other transcription regulators in the cell, in which the wrong transcription factor binds to the site (Gerland et al., 2002; Sengupta et al., 2002). This cross-talk is difficult to prevent, because the concentration and activity of the regulators in the cell changes in response to varying conditions, leading to an ever-changing set of cross-reacting affinities to the site. This cross-talk can act to reduce or increase expression, leading to errors. A second source of error arises from residual binding of the designated regulator in its inactive form to its own site: in many cases, the affinity of the inactive regulator is only about one to two orders of magnitude lower than its affinity in the active state. Since regulator levels fluctuate from cell to cell (see Appendix D), there will be a varying degree of residual binding to the free site, causing cell-cell fluctuations in expression. These errors in expression deviate from the optimal level, leading to a reduction in fitness.

Let us now compare the error-loads of the positive and negative modes of regulation. Consider a gene regulated by an activator, and the same gene regulated by a repressor, such that the two regulatory mechanisms lead to the same input-output relationship.[1] The regulated gene has a demand p, defined as the fraction of time that full expression of the gene product is needed in the environment. Now, for either mode of control, errors occur mainly when the DNA site is free and exposed; The two modes differ in the expression state that is associated with a free site. In the case of a repressor, a free site corresponds to high expression, which occurs a fraction p of the time. Errors in expression lead to a relative fitness reduction of Δf_1, where the subscript '1' denotes the high expression state. The

---

[1] The regulators are assumed to be equivalent in terms of design criteria such as the sharpness of the response function, its dynamic range etc., as well as the production cost of the regulatory proteins themselves. Indeed, sharp and high gain switches are known with both activators and repressor.

average reduction in fitness for a repressor, taking into account only errors from the free site, is therefore

$$E_R = p\Delta f_1 \qquad (11.1)$$

For an activator, the free site corresponds to low expression, which occurs a fraction $1-p$ of the time (the fraction of time that the gene is not in demand). Errors in the expression level lead to a relative fitness reduction of Δf_0, where the subscript '0' denotes the low expression state. The average reduction in fitness for an activator, taking into account only errors from the free site, is therefore

$$E_A = (1 - p)\Delta f_0 \qquad (11.2)$$

In this simplest case, a repressor will have a fitness advantage over an otherwise equivalent activator when it has a lower error-load

$$E_R < E_A \qquad (11.3)$$

Using Equation 11.1 and Equation 11.2 in Equation 11.3, we see that repressors are advantageous when the demand is lower than a threshold determined by the ratio of the relative reductions in fitness:

$$p < 1 / (1 + \Delta f_1/\Delta f_0) \qquad (11.4)$$

Thus, repressors are advantageous for low-demand genes, and activators for high-demand genes (Figure 11.2). The reason for this is that repressors in low-demand genes and activators in high-demand genes ensure that the site is bound to its designated regulator most of the time. The demand rule therefore minimizes the fraction of time that the site is exposed to errors.

11.4 THE SELECTION PRESSURE FOR OPTIMAL REGULATION

Can error-load create a selection pressure sufficient to cause a regulatory system to be replaced by a system with the opposite mode of control? Consider a wild-type population with a regulatory system in place. Suppose that conditions vary, leading to a permanent change in the demand for the gene, so that the opposite mode of control becomes optimal. Mutants with the opposite mode of control arise in the population, by genomic mutation or lateral gene transfer from other organisms[1]. These mutants have a lower error-load, and hence a relative fitness advantage, which is equal to $E_A - E_R$ in the case of a repressor and $E_R - E_A$ in the case of an activator. The mutants can become fixed if their relative

---

[1] There are well characterized examples where the same regulatory protein can act either as a repressor or as an activator depending on the position and strength of its regulatory site (Collado-Vides et al., 1991, Choy et al., 1995, Monsalve et al., 1996); Other cases are known where mutations in the regulator coding region can cause a repressor to become an activator and vise versa (Bushman et al., 1988, Bushman et al., 1989, Lamphier et al., 1992, Ptashne et al., 2002).

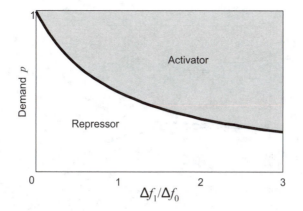

FIGURE 11.2 Selection diagram for error-load minimization. Each region corresponds to the mode of control with the smaller error-load. The vertical axis is the demand p, defined as the fraction of the time that the gene product is needed at full expression. The horizontal axis is the ratio of the fitness reductions arising from errors in the free sites of the positive and negative control mechanisms.

fitness advantage exceeds a minimal selection threshold, s_{min}. The selection threshold s_{min} has been estimated in bacteria and yeast to be in the range 10^{-8}-10^{-7} (Hartl et al., 1994, Wagner, 2005a).

The condition for fixation of a repressor mutant is thus $E_A - E_R > s_{min}$, whereas the condition for fixation of an activator mutant is $E_R - E_A > s_{min}$. These inequalities lead to a selection diagram (Figure 11.3), in which the error-minimizing regulatory mechanism becomes fixed at a given demand p only if the ratio $s_{min} / (\Delta f_0 + \Delta f_1)$ is sufficiently small. In cases where the fitness advantage is smaller than s_{min}, there exists a region in parameter space where historical precedent determines the mode of control.

One can estimate whether the fitness reductions caused by expression errors, Δf_0 and Δf_1, lead to selectable error-load differences. The fitness as a function of protein expression was described in the previous chapter for the *lac* system of *E. coli*. The fitness function indicates that a 1% error in expression leads to relative fitness reductions Δf_0 and Δf_1 on the order of 10^{-3}, which is four orders of magnitude higher than the selection threshold s_{min}. Similarly, Wagner estimated that in yeast, a 2% change in the expression level of any protein is sufficient to cause fitness differences that exceed the selection threshold (Wagner, 2005a). These considerations suggest that even minute expression errors lead to error-load effects that can dominate over historical precedent in determining the choice of regulatory system.

11.5 DEMAND RULES FOR MULTI-REGULATOR SYSTEMS

So far, we have analyzed the demand rules for a gene with a single regulator. We have seen that activators are better than repressors for regulating high demand genes and that repressors are better than activators for regulating low demand genes. In both cases, the DNA site is bound for most of the time, minimizing errors. Let us now turn to systems with multiple regulators. For clarity, we will consider in detail the *lac* system of *E. coli*, even though the present considerations can be generally applied to other systems.

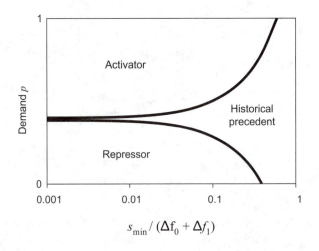

FIGURE 11.3 Selectability of an error-minimizing mode of control in the face of an existing mechanism with the opposite mode. The region marked activator (repressor) is a region in which a mutant bearing an activator (repressor) regulatory system can become fixed in a wild-type population carrying the opposite mode of control. In the region marked historical precedent, mutants with optimal regulatory mehanisms do not have sufficient fitness advantage to take over a population with an existing, suboptimal mode of regula-tion. The x-axis is $s_{min}/(\Delta f_0 + \Delta f_1)$, where s_{min} is the minimal selection advantage needed for fixation, Δf_0 is the reduction in fitness due to errors in the free state of an activator site and Δf_1 is the reduction in fitness due to errors in the free state of a repressor site. In this plot, the ratio $\Delta f_1/\Delta f_0$ is constant, equal to 1.5.

As mentioned in previous chapters, the Lac proteins transport the sugar lactose into the cell, and participate in its degradation. The system has two input stimuli, lactose and glucose. Expression of the Lac proteins is induced in the presence of lactose to allow uti-lization of the sugar. Expression is inhibited in the presence of glucose, which is a better energy source than lactose. The input-output relationship, which maps the levels of the two input sugars onto the expression levels of the Lac proteins, is shown in Figure 11.4a.

This input-output relationship is implemented by two regulators, the repressor LacI and the activator CRP. When lactose enters the cell, the repressor LacI does not bind its DNA sites, causing increased expression of the Lac proteins. When glucose enters the cells, the activator CRP does not bind its DNA site, leading to a reduction in expression. The *lac* system has an additional mechanism that inhibits expression in the presence of glucose, which is called **inducer-exclusion**: when glucose is pumped into the cell, lactose entry is blocked, preventing the induction of the *lac* system (Postma et al., 1993, Thattai and Shraiman, 2003).

The relation between the input-states, the DNA binding-states and the output-levels of the *lac* system is shown in Figure 11.4b. There are four possible binding-states, depending on whether the CRP and LacI sites are bound or free. These binding-states are denoted [CRP,LacI] = [0,0], [0,1], [1,0] and [1,1], where 1/0 correspond to bound/free. One of the four binding-states, [CRP, LacI] = [0,0], is not reached by any input-state, because inducer-exclusion prevents lactose and glucose from being present in the cell at the same time. Thus, in the presence of glucose, the LacI site is bound even if lactose is present in the environment. As a result, the binding-state [0,0] does not correspond to any input-state, and may therefore be called an **excluded state**. Its expression level was experimentally determined by using artificial inducers such as IPTG, which are not subject to inducer-exclusion.

The naturally occurring mechanism, with a glucose-responsive activator and a lactose-responsive repressor, is only one of the four possible mechanisms in which the two regulators can have either mode of control (Figure 11.5). The four mechanisms can be denoted RR, RA, AR and AA where the first letter denotes the mode of the glucose regulator, the second letter denotes the mode of the lactose regulator, and the designation A/R corresponds to activator/repressor. The wild-type *lac* system has the AR mechanism, with activator CRP and repressor LacI (Figure 11.5a).

These four mechanisms all map the input-states onto the expression levels in the same way. The mechanisms differ only in the promoter binding-states that correspond to each input- and output-state. All four mechanisms have inducer exclusion, and thus have an excluded state, although the identity of the excluded state differs between the mechanisms (Figure 11.5): The excluded state is [CRP,LacI]=[0,0], [0,1], [1,0], [1,1] in the AR, AA, RR and RA mechanisms respectively.

Let us now consider the errors in this system. As before, errors are assumed to be associated with free binding sites. Table 11.4 lists the fitness reductions resulting from errors that occur when one or both of the regulator sites are free. For example, consider the AR mechanism with the input-state (glucose, lactose) = (0,1). This input-state, which corresponds to the highest expression level Z_4, is mapped onto binding-state [CRP, LacI] = [1,0], where the glucose-responsive regulator site is bound and the lactose-responsive regulator site is free. Thus, only the latter site is exposed to errors, contributing a reduction $\Delta f_4'$ to fitness.

The average error-load is calculated by multiplying the probability of each input-state by the appropriate fitness reductions and summing over all input-states (Table 11.4). It turns out that in this system, the probabilities of the input states can be fully described by two numbers: p_{00}, the probability that neither glucose nor lactose are present in the

TABLE 11.4 Error-Load of Four Possible Mechanisms for the lac System

INPUT-STATE (GLUCOSE, LACTOSE)	(0,0)	(0,1)	(1,0) / (1,1)
REGULATORY MECHANISM			
AA	$\Delta f_2'$	0	$\Delta f_1 + \Delta f_1'$
AR	0	$\Delta f_4'$	Δf_1
RA	$\Delta f_2 + \Delta f_2'$	Δf_4	$\Delta f_1'$
RR	Δf_2	$\Delta f_4 + \Delta f_4'$	0

Table 11.4: Fitness reduction due to errors from free DNA sites in the four possible regulatory mechanisms of the *lac* system. Columns correspond to input-states (glucose, lactose), where 1/0 means saturating/no input. The reductions in fitness due to errors from the glucose-responsive (lactose-responsive) regulator binding site are Δf_i ($\Delta f_i'$), where i = 1...4 corresponds to the output-level. The input-states (glucose, lactose) = (1,0) and (1,1) are both mapped onto the same binding-state due to inducer-exclusion.

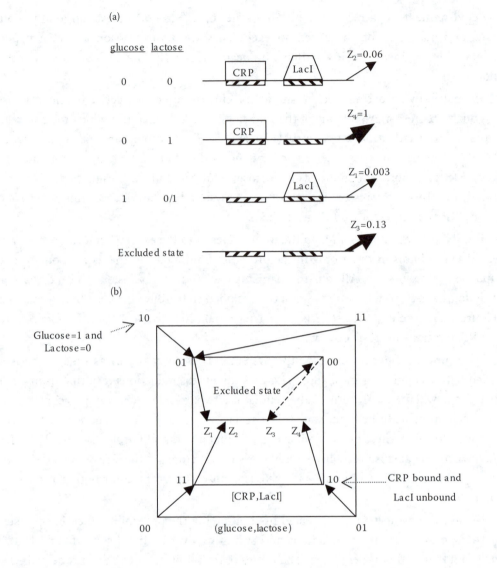

FIGURE 11.4 The *lac* system of *E. coli* and its regulatory mechanism. (a) Input-output relationship of the *lac* system. Glucose and lactose levels are 1/0 corresponding to saturating/no sugar in the environment. Z_1, Z_2, Z_3 and Z_4 are the relative protein expression levels from the *lac* promoter. The binding-states of CRP and LacI are shown. (b) Diagram mapping the input-states (glucose, lactose) onto the binding-states [CRP, LacI] and finally onto the output-states Z_1, Z_2, Z_3 and Z_4. The dashed arrow indicates the excluded state, which is not reached by any of the input-states due to inducer exclusion.

environment, and p_{01}, the probability that glucose is absent but lactose is present. For example, the error-load of the AR mechanism is

$$E_{AR} = p_{01}\Delta f_4' + (1 - p_{00} - p_{01})\Delta f_1$$

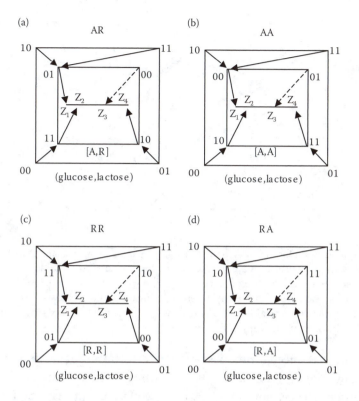

FIGURE 11.5 The four possible regulatory mechanisms of the *lac* system. The mechanisms are labeled by the mode of regulation of the glucose-responsive and lactose-responsive regulators, where A/R means activator/ repressor. Excluded states map to output states with dashed arrows.

We can now compare the different mechanisms, and find which one has the lowest error-load in a given environment (p_{00}, p_{01}). This results in the selection diagram shown in Figure 11.6. The diagram is triangular because $p_{00} + p_{01} \leq 1$. We can see that three of the four mechanisms minimize the error-load, each in a different region of the diagram. One of the four possible mechanisms (RA) never minimizes the error-load. The wild-type mechanism of the *lac* system, AR, minimizes the error-load in a region of the diagram that includes environments where lactose and glucose are present with low probability, namely $p_{01} \ll 1$ and $p_{00} \approx 1$. This is consistent with the empirical observation that in the natural environment of *E. coli*, both glucose and lactose are rare.

It is easy to see why the AR mechanism minimizes the error-load in the natural environment of *E. coli*. The most frequent input-state in the environment is that both sugars are absent, (glucose, lactose) = (0,0). This input state maps onto the second lowest output-level Z_2, in which the lac system is induced to a low but non-zero level, in order to maintain Lac proteins to sense lactose in the case that the sugar appears. In the AR mechanism, this most frequent state corresponds to the binding-state [CRP,LacI] = [1,1], where both regulators bind their DNA sites. Thus, the AR mechanism keeps the DNA sites protected from errors most of the time.

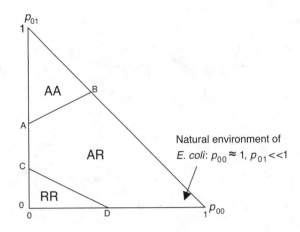

FIGURE 11.6 Selection diagram for the *lac* system, indicating the regulatory mechanism that minimizes the error-load in each environment. The axes are the probability for neither glucose nor lactose in the environment (p_{00}), and for lactose in the absence of glucose (p_{01}). The wild-type mechanism, AR, minimizes the error-load in a region that includes environments where both sugars are rare. Line AB has slope $(\Delta f_2' - \Delta f_1') / (\Delta f_1' + \Delta f_4')$ and intercepts the vertical axis at $\Delta f_1' / (\Delta f_1' + \Delta f_4')$. Line CD has slope $-(\Delta f_1 + \Delta f_2) / (\Delta f_1 + \Delta f_4)$ and intercepts the vertical axis at $\Delta f_1 / (\Delta f_1 + \Delta f_4)$.

In addition, the AR mechanism has another error-minimizing feature: the most error-prone binding-state [CRP, LacI] = [0,0], in which both sites are free, is concealed by inducer exclusion. Hence, not only does the AR mechanism map the most frequent input-state onto the error-free binding-state [1,1], but also it excludes the most error-prone binding-state [0,0] and prevents it from ever being reached by any input-state. This is in contrast to the three other possible mechanisms that make the error-prone binding-state [0,0] accessible to environmental conditions.

11.6 SUMMARY

We have seen that one can formulate rules to understand the selection of regulatory mechanism in transcription. In the realm of bacteria, accumulated evidence points to the following tendency, called the Savageau demand rule: Genes that are often needed at full expression in the natural environment tend to have activator control, whereas genes rarely needed at full expression tend to have repressor control.

We saw that the demand rule can be understood in terms of error minimizing strategies. In such strategies, biological regulation systems in which open sites are error-prone will tend to evolve mechanisms that keep the sites bound for most of the time, thus minimizing errors. Hence, genes whose product is required at full expression for a small fraction of the time (low-demand genes) will tend to have repressor control, so that the repressor binds and protects the site most of the time. Genes needed at full expression a large fraction of the time (high-demand genes) will tend to have activator control, so that the activator binds and protects its site most of the time. The expected selective advantage of error-load minimization appears to be sufficient to overcome historical precedent in the choice of regulatory mechanism.

We also saw that this approach can be generalized to multi-input systems, as demonstrated for the *E. coli lac* system. We saw how the most common environmental state maps onto an internal state in which both regulators bind their sites in the promoter, protecting it from mis-binding errors. The most error-prone state, in which the promoter is free and exposed to mis-binding, is kept hidden and inaccessible to normal environmental conditions by inducer exclusion.

These conclusions for transcription regulation assume that free DNA sites are more error-prone than sites bound to their cognate regulators. In cases where the reverse is true, that is when bound sites are more error-prone, the predictions are opposite, namely that repressors (activators) correspond to high (low) demand genes. One possible scenario in which bound sites might be more error-prone than free sites may occur in eukaryotic genes. In eukaryotes, DNA regions that bear genes that are not expressed are often packed into a compact conformation studded with protective proteins, called closed chromatin. When the gene needs to be expressed, the chromatin is changed to open conformations that allow better access to regulatory proteins. In this case, closed chromatin may protect the free sites from errors, whereas regulator binding may require opening of chromatin allowing increased exposure to mis-binding errors. Thus, some eukaryotic genes may have opposite demand rules. Further study is needed to assess the error-loads associated with such chromatin states.

Although this chapter addressed transcription regulation, note that the same considerations can be applied to other biological systems in which regulation involves the binding of bio-molecules. One example is protein-protein interactions mediated by specific protein binding domains. In this case, positive and negative modes of regulation correspond to the activation of proteins either by the binding or the unbinding of a regulatory domain. Indeed, experiments suggest that cross-talk between different members of the same class of binding domains sets a selectable constraint on the fitness of the organism (Zarrinpar et al., 2003). Hence, selection of mode of control (positive or negative) according to demand rules is a possible way for the cells to evolve increased specificity, by minimizing the time that a given site is free and exposed to cross-reactivity. These considerations might help us to make sense of what may otherwise appear to be an arbitrary choice of mechanism in each instance.

In conclusion, we have examined rules for biological regulation based on minimizing errors. In systems where a free site is more exposed to errors than a site bound to its cognate interaction partner, it is predicted that mechanisms that keep the site bound most of the time will have a lower error-load and hence a selectable advantage. In the context of transcription, this explains the Savageau demand rule, as well as offers rules for multi-regulator systems.

Biological mechanisms are rich in variety, and different systems often have different biochemical details. How can we understand this variety? We end with the words of Savageau (Savageau, 1989):

> "Such differences [in biochemical details] might be the result of historical accidents
> that are functionally neutral, or they might be governed by additional rules that

have yet to be determined. One can always assume that certain differences are the result of historical accident, but such an explanation has no predictive power and tends to stifle the search for alternative hypotheses. It generally tends to be more productive if one starts with the working hypothesis that there are rules. One may end up attributing differences to historical accident, but in my opinion it is a mistake to start there."

FURTHER READING

Savageau, M.A. (1989). Are There Rules Governing Patterns of Gene Regulation? In *Theoretical Biology*, Goodwin, B.C. and Saunders, P.T., Eds. Edinburgh University Press. pp. 42–66.

Savageau, M.A. (1977). Design of molecular control mechanisms and the demand for gene expression. *Proc. Natl. Acad. Sci. U.S.A.* 74: 5647-51.

Shinar, G., Dekel, E., Tlusty, T., and Alon, U. (2006). Rules for Gene Regulation Based on Error Minimization. *Proc. Natl. Acad. Sci. U.S.A.* 103: 3999–4004.

EXERCISES

11.1. *Optimization versus historical precedent*: Imagine a population of organisms with a regulatory mechanism in place for a certain gene. Conditions change, and the opposite mode of regulation is now more optimal for that gene, in the sense that it has a lower error load. The demand for the gene in the new environment is p, and the error loads associated with errors in the high and low expression states are Δf_1 and Δf_0. Mutants with the opposite mode arise in the population, but they can only fully replace the original population if their fitness advantage exceeds a minimal value s_{min}.

(a) Calculate the conditions on the demand p so that the mutants with the optimal mode can take over the population.

(b) When is the mode of regulation determined by historical precedent? Explain.

11.2. *Error-load of variability in protein expression*: The expression of proteins varies from cell to cell. This means that different cells deviate from the optimal expression level. In this exercise we will calculate the average reduction in fitness due to such variations, for the case of the *lac* system. The fully induced *lac* promoter has a cell-cell variation in expression with a coefficient of variation (standard deviation of protein level Z divided by the mean) of about V=0.1 in the fully induced state (Elowitz et al., 2002). The fitness function for this exercise, similar to the function we saw in chapter 10, is $f(Z) = \eta \, Z/(1-Z/M) + \delta \, Z$, with $\delta = 0.17Z_1^{-1}$, $\eta = 0.02Z_1^{-1}$ and $M = 2Z_1$, where Z_1 is the fully induced expression level.

(a) Show that the mean reduction in fitness due to small cell-cell variations Z + ΔZ is $\Delta f = C \langle \Delta Z^2 \rangle$, where C is the curvature of the fitness function near its maximum $C = 1/2 \, d^2f/dZ^2$ and the brackets $\langle \rangle$ denote a population average. Hint: use a Taylor expansion of f near its maximum.

(b) Compute the mean reduction in fitness due to variations in Z in the fully induced state (the average value of Z in the population is Z_1). Answer: $\Delta f_1 \sim 0.1\%$.

11.3. *Demand rules for developmental genes*: Consider a cell which, during the developmental process of the organism, can assume either fate A or fate B. A set of genes G_A is expressed in fate A and not in fate B, and a different set G_B is expressed in fate B and not in A. This cell-fate decision is regulated by two transcription-factors X and Y. X activates its own transcription and represses the transcription of Y, whereas Y activates its own transcription and represses the transcription of X. Furthermore, X transcriptionally activates G_A and represses G_B, and Y has the opposite effect, activating G_B and repressing G_A.

(a) Draw the transcription network in this case.

(b) Explain the mode of regulation of each gene in terms of the demand rule.

11.4. *Error-load of two-input genes*: Compute the error-loads of all possible four regulatory mechanisms for the lac system, according to Table 11.4. Compute the conditions (the range of p_{00} and p_{01}) in which each of the four mechanisms is optimal. Compare your results to Fig 11.6.

11.5. *Error-load of a feed-forward loop (FFL)*: Consider a type-1 coherent FFL (Chapter 4). In this gene circuit, activators X and Y activate gene Z, and X also activates Y so that at steady state, Y levels are zero unless X is transcriptionally active. The regulators X and Y respond to input signals S_x and S_y. The promoter of gene Z is activated in an additive fashion by X and Y, such that at steady-state the expression level is $Z = a X^* + b Y^*$.

(a) Plot the steady-state relationship between input states (S_x, S_y), internal states $[X^*, Y^*]$ and output states (Z_1, Z_2, Z_3, Z_4) for all four combinations of S_x, S_y =0 or 1 (Similar to Fig 11.4).

(b) Is there an excluded-state in this case (at steady-state)?

(c) Repeat this for an incoherent type-1 FFL in which the output is $Z = a X^* - b Y^*$. Assume that $b < a$, and that at steady-state, Y levels are zero unless X is transcriptionally active.

CHAPTER 12

Epilogue: Simplicity in Biology

I have almost finished writing this book and have gone over many drafts. I am happy to still have a sense of wonder when reading the chapters. The wonder comes because networks of thousands of interacting components are generally incomprehensible. There is no *a priori* reason that immensely complex biological systems would be understandable. But despite the fact that biological networks evolved to function and not to be comprehensible, simplifying principles can be found that make biological design understandable to us.

This epilogue will discuss simplicity in several aspects of biological networks. We will review simplicity in structure and timescales, in the ability to form simplified models of regulatory networks, and in recurring design principles.

One level of simplicity occurs in the structure of biological networks. In large networks there are a huge number of possible interaction patterns. The surprise is that biological networks are built, to a good approximation, from only a few types of recurring interaction patterns called network motifs.

Each of the network motifs can perform defined information processing functions. The main network motifs found in sensory transcription networks and their functions are summarized in Figure 5.15. For example, feed-forward loops can act as sign-sensitive filters, pulse generators, and response accelerators. Negative autoregulation can speed and stabilize responses. Single-input modules can generate temporal programs of expression. No doubt additional functions of these motifs will be discovered as our knowledge of networks becomes more complete.

The same network motifs appear in the sensory transcription networks of diverse organisms. It is important to stress that the similarity in circuit patterns does not necessarily stem from circuit duplication. Evolution appears to have *converged* on the same network motifs again and again in different systems, suggesting that they are selected because of their function. These functions can be readily tested experimentally in each system.

Network motifs are embedded in the network and are connected to each other. Importantly, the motifs often appear to be embedded in a way that allows them to carry out their functions even in the presence of additional interactions. This property is due to the particular ways that the motifs are wired together. In many systems, network motifs appear to be connected to each other in ways that do not spoil the independent functionality of each motif, allowing us to understand the network, at least partially, based on the functions of individual motifs. Simple examples include the way that three-node FFLs are connected to each other to form a multi-output FFL. This pattern preserves the functionality of each three-node FFL (sign-sensitive filtering, etc.). In addition, the multi-output FFL can generate rather elaborate programs of expression timing between output genes, as we saw in Chapter 5. As a result of the way the motifs are embedded into the network, they can, at least in many cases, be considered as elementary circuit elements.

In addition to the simple ways in which motifs are wired together, motifs can act as elementary circuit elements due to the separation of timescales of different interactions. The strong separation of timescales between different biological processes is a general principle that is found in virtually all of the networks in the cell. It allows us to understand the dynamics on the slow timescale by using steady-state approximations for the interactions on fast timescales. For example, transcriptional motifs that carry out their computations on a slow timescale of minutes to hours can be understood, at least schematically, as if they acted in isolation, despite the fact that they are embedded in additional feedback loops on the level of protein–protein interactions that function on the timescale of seconds. In short, biological networks can be understood, to a first approximation, in terms of a rather limited set of recurring circuit patterns, each carrying out computations on a different timescale.

In addition to the reuse of network motifs, biological networks have an additional important structural feature: **modularity** (Hartwell et al., 1999; Ihmels et al., 2002; Segal et al., 2003; Wolf and Arkin, 2003; Schlosser and Wagner, 2004). Most biological functions are carried out by specific groups of genes and proteins, so that one can separate the structure into functional modules. For example, proteins work in coregulated groups such as pathways and complexes. Transcription networks are nearly decomposable into single-input modules and multi-input dense overlapping regulons (DORs), as described in Chapter 5. Signal transduction networks display distinct signaling pathways shaped as multi-layer perceptron modules, discussed in Chapter 6, etc. The modules in biological networks can be compared on the metaphorical level to the modules used in engineering, such as subroutines in software and replaceable parts in machines.

A working definition of a module is a set of nodes that have strong interactions and a common function. A module has defined input and output nodes that control the interaction with the rest of the network. A module also has internal nodes that do not significantly interact with the nodes outside the module. Modules in engineering, and probably also in biology, have special features that make them easily embedded in almost any system. For example, output nodes should have low impedance, so that adding on additional downstream clients should not drain the output to existing clients (up to some limit).

Why does modularity exist in biological networks? It is important to realize that not all evolved networks are modular. The opposite is true: nonmodular solutions are the norm in simple computer simulations of evolution. In evolutionary simulations, a population of

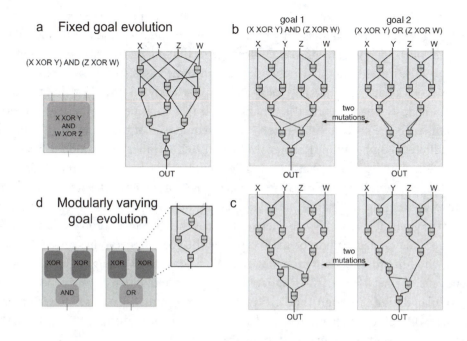

FIGURE 12.1 Electronic circuits evolved under a fixed goal and modularly varying goals. Networks made of logic gates (NAND gates that compute a NOT-AND function on two inputs) were evolved to attain a goal, defined as a logic function of four inputs: X, Y, Z, and W. (a) A typical network evolved towards a fixed goal, the goal G_1 = (X XOR Y) AND (Z XOR W), where XOR is the exclusive-OR function. The network is a perfect solution for this goal. Similar nonmodular solutions were found for goal G_2 = (X XOR Y) OR (Z XOR W). (b) Networks evolved with modularly varying goals, in which the goal switched between G_1 and G_2 every 20 generations. Note that G_1 and G_2 share the same subproblems (namely, X XOR Y, Z XOR W), but in different combinations. Connections that are rewired when the goal is switched are marked in light lines. Each network is a perfect solution for its corresponding goal. (C) Another example of networks evolved with modularly varying goals toward G_1 and G_2. (d) Modular structure of the networks evolved under modularly varying goals: the networks are composed of two XOR modules that input into a third module that implements an AND/OR function, depending on the goal. (From Kashtan and Alon, 2005.)

networks is evolved by randomly adding, removing, and changing connections between nodes — and even duplicating and recombining parts of the networks — until the networks perform a given computation goal, that is, until the networks give the correct output-to-input signals. Unlike biological networks, simulated networks evolved in this way are usually nonmodular (Figure 12.1a). They typically have a highly interconnected structure that cannot be decomposed into nearly independent subsystems (Thompson, 1998).

Viewed in this perspective, the modularity of biological networks is puzzling. The evolutionary simulations make it clear that modular structure is usually less optimal than fully wired, nonmodular structures. After all, modules greatly limit the number of possible connections in the network, and usually a connection can be added that reduces modularity and increases the fitness of the network. This is the reason that the evolutionary simulations almost always display a nonmodular solution.

A clue to the reason why modules evolve in biology can be found in engineering. Modules in engineering convey an advantage in situations where the design specifications change from time to time. New devices or software can be easily constructed from existing, well-tested modules (Lipson et al., 2002). Similarly, modular biological networks might offer an advantage in real environments that change over time (Gerhart and Kirschner, 1997). Indeed, modular networks can, in some cases, evolve in simulations in which the evolutionary goal changes over time. Importantly, in order for modularity to arise spontaneously, the goals need to change such that each new goal shares the same subproblems with the previous goals: Each goal is composed of the same set of subproblems in a different combination (Kashtan and Alon, 2005). Under such modularly varying goals, networks rapidly evolve that have high fitness to the current goal and, every time the goal changes, rapidly rewire to satisfy the new goal. These networks are highly modular in structure (Figure 12.1b and c). They include a module for each of the subproblems shared by the goals. It is as though the network learns the shared subproblems and creates a specific module for each subproblem. Every time the goal changes, these modules need only be rewired in order to satisfy the new goal. If the goal stops changing for a sufficient length of time, the networks in the simulation begin to lose modularity and evolve toward a nonmodular design, a design that is typically more optimal (e.g., uses fewer components). Thus, the ability to reconfigure and adapt to new conditions may be one force that helps to maintain modular structure in biological systems.

In addition to the structural simplicity associated with modularity and the small number of network motifs that make up the networks, there is a second level of simplicity. This simplicity is found in the realm of models of biological interactions, in the ability to treat regulatory circuits with simplified mathematical models that capture the essence of the behavior and have a certain degree of universality. This abstract mode of description is surprising because it contrasts with the complex and idiosyncratic biochemical mechanisms by which each protein carries out its function. These biochemical particulars are astoundingly rich, Baroque, but on the level of the dynamics of transcription circuits, we saw that one can use rather simple mathematical models that do not require precise knowledge of most of the molecular details. These molecular details are instead chunked into systems-level parameters. The models include only information on whether X activates or inhibits Y, and at what activity threshold.

In these models, logic input functions can be used to gain a back-of-the-envelope sketch of the behaviour of diverse circuits, aided by the graphic intuition of piecewise-exponential dynamics that cross activity thresholds. Simplified models are particularly useful for schematically deducing dynamical behavior and its qualitative dependence on biochemical parameters. This analysis can readily be experimentally tested, by using controlled experiments that keep many parameters relatively constant, and thus approach the idealized situations described by the models.

Furthermore, the same mathematical models often apply to different networks. For example, production–degradation equations describe gene expression dynamics in transcription networks (Chapters 3 to 5), kinase activities in signal transduction networks (Chapter 6), and even simple models of neurons (Chapter 6). Thus, simplified models can connect motifs that work in different networks on different timescales.

A third level of simplicity is the conceptual similarity of seemingly unrelated systems, a similarity expressed in terms of unifying design principles. One such design principle is robustness to component fluctuations: A biological system must work under all possible insults and interferences that come with the inherent properties of the components and the environment. Thus, *E. coli* needs to be robust with respect to temperature changes over a few tens of degrees, and no circuit in the cell should depend on having precisely 100 copies of protein X and not 103. The fact that a gene circuit must be robust to such perturbations imposes severe constraints on its design: only a small percentage of the possible circuits that perform a given function can perform it robustly. Since most possible mechanisms are not robust, robustness can help theorists to recognize the correct model.

We examined specific examples of robustness in bacterial chemotaxis and in embryonic patterning in Chapters 7 and 8.

There are several ways to achieve robustness. For example, integral feedback can provide robust adaptation: it can lead the output of a system to a desired goal in the face of wide variations in the environment or the internal parameters of the system (Chapter 7). Integral feedback employs a negative feedback signal proportional to the time integral over the difference between the actual output and the goal. Integral feedback can be shown in many cases to be a unique solution to the problem of robust adaptation in the context of engineering control theory.

Biological systems can readily implement integral feedback because integration over time is an inherent feature of the production–degradation equations mentioned above, which describe many of the biochemical interactions in the cell. We saw how integral feedback is implemented in the chemotaxis system by means of accumulating methylation that regulates the activity of the chemotaxis receptors. In other systems, accumulating protein levels or protein modifications can play the role of the integrator to achieve integral feedback and robust adaptation.

An additional principle of robustness in spatial patterning systems is self-enhanced degradation of the morphogen, as we saw in the case of fruit fly development (Chapter 8). Self-enhanced degradation allows long-range pattern formation that is nearly independent of the production rate of the morphogen. This principle appears to be used again and again in different patterning systems with different morphogens.

A third general principle that confers robustness is kinetic proofreading, a mechanism that allows molecular detection of a specific molecule despite the background signal of chemically similar molecules in the cell (Chapter 9). Kinetic proofreading relies on time delays that can be implemented by diverse kinds of nonequilibrium reactions. The same principle seems to appear in systems ranging from translation in the ribosome to antigen detection in the immune system.

All of these design principles allow us to understand subtle biochemical details that might otherwise appear as wasteful side reactions. The few ways to achieve robustness help to limit the number of possible circuits in biology, and to give the circuits that do appear a defined style. It is likely that additional general ways of achieving robust designs will be discovered that can unite our understanding of diverse systems (Kitano, 2004; Wagner, 2005b).

Throughout this book, we have used engineering metaphors and gained inspiration from engineering principles. One example is the consideration of response time and stability in autoregulatory feedback loops (Chapter 3) and other network motifs. Stability and response time trade-offs are of the essence in electronic and mechanical engineering design. Another point of similarity is the principle of robustness to component fluctuations. Robustness is a guiding principle in engineering; for example, electronic circuits must work despite variations in the resistance of each resistor. Engineering textbooks are filled with robust designs, and many alternative nonrobust designs for the same circuits are avoided. Good engineering uses modularity and recurring circuit elements (network motifs) to build reliable and scalable devices from simpler subsystems. An additional similarity is optimal design, with cost and benefit trade-offs, which we saw guides evolutionary selection in simple systems (Chapter 10). Such cost–benefit trade-offs are ubiquitous in engineering.

In addition to the similarities with engineering, biological circuitry also has fundamental conceptual differences from engineered devices. One important difference is the stochastic nature of biological function. Here, I mean that genetically identical cells in the same environment respond in a probabilistic way to a given stimulus (see Appendix D). Often, the response of each cell is not predictable, but the proportion of cells with a given response is predetermined and is regulated according to environmental signals. For example, we saw that swimming bacteria have individual characters and perform chemotaxis with different tumbling rates and adaptation times. More dramatic examples include cellular decisions such as differentiation or cell death, in which a fraction of the cells assume one fate and other cells a completely different fate. Evolution appears to select for a probabilistic outcome.

An element of randomness in behavior is one of the most familiar features of living organisms. In contrast, most engineered designs are made to try and avoid stochastic outcomes. Engineered devices such as a radio are designed to function with 100% probability if, say, the ON button is pressed. We would say that a probabilistic radio is a malfunctioning radio.

Stochastic designs are not normally found in many areas of engineering, but they are common in the fields of game theory and economics. In game theory, it can sometimes be shown that a probabilistic strategy is optimal when competing with other organisms. A deterministic response would be easily exploited by competitors. The stochastic nature of cellular responses broadens the range of possible responses in an unpredictable future, increasing the probability that at least a fraction of the cells will be able to cope with sudden unforeseen changes in circumstances. Study of the role of probabilistic design in biology is only at its beginnings.[1]

An additional intriguing set of questions concerns the interplay between the ecology and the biological design of the organism. We currently have more information on the detailed structure of biological circuits than on the environment in which they evolved. We know little about the constraints and functional goals of cells within complex organ-

---

[1] See Kerr et al., 2002; Nowak and Sigmund, 2002, 2004; Wolf and Arkin, 2003; Balaban et al. 2004; Thattai and van Oudenaarden, 2004; and Kussell and Leibler, 2005 for interesting theoretical and experimental studies.

isms, and are only beginning to understand the optimizations and trade-offs that underlie their design. It is an interesting question whether it would be possible to form a theory of biological design that can help unify aspects of ecology, evolution, and molecular biology.

We are almost at the end of this book, which tried to present an introduction to systems biology. At this time, one can only write an introduction, since we are only at the beginning of the adventure to find the design principles of biological systems. In this book I have tried to emphasize simplicity in biology, within its undoubted complexity, to encourage the optimistic point of view that general principles can be discovered. Without such principles, it is difficult to imagine how we can make sense of biology on the level of an entire cell, tissue, or organism.

Will a complete description of the biological networks of an entire cell or organism ever be available? The task of mapping an unknown network is known as reverse engineering. Much of engineering is actually reverse engineering, because prototypes often do not work and need to be understood in order to correct their design. The program of biology is reverse engineering on a grand scale. There are many difficulties to be overcome in this project. Reverse engineering of a nonmodular, highly wired network of a few thousand components and their nonlinear interactions is virtually impossible. However, the simplifying features that we have discussed give hope that biological networks are structures that human beings can understand. Modularity, for example, is at the root of our ability to separate the problem into smaller bits that can be studied nearly independently, to assign functions to genes, proteins, and pathways, and so on. The principle of robustness limits the range of possible circuits that function on paper to only a few designs that can work in the cell. This can help theorists to home in on the correct design with limited data. Network motifs define the few basic patterns that make up a network, and the dictionary of elementary dynamical functions that the network can perform. These concepts, together with the current technological revolution in biology, may eventually allow characterization and understanding of cell-wide networks, with great benefit to medicine. The similarity between the creations of evolution and engineering also raises a fundamental scientific challenge: understanding the laws of nature that unite evolved and designed systems.

The Input Function of a Gene: Michaelis–Menten and Hill Equations

A.1   BINDING OF A REPRESSOR TO A PROMOTER

This appendix provides a simplified introduction to basic models in biochemistry. We will begin with understanding the interaction of a repressor protein with DNA and with its inducer.[1] We will then turn to activator proteins. The repressor X recognizes and binds to a specific DNA site, D, in a promoter: X and D bind to form a complex, [XD]. Transcription of the gene occurs only when the repressor is *not* bound, that is, when D is free. The DNA site can thus be either free, D, or bound, [DX], resulting in a conservation equation:

$$D + [XD] = D_T \tag{A.1.1}$$

where D_T is the total concentration of the site. For example, a single DNA binding site per bacterial cell means that D_T = 1/cell volume ~ $1/\mu m^3$ ~ 1 nM. In eukaryotic cells, the volume of the nucleus is on the order of 10–100 μm^3.

The repressor X and its target D diffuse in the cell and occasionally collide to form a complex, [XD]. This process can be described by mass-action kinetics: X and D collide and bind each other at a rate k_{on}. The rate of complex formation is thus proportional to the collision rate, given by the product of the concentrations of X and D:

$$\text{rate of complex formation} = k_{on} XD$$

---

[1] The theoretical treatment for the input function of simple gene regulation was initiated by Gad Yagil in the context of the *lac* system of *Escherichia coli* (Yagil and Yagil, 1971).

The complex [XD] falls apart (dissociates) at a rate k_{off}. The rate of change of [XD] based on these collision and dissociation processes is described by

$$d[XD]/dt = k_{on} \, XD - k_{off} \, [XD] \tag{A.1.2}$$

The rate parameter for the collisions, k_{on}, describes how many collision events occur per second per protein at a given concentration of D, and thus has units of 1/time/concentration. It is useful to remember that k_{on} in biochemical reactions is often limited by the rate of collisions of a diffusing molecule hitting a protein-size target, and has a **diffusion-limited** value of about $k_{on} \sim 10^8 - 10^9 \, M^{-1} \, sec^{-1}$, independent of the details of the reaction. For the case of a transcription factor and DNA, the diffusion limit is usually higher because of one-dimensional diffusion effects due to sliding of the transcription factor along the DNA, $k_{on} \sim 10^{10} - 10^{11} \, M^{-1} \, sec^{-1}$ (Berg et al., 1981). The off-rate k_{off}, on the other hand, has units of 1/time and can vary over many orders of magnitude for different reactions, because k_{off} is determined by the strength of the chemical bonds that bind X and D.

The kinetics of Equation A.1.2 approach a steady-state in which concentrations do not change with time, $d[XD]/dt = 0$. Solving Equation A.1.2 at steady-state, we find that the balance between the collision of X and D and the dissociation of [XD] leads to the **chemical equilibrium** equation:

$$K_d \, [XD] = XD \tag{A.1.3}$$

where K_d is the **dissociation constant**,

$$K_d = k_{off}/k_{on}$$

The dissociation constant K_d has units of concentration. The larger the dissociation contant, the higher the rate of dissociation of the complex, that is, the weaker the binding of X and D.

Solving for the concentration of free DNA sites, D, using Equations. A.1.1 and A.1.3, we find $K_d \, (D_T - D) = XD$, which yields

$$\frac{D}{D_T} = \frac{1}{1 + X/K_d} \tag{A.1.4}$$

For many repressors, [XD] complexes dissociate within less than 1 sec (that is, $k_{off} > 1$ sec^{-1}). Therefore, we can average over times much longer than 1 sec and consider D/D_T as the probability that site D is free, averaged over many binding and unbinding events.

The probability that the site is free, D/D_T, is a decreasing function of the concentration of repressor X. When there is no repressor, $X = 0$, the site is always free, $D/D_T = 1$. The site has a 50% chance of being free, $D/D_T = 1/2$, when $X = K_d$.

When site D is free, RNA polymerase can bind the promoter and transcribe the gene. The rate of transcription (number of mRNAs per second) from a free site is given by the **maximal transcription rate** β. (Note that in the main text we used β to denote the rate of protein production. This rate is proportional to the transcription rate times the

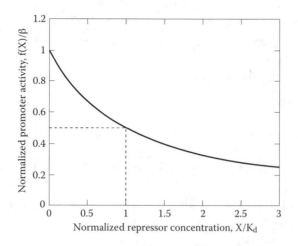

FIGURE A.1 Normalized promoter activity vs. repressor concentration in units of its K_d. Half-maximal activity occurs when $X = K_d$ (dashed lines). Here X corresponds to the concentration of repressor in its active form, X^*.

number of proteins translated per mRNA provided that there is a constant mRNA lifetime and translation rate.) The maximal transcription rate depends on the DNA sequence and position of the RNA polymerase binding site in the promoter and other factors. It can be tuned by evolutionary selection, for example, by means of mutations that change the DNA sequence of the RNAp binding site. In different genes, β ranges over several orders of magnitude, $\beta \sim 10^{-4} - 1$ mRNA/sec. The rate of mRNA production, called the **promoter activity**, is β times the probability that site D is free:

$$\text{promoter activity} = \frac{\beta}{1 + X/K_d} \tag{A.1.5}$$

Figure A.1 shows the promoter activity as a function of X. When X is equal to K_d, transcription is reduced by 50% from its maximal value. The value of X needed for 50% maximal repression is called the **repression coefficient**.

For efficient repression, enough repressor is needed so that site D is almost always occupied with repressor. From Equation A.1.4, this occurs when repressor concentration greatly exceeds the dissociation constant, such that $X/K_d \gg 1$. This is the case for many repressors, including the *lac* repressor LacI.

So far we discussed how the repressor binds the promoter and inhibits transcription. To turn this gene system ON, a signal must cause X to unbind from the DNA. We will treat the simplest case, in which a small molecule (an inducer) is the signal. The inducer directly binds to protein X and causes it to assume a molecular conformation where it does not bind D with high affinity. Typically, the affinity of X to its DNA sites is reduced by a factor of 10 to 100. Thus, the inducer frees the promoter and allows transcription of the gene. We now consider the binding of inducer to X.

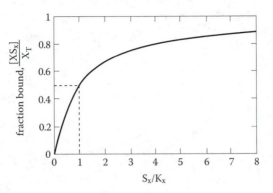

FIGURE A.2 Fraction of X bound to S_x as a function of S_x concenration. Half of X is bound when $S_x = K_x$.

A.2 BINDING OF A REPRESSOR PROTEIN TO AN INDUCER: MICHAELIS–MENTEN EQUATION

The repressor protein X is designed to bind a small molecule inducer S_x, which can be considered as its input signal. The two can collide to form a bound complex, $[XS_x]$. The repressor is therefore found[1] in either free form, X, or bound form, $[XS_x]$, and a conservation law states that total concentration of repressor protein is X_T:

$$X_T = X + [XS_x] \qquad (A.2.1)$$

X and S_x collide to form the complex $[XS_x]$ at a rate k_{on}, and the complex $[XS_x]$ falls apart (dissociates) at a rate k_{off}. Thus, the mass-action kinetic equation is:[2]

$$d\,[XS_x]/dt = k_{on}\, X\, S_x - k_{off}\, [XS_x] \qquad (A.2.2)$$

At steady state, $d[XS_x]/dt = 0$, and we find the chemical equilibrium relation:

$$K_x\, [XS_x] = X\, S_x \qquad (A.2.3)$$

where K_x is the dissociation constant (for the *lac* repressor, $K_x \sim 1\ \mu M \sim 1000$ inducer (IPTG) molecules/cell).[3] Using the conservation of total repressor X (Equation A.2.1), we arrive at a useful equation that recurs throughout biology (this equation is known as the **Michaelis–Menten equation** in the context of enzyme kinetics; we use the same name in the present context of inducer binding):

$$[XS_x] = \frac{X_T S_x}{S_x + K_x} \qquad \textit{Michaelis–Menten equation} \quad (A.2.4)$$

[1] We assume that S_x can bind X regardless of whether it is bound to D or not.
[2] Usually the number of S_x molecules is much larger than the number of X molecules, and so we need not worry about conservation of S_x, $S_{x,total} = S_x + [XS_x]$. For example, in the *lac* system, the number of LacI repressors, each made of a tetramer of LacI proteins, is $X_T \sim 10$ units/cell, which is negligible relative to S_x, which is at least 1000/cell for a detectable response.
[3] In the case of the *lac* repressor, $K_x \sim 1\ \mu M$. Using the diffusion-limited value for $k_{on} \sim 10^9$/M/sec, we find the lifetime of the complex is $1/k_{off} \sim 1$ msec.

The Michaelis–Menten function (Figure A.2) has three notable features:

1. It reaches saturation at high S_x.

2. It has a regime where $[XS_x]$ increases linearly with S_x, when $S_x \ll K_x$.

3. The fraction of bound protein reaches 50% when $S_x = K_x$.

The dissociation constant thus provides the scale for detection of S_x: S_x concentrations far below K_x are not detected; concentrations far above K_x **saturate** the repressor at its maximal binding. The saturated regime ($S_x \gg K_x$) is known as **zero-order** because $[XS_x] \sim S_x^0$, and the linear regime ($S_x \ll K_x$) is known as **first-order** since $[XS_x] \sim S_x^1$.

Recall that in cases like LacI, only X unbound to S_x, is active in the sense that it can bind the promoter D to block transcription. Because free X is active, we denote it by X^*. Active repressor, $X^* = X_T - [XS_x]$, decreases with increasing inducer levels:

$$X^* = \frac{X_T}{1+S_x / K_x} \qquad \textit{concentration of X not bound to } S_x \qquad (A.2.5)$$

A.3   COOPERATIVITY OF INDUCER BINDING AND THE HILL EQUATION

Before returning to the input function, we comment on a more realistic description of inducer binding. Most transcription factors are composed of several repeated protein subunits, for example, dimers or tetramers. Each of the protein subunits can bind inducer molecules. Often, full activity is only reached when multiple subunits bind the inducer. A useful phenomenological equation for this process can be derived by assuming that n molecules of S_x can bind X.

To describe the binding process, we need to describe the binding of n molecules of S_x to X. The protein (protein multimer) X can either be bound to n molecules of S_x, described by the complex $[nS_x X]$, or unbound, denoted X_o (in this simple treatment, intermediate states where fewer than n molecules are bound are neglected). The total concentration of bound and unbound X is X_T, and the conservation law is

$$[nS_x X] + X_o = X_T \qquad (A.3.1)$$

The complex $[nS_x X]$ is formed by collisions of X with n molecules of S_x. Thus, the rate of the molecular collisions needed to form the complex is given by the product of the concentration of free X, X_o, and the concentration of S_x to the power n (the probability of finding n copies of S_x at the same place at the same time):

$$\text{collision rate} = k_{on} X_o S_x^n \qquad (A.3.2)$$

where the parameter k_{on} describes the on-rate of complex formation. The complex $[nS_x X]$ dissociates with rate k_{off}:

$$\text{dissociation rate} = k_{off} [nS_x X] \qquad (A.3.3)$$

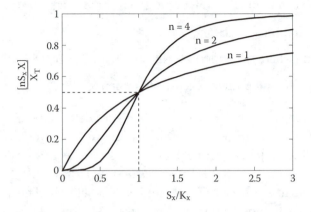

FIGURE A.3 Hill curves for ligand binding with Hill coefficient n = 1, 2, and 4. Note that the curve is steeper the higher the Hill coefficient n. Half maximal binding occurs at $S_x = K_x$.

The parameter k_{off} corresponds to the strength of the chemical bonds between S_x and its binding sites on X. The total rate of change of the concentration of the complex is thus the difference between the rate of collisions and dissociations:

$$d[nS_x X]/dt = k_{on} X_o S_x^n - k_{off} [nS_x X] \tag{A.3.4}$$

This equation reaches equilibrium within milliseconds for typical inducers. Hence, we can make a steady-state approximation, in which $d[nS_x X]/dt = 0$, to find that dissociations balance collisions:

$$k_{off} [nS_x X] = k_{on} X_o S_x^n \tag{A.3.5}$$

We can now use the conservation equation (Equation A.3.1) to replace X_o with $X_T - [nS_x X]$, to find

$$(k_{off}/k_{on}) [nS_x X] = (X_T - [nS_x X]) S_x^n \tag{A.3.6}$$

Finally, we can solve for the fraction of bound X, to find a binding equation known as the **Hill equation**:

$$\frac{[nS_x X]}{X_T} = \frac{S_x^n}{K_x^n + S_x^n} \qquad \textit{Hill equation} \tag{A.3.7}$$

where we have defined the constant K_x such that

$$K_x^n = k_{off}/k_{on} \tag{A.3.8}$$

Equation A.3.7 can be considered the probability that the site is bound, averaged over many binding and unbinding events of S_x.

The parameter n is known as the **Hill coefficient**. When n = 1, we obtain the Michaelis–Menten equation (Equation A.2.4). As shown in Figure A.3, both the Michaelis–Menten and Hill equations reach half-maximal binding when $S_x = K_x$.

The steepness of the Hill curve is greater the larger the Hill coefficient n (Figure A.3). In the *lac* system, n = 2 with the inducer IPTG (Yagil and Yagil, 1971). Reactions described by Hill coefficients n > 1 are often termed **cooperative** reactions.

The concentration of unbound repressor X is given by:

$$\frac{X^*}{X_T} = \frac{1}{1+(S_x / K_x)^n} \tag{A.3.9}$$

A.4   THE MONOD, CHANGEUX, AND WYMANN MODEL

We note that a more rigorous and elegant analysis of cooperative binding based on symmetry principles is due to Monod, Changeux, and Wymann, in a paper well worth reading (Monod et al., 1965), usually also described in biochemistry textbooks. In this model X switches to an active state X^* and back. The signal S_x binds X with dissociation constant K_x, and binds X^* with a lower dissociation constant K_x^*. Up to n molecules of S_x can bind to X. The two states, X and X^* spontaneously switch such that in the absence of S_x, X is found at a probability larger by L than X^*. The result is:

$$\frac{X^*}{X_T} = \frac{(1+S_x / K_x^*)^n}{L(1+S_x / K_x)^n +(1+S_x / K_x^*)^n}$$

Interesting extensions to this model make analogies to Ising models in physics (Duke et al., 2001). One difference between the rigorous models and the Hill curve is that binding at low concentrations of S_x is linear in S_x rather than a power law with coefficient n, as in Equation A.3.7. This linearity is due to the binding of a single site on X, rather than all sites at once.

A.5   THE INPUT FUNCTION OF A GENE REGULATED BY A REPRESSOR

We can now combine the binding of inducer to the repressor (Equation A.2.5) and the binding of the repressor to the DNA (Equation A.1.4) to obtain the input function of the gene. The input function in this case describes the rate of transcription as a function of the input inducer concentration S_x:

$$f(S_x) = \frac{\beta}{1+X^* / K_d} = \frac{\beta}{1+X_T / K_d /(1+(S_x / K_x)^n)} \tag{A.5.1}$$

Figure A.4 shows how the transcription rate of a gene repressed by X increases with increasing inducer concentration S_x. Note, when no inducer is present, there is a **leakage** transcription rate, $f(S_x = 0) = \beta/(1 + X_T/K_d)$, also called the **basal promoter activity**. This leakage is smaller. The stronger X binds its DNA site.

The input function reaches half-maximal value at inducer concentration $S_x = S_{1/2}$. This halfway induction point is approximately (when $X_T \gg K_d$)

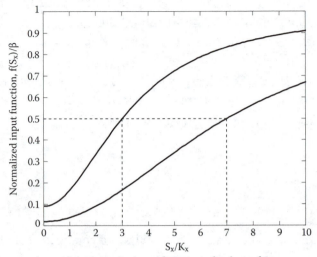

FIGURE A.4 Input function (rate of transcription) as a function of inducer for a gene regulated by a repressor. Shown are $X_T/K_d = 10$ (top curve) and $X_T/K_d = 50$ (bottom curve), both with $n = 2$. There is a leakage transcription at $S_x = 0$, and half-maximal induction is reached at $S_x = 3\ K_x$ and $S_x = 7\ K_x$.

$$S_{1/2} \sim (X_T/K_d)^{1/n}\ K_x \tag{A.5.2}$$

The halfway inducer concentration $S_{1/2}$ can be significantly larger than K_x (Figure A.4). For LacI, for example, $X_T/K_d \sim 100$ and $n = 2$, so that $S_{1/2} \sim 10\ K_x$.

We now turn to describe transcription activators.

A.6   BINDING OF AN ACTIVATOR TO ITS DNA SITE

In the decade following the discovery of the *lac* repressor, other gene systems were found to have repressors with a similar principle of action. It is interesting that it took several years for the scientific community to accept evidence that there also existed transcriptional **activators**.

An activator protein increases the rate of transcription when it binds to its DNA site in the promoter. The rate of transcription is thus proportional to the probability that the activator X is bound to D. Using the same reasoning as above, the binding of X to D is described by a Michaelis–Menten function:

$$\text{promoter activity} = \frac{\beta X^*}{X^* + K_d} \tag{A.6.1}$$

Many activators have a specific inducer, S_x, such that X is functional (in the sense that it can bind DNA to activate transcription) only when it binds S_x.[1] Thus, we obtain

---

[1] In some systems, the activator is active when it is unbound to S_x and inactive when it is bound. In such cases S_x is an inhibitor of X. Similarly, some repressors can be activated by binding S_x. These cases can be readily described using the reasoning in this appendix.

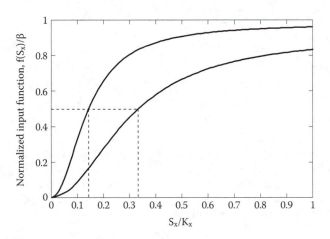

FIGURE A.5 Input function for a gene regulated by an activator as a function of the inducer level. Shown are the curves for $X_T/K_d = 10$ (bottom curve) and $X_T/K_d = 50$ (top curve), both with $n = 2$. There is no basal transcription at $S_x = 0$, and half-maximal induction is reached at $S_x \sim 1/3\ K_x$ and $S_x \sim 1/7\ K_x$.

$$X^* = [XS_x] = \frac{X_T S_x^n}{K_x^n + S_x^n} \tag{A.6.2}$$

The genes input function is

$$f(S_x) = \beta\, X^*/(K_d + X^*) \tag{A.6.3}$$

This function, shown in Figure A.5, is an increasing function. The basal transcription level is zero in this regulation function, $f(S_x = 0) = 0$. Simple activators thus can have lower leakage than repressors. If needed, however, a nonzero basal level can be readily achieved by allowing RNAp to bind and activate the promoter to a certain extent even in the absence of activator.

The inducer level needed for half-maximal induction of an activator can be much smaller than K_x:

$$S_{1/2} \sim (K_d/X_T)^{1/n}\, K_x \tag{A.6.4}$$

in contrast to the repressor case (Equation A.5.2). Overall, however, similar input function shapes as a function of inducer S_x can be obtained with either activator or repressor proteins. Rules that seem to govern the choice of activator or repressor for a given gene are discussed in Chapter 11.

In this appendix we described a simplified model that captures the essential behavior of a simple gene regulation system, in which proteins are transcribed at a rate that increases with the amount of inducer S_x. Many real systems have additional important details that

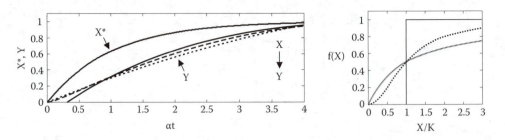

FIGURE A.6 Activation of Y by X with logic and Hill input functions. Three forms of the input function f(x) are compared, where $dY/dt = f(X^*) - \alpha Y$. The full curve results from a logic input function, $f(X) = \theta(X > K)$. The dotted line results from the Hill input function, with Hill coefficient $n = 1$ (a Michaelis–Menten function), $f(X) = X/(K + X)$, and the dashed curve from Hill kinetics, with $n = 2$, $f(X) = X^2/(K^2 + X^2)$. In all cases, $K = 1/3$, $\alpha = 1$. Right panel: the normalized input functions.

make them tighter and sharper switches.[1] The present description is sufficient, however, to understand basic circuit elements in transcription networks.

A.6.1 Comparison of Dynamics with Logic and Hill Input Functions

How good is the approximation of using logic input functions (see Section 2.3.4) instead of graded functions like Hill functions? In Figure A.6, the dynamics of accumulation of a simple one-step transcription cascade are shown, using three different forms of the input function f(X). The input functions are Hill functions with $n = 1$ and $n = 2$, and a logic input function. At time $t = 0$, X^* starts to be produced, and its concentration increases gradually with time. The graded input functions show expression as soon as X^* appears, whereas the logic input function shows expression only when X^* crosses the threshold K. Overall, the dynamics in this cascade are quite similar for all three input functions.

A.7 MICHAELIS–MENTEN ENZYME KINETICS

We now briefly describe a useful model of the action of an enzyme X on its substrate S, to catalyze formation of product P. Enzyme X and substrate S bind with rate k_{on} to form a complex [XS], which dissociates with rate k_{off}. This complex has a small rate v to form product P, so that

$$X + S \underset{k_{off}}{\overset{k_{on}}{\rightleftarrows}} [XS] \xrightarrow{v} X + P \qquad (A.7.1)$$

The rate equation for [XS], taking into account the dissociation of [XS] into X + S, as well as into X + P, is

$$d[XS]/dt = k_{on} X S - k_{off} [XS] - v[XS] \qquad (A.7.2)$$

At steady-state, we obtain

---

[1] The model is actually a reasonable description of genetically engineered *lac* promoters that include a single LacI site in a bacterium lacking the LacY pump, used as a general tool for expressing proteins under control of the inducer IPTG.

$$[XS] = k_{on}/(v + k_{off}) \, X \, S \tag{A.7.3}$$

If substrate S is found in excess, we need only worry about the conservation of enzyme X:

$$X + [XS] = X_T \tag{A.7.4}$$

Using this in Equation A.7.3, we find the Michaelis-Menten equation:

$$\text{rate of production} = v[XS] = v \, X_T \, S/(K_m + S) \qquad \textit{Michaelis–Menten enzyme kinetics} \tag{A.7.5}$$

where the Michaelis–Menten coefficient of the enzyme is:

$$K_m = (v + k_{off})/k_{on} \tag{A.7.6}$$

This constant has units of concentration and is equal to the concentration of substrate at which the production rate is half maximal. When substrate is saturating, $S \gg K_m$, production is at its maximal rate, equal to $v \, X_T$. Thus, the production rate does not depend on S (that is, it depends on S to the power zero) and is known as zero-order kinetics:

$$\text{production rate} = v \, X_T \qquad \textit{zero-order kinetics} \tag{A.7.7}$$

In the main text we will sometimes make approximations to this function, in which the substrate S is found in low concentrations, $S \ll K_m$. In this case, the production rate becomes linear in S, as can be seen from Equation A.7.5 by neglecting S in the denominator. This regime is known as first-order kinetics:

$$\text{production rate} = v \, X_T \, S/K_m \qquad \textit{first-order kinetics}$$

FURTHER READING

Ackers, G.K., Johnson, A.D., and Shea, M.A. (1982). Quantitative model for gene regulation by lambda phage repressor. *Proc. Natl. Acad. Sci. U.S.A.*, 79: 1129–1133.

Monod, J., Wyman, J., and Changeaux, J.P. (1965). On the nature of allosteric transitions: a plausible model. *J. Mol. Biol.*, 12: 88–118.

Ptashne, M. (2004). *Genetic Switch: Phage Lambda Revisited*, 3rd ed. Cold Spring Harbor Laboratory Press.

Setty, Y., Mayo, A.E., Surette, M.G., and Alon, U. (2003). Detailed map of a *cis*-regulatory input function. *Proc. Natl. Acad. Sci. U.S.A.*, 100: 7702–7707.

Stryer, L. (1995). *Biochemistry Enzymes: Basic Concepts and Kinetics*. W.H. Freeman & Co., Chap. 8.

EXERCISES

1.A.1.   Given a simple repressor with parameters β, X_T, K_d, K_x, and n, design an activator that best matches the performance of the repressor. That is, assign values to β, X_T, K_d, and K_x for the activator so its input function will have the same maximal expression, and the same $S_{1/2}$, and the same slope around $S_{1/2}$ as the repressor input function.

1.A.2. Derive the approximate value of diffusion-limited k_{on} based on dimensional analysis. Dimensional analysis seeks a combination of the physical parameters in the problem that yields the required dimensions. If only one such combination exists, it often supplies an intuitive solution to otherwise complicated physical problems. Assume a target protein with a binding site of area $a = 1$ nm^2, and a small molecule ligand that diffuses with diffusion constant $D = 1000$ μm^2/sec. The affinity of the site is so strong that it binds all ligand molecules that collide with it.

Solution:

To study the on-rate k_{on}, place a single protein in a solution of 1 M ligand L (concentration of ligand is $\rho = 1M = 6 \cdot 10^{23}$ molecules/l $\sim 10^9$ mol/μm^3). The number of L molecules colliding with the binding site of the protein has dimensions of molecules/sec and should be constructed from ρ, D, and a. The combination with the desired dimensions is $k_{on} \sim \rho D \sqrt{a}$, because D has units of $[x]^2/[t]$ and a has units of $[x]^2$. This combination makes sense: it increases with increasing ρ, a, and D as expected. Inserting numbers, we find $k_{on} \sim \rho D \sqrt{a} \sim 10^9$ mol/μm$^3 \cdot 1000$ μm^2/sec$\cdot 10^{-3}$ μm $= 10^9$ mol/sec, hence $k_{on} \sim 10^9/M$/sec. Note that dimensional analysis neglects dimensionless prefactors and is often only accurate to within an order of magnitude.

1.A.3. What is the expected diffusion-limited k_{on} for a protein sliding along DNA to bind a DNA site. The protein is confined to within $r = 1$ nm of the DNA. The total length of DNA in a bacterium such as *E. coli* is on the order of 1 mm, and the volume of the *E. coli* cell is about ~1 μm^3. Discuss the biological significance of the increase in k_{on} relative to free diffusion in space.

1.A.4. Estimate the off-time $(1/k_{off})$ of a diffusion-limited repressor that binds a site with $K_d = 10^{-11}$ M. What is the off-time of a small-molecule ligand from a receptor that binds it with $K_d = 10^{-6}$ M, $K_d = 10^{-9}$ M?

Multi-Dimensional Input Functions

Many genes are regulated by more than one transcription factor. The combined effects of these regulators can be described by a multi-dimentional input function. As an example, let us examine one simple case and then discuss the more general forms of the input function.

B.1 INPUT FUNCTION THAT INTEGRATES AN ACTIVATOR AND A REPRESSOR

Let us take a look at an input function that integrates an activator X and a repressor Y at a promoter. How can an activator and repressor work together?

A common situation is that the activator and repressor bind the promoter independently on two different sites. Thus, there are four binding states of promoter D (Figure 4.11b): D, DX, DY, and DXY, where DXY means that both X and Y bind to D. Transcription occurs mainly from the state DX, in which the activator X but not the repressor Y bind. In the following we use X and Y to denote the active forms X* and Y*.

The probability that X is bound is given by the (now familiar) Michaelis–Menten function (Appendix A):

$$P_{X\,bound} = \frac{X}{K_1 + X} = \frac{X/K_1}{1 + X/K_1} \tag{B.1.1}$$

The probability that Y is not bound is given by the Michaelis–Menten term equal to 1 minus the probability of binding:

$$P_{Y\,not\,bound} = 1 - \frac{Y}{K_2 + Y} = \frac{1}{1 + Y/K_2} \tag{B.1.2}$$

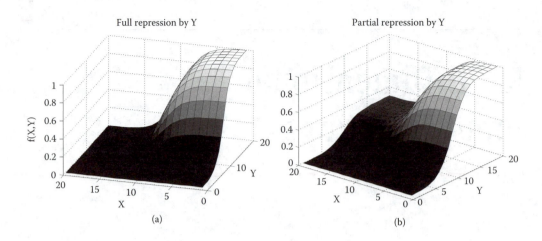

FIGURE B.1 Input function of a gene regulated by activator X and repressor Y. (a) full repression $\beta_z' = 0$, and (b) with partial repression $\beta_z' = 0.3$ (b). In both cases, $K_1 = K_2 = 10$.

Since the two binding events are *independent*, the probability that the promoter D is bound to X and not to Y is given by the product of the two probabilities:[1]

$$P_{\text{X bound AND Y not bound}} = P_{\text{X bound}} \cdot P_{\text{Y not bound}} = \frac{X / K_1}{1 + X / K_1 + Y / K_2 + XY / K_1 K_2} \qquad (B.1.3)$$

and the output promoter activity is given by the production rate β_z times this probability:

$$P_z = \beta_z X/K_1/(1 + X/K_1 + Y/K_2 + XY/K_1 K_2) \qquad (B.1.4)$$

This results in an X AND NOT Y input function, shown in Figure B.1a.

In many promoters, when the repressor binds, repression is only partial and there is basal transcription (leakage). In such cases, the state in which both X and Y bind, DXY, also contributes a transcription rate, $\beta_z' < \beta_z$, to the promoter activity of Z:

$$P_z = \frac{\beta_z X / K_1 + \beta_z' XY / K_1 K_2}{1 + X / K_1 + Y / K_2 + XY / K_1 K_2} \qquad (B.1.5)$$

This results in an input function with three plateau levels: zero when X = 0, β_z when X is high but Y is low, and β_z' when both are high (Figure B.1b). This continuous input function can be approximated by a logic function:

$$P_z = \theta (X > K_1)(\beta_z [1 - \theta (Y > K_2)] + \beta_z' \theta (Y > K_2)) \qquad (B.1.6)$$

where we remember that θ is the step function, equal to 0 or 1.

These results have some generality. The input functions can often be described by the ratio of polynomials of the active concentrations of the input transcription factors X_i, i = 1, …, n, for example,

---

[1] Fans of statistical physics will recognize the partition function in this expression. The relation of partition functions to promoters has been worked out by Ackers et al., 1982 and Gerland et al., 2002.

$$f(X_1, ..., X_m) = \frac{\sum_i \beta_i (X_i / K_i)^{n_i}}{1 + \sum_i (X_i / K_i)^{m_i}} \qquad (B.1.7)$$

The parameter K_i is the activation or repression coefficients for transcription factor X_i, β_i is its maximal contribution to expression, and the Hill coefficients are $n = m$ for activators and $n = 0$, $m > 0$ for repressors. These types of functions have been found to describe experimentally determined input functions (Setty et al., 2003). More complicated expressions are possible if the different transcription factors interact with each other on the protein level (Buchler et al., 2003).

EXERCISES

B.1. *This promoter ain't big enough.* Activator X and repressor Y bind a promoter. The repressor and activator sites overlap so that X and Y cannot both bind at the same time. What is the resulting input function? How does it differ from the input function obtained from independent binding?

Graph Properties of Transcription Networks

C.1 TRANSCRIPTION NETWORKS ARE SPARSE

What is the maximal number of edges in a network with N nodes? Each node can have an outgoing edge to each of the N – 1 other nodes, for a total of $E_{max} = N(N - 1)$ edges. If we also allow self-edges, there are an additional N possible edges, for a total of $E = N^2$. Note that a maximally connected network has a pair of edges in both directions (mutual edges) between every two nodes.

The number of edges actually found in transcription networks, E, is much smaller than the maximum possible number of edges. The networks are **sparse**, in the sense that $E/E_{max} \ll 1$. Typically, less than 0.1% of possible edges are found in the network.

Transcription networks are the product of evolutionary selection. It is important to note that it is very easy to lose an edge in the network: a single mutation in the binding site of X in the promoter of Y can cause the loss of the interaction. Therefore, every edge in the network is under evolutionary selection The sparse nature of the network reflects the fact that only very few and specific interactions, with useful function, are selected and appear in the network.

C.2 TRANSCRIPTION NETWORKS HAVE LONG-TAILED OUTPUT DEGREE SEQUENCES AND COMPACT INPUT DEGREE SEQUENCES

We saw that nodes in the transcription network correspond to genes. Incoming edges to a node in the network correspond to transcription factors that regulate the gene. The number of edges that point into a node is called the node's **in-degree**. The **out-degree** is the number of edges pointing out of a node, corresponding to the number of genes regulated by the transcription factor protein that is encoded by the gene (or operon) that corresponds to the node.

The mean number of edges per node, called the mean connectivity of the network, is λ = E/N. Typically λ is on the order of 2 to 10 edges/node.

Do all nodes have similar degrees? Transcription networks almost always have nodes that show much higher out-degrees than the average node. Transcription networks often have many transcription factors that regulate a few genes, fewer nodes that regulate tens of genes, and even fewer that regulate hundreds of genes. The latter are called **global regulators** and usually respond to key environmental signals to control large ensembles of genes (examples of global regulators in bacteria include CRP, which responds to glucose starvation, and RpoS, which responds to general stresses). Thus, the out-degree distribution has a long tail and can be roughly described as a power law, at least over a certain range (Barabasi and Oltvai, 2004). That is, the number of nodes with out-degree k is roughly $P(k) \sim k^{-\gamma}$, with $\gamma \sim 1$ to 2. Note that the out-degree distribution is only approximately power law; for example, it is bounded by the total number of genes N.

The long-tailed distribution is sometimes called "scale-free" because there are sets of regulated genes of many different sizes with no typical scale. Nodes with many more connections than the average are called **hubs**. Hubs are found in many types of natural and engineered systems. The question of their origin in biological networks is an interesting one.[1]

In contrast to the long tail of the out-degree distribution, the **in-degree** distribution is concentrated around its average value[2] (Thieffry et al., 1998; Guelzim et al., 2002; Shen-Orr et al., 2002). The in-degrees range between zero and a few times the mean connectivity, λ. There is little chance of finding a node regulated by 10 or 100 times more inputs than the average node. In other words, the in-degree distribution does not have a long tail, and instead resembles compact distributions such as the Poisson distribution, whose standard deviation is about the same as the mean.

The compact distribution of in-degrees may correspond in part to a physical limitation. In simple organisms, promoters are short. The region near the RNAp binding site that participates in regulation is on the order of a few hundred base-pairs (DNA letters). There is no space in the promoter region to accommodate more than a few binding sites for transcription factors (each on the order of 10 base-pairs). In more complex organisms, transcription factors can affect a gene even if bound far away on the DNA, through DNA-looping

---

[1] Many natural and engineered networks have hubs and degree distributions that appear to be power laws over a certain range (Barabasi and Oltvai, 2004). This power law behavior can stem from multiple different reasons, and probably has a different origin in each type of network. A general mechanism for generation of power law connectivity was proposed in the context of networks by Barabasi and Albert, 1999. In this model, called preferential attachment, new nodes are added to a growing network and connect with higher probability to nodes that already have many connections. This process generates networks with scale-free degree distributions. However, this is not a reasonable model of the evolution of transcription networks in which edges are continually selected for function. In some communication networks, scale-free distributions have been proposed to afford robustness of network connectivity with respect to the deletion of nodes. However, robustness to node removal does not appear to be the function of the degree distribution in transcription networks. These networks are often not robust to mutations (deletion of nodes), especially in bacteria. We believe that the origin of long-tailed degree distribution lies in a broad distribution of the benefit of the functions that need to be performed by the cells, and which require partitioning of gene resources into coregulated modules of widely differing sizes. An interesting theory on the origin of power laws in designed or optimized systems along these lines has been suggested by Carlson and Doyle, 1999.

[2] The average in-degree is equal to the average out-degree, because the sum of the in-degrees is equal to the total number of edges, as is the sum of the out-degrees.

interactions and other effects. Such action at a distance can increase the number of input transcription factors to a given gene. Higher organisms often display larger in-degrees than microorganisms, accommodating the complex computations needed during development.

C.3 CLUSTERING COEFFICIENTS OF TRANSCRIPTION NETWORKS

An additional statistical property of graphs is the clustering coefficient, which corresponds to whether the neighbors of a given node are connected to each other. Let us consider the network as nondirected; that is, disregard the direction of the edges. A node with k neighbors can be a part of at most $k(k-1)/2$ triangles, one for each possible pair of neighboring nodes. The clustering coefficient C is the average number of triangles that a node participates in, divided by this maximal number. Transcription networks have average clustering coefficients larger than those of randomized networks.

As described in Chapters 4 through 6, network motifs in sensory transcription networks generally include one main type of triangle, the feed-forward loop. The major contribution to the clustering coefficient of transcription networks therefore stems from feed-forward loops. This pattern appears to be selected due to its functions, such as filtering and response acceleration.

The clustering coefficient can also be measured as a function of the number of neighbors that each node has, resulting in a clustering sequence C(k). Often, $C(k) \sim 1/k$ over a certain range, so that the more neighbors a node has, the lower its clustering coefficient (Barabasi and Oltvai, 2004). In transcription networks, this tendency appears to correspond to the way that feed-forward loops connect to each other. The chief arrangement of feed-forward loops in sensory transcription networks is the multi-output FFL, discussed in Chapter 5. In the multi-output FFL, node X regulates (and is thus a neighbor of) Y, and both X and Y regulate k output nodes. These output nodes are typically not neighbors. Thus, node X has $k + 1$ neighbors (Y and the k output nodes), with only k connections between these neighbors (the connections of Y to the outputs), resulting in a clustering coefficient $C \sim k/k^2 \sim 1/k$.

Generally, it appears that global statistical properties of biological networks such as degree sequences and clustering sequences are the result of selection working on the detailed circuit patterns in each individual system. Different networks have different selection constraints, which must be understood in order to understand their graph properties.

C.4 QUANTITATIVE MEASURE OF NETWORK MODULARITY

Network modularity is the degree to which it can be separated into nearly independent sub-networks. A quantitative measure of modularity was developed by Newman and Girvan (Newman 2004; Newman and Girvan, 2004). Briefly, the Newman and Girvan algorithm finds the division of the nodes into modules that maximizes a measure Q. This measure is defined by the fraction of the edges in the network that connect between nodes in a module minus the expected value of the same quantity in a network with the same assignment of nodes into modules but random connections between the nodes:

$$Q = \sum_{s=1}^{K} \left[\frac{l_s}{L} - \left(\frac{d_s}{2L} \right)^2 \right]$$

(C.1)

where K is the number of modules, L is the number of edges in the network, l_s is the number of edges between nodes in module s, and d_s is the sum of the degrees of the nodes in module s. The rationale for this modularity measure is as follows (Guimera and Amaral, 2005): a good partition of a network into modules must comprise many within-module edges and as few as possible between-module edges. However, if we try to minimize the number of between-module edges (or equivalently maximize the number of within-module edges), the optimal partition consists of a single module and no between-module edges. Equation C.1 addresses this difficulty by imposing $Q = 0$ if nodes are placed at random into modules or if all nodes are in the same module.

This measure can be further refined by normalizing it with respect to randomized networks. The normalized measure Q_m is (Kashtan and Alon, 2005):

$$Q_m = (Q_{real} - Q_{rand}) / (Q_{max} - Q_{rand})$$

(C.2)

where Q_{real} is the Q value of the network, Q_{rand} is the average Q value of randomized networks, and Q_{max} is the maximal possible Q value of a network with the same degree sequence as the real network. The values of Q_{real}, Q_{rand}, and Q_{max} can be calculated by efficient algorithms (Kashtan and Alon, 2005).

The Q_m measure of modularity normalizes out the effects of network size and connectivity. Biological networks show high modularity according to this measure: The transcription network of the bacterium *Escherichia coli* has $Q_m = 0.54$, the neuronal synaptic network of the nematode *Caenorhabditis elegans* has $Q_m = 0.54$, and a human signal transduction network has $Q_m = 0.58$.

Cell–Cell Variability in Gene Expression

The concentration of a protein X in a population of genetically identical cells varies from cell to cell due to stochastic processes (reviewed in McAdams and Arkin, 1999; Kaern et al., 2005). The concentration of a given protein often has a coefficient of variation (standard deviation divided by the mean) in the range CV = 0.1 to 1 (Elowitz et al., 2002; Ozbudak et al., 2002; Blake et al., 2003; Raser and O'Shea, 2004). That is, the cell–cell variations are on the order of tens of percents of the mean. The dynamics of protein levels thus have a stochastic component (Figure D.1).

One important source of noise is extrinsic noise, in which the cellular capacity to produce proteins, and the regulatory systems that regulate a gene, fluctuate over time. For example, fluctuations in a transcription factor concentration can affect the expression rate of its targets. The correlation time of these variations in production rates is often on the scale of a cell generation: that is, a cell with high production levels often tends to stay high for a cell cycle or more (Rosenfeld et al., 2005).

In addition to extrinsic noise, there is also intrinsic noise, which is due to stochastic variations in the transcription and translation events of the gene. An elegant experiment by Michael Elowitz and colleagues (Elowitz et al., 2002) measured the relative level of intrinsic and extrinsic noise, by measuring the levels of two fluorescent proteins expressed by identical promoters (Figure D.2). Intrinsic noise appears to fluctuate on a timescale of minutes in bacteria (Rosenfeld et al., 2005).

The cell–cell distribution of protein numbers is often similar to lognormal (a Gaussian distribution in the variable log(X)). Whereas Gaussian distributions describe processes that are a sum of random variables with finite mean and variance, lognormal distributions characterize processes with several *multiplicative* stochastic steps (because log(X) is then a sum of random variables). Examples of multiplicative steps in the production of a protein are transcription and translation.

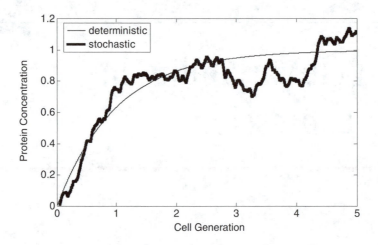

FIGURE D.1 Stochastic dynamics of protein concentration. A stable protein is produced with a stochastically fluctuating production rate. The resulting dynamics show fluctuations on both slow and fast time-scales. The dynamics according to the deterministic model (Equation 2.4.2) is also shown (light curve).

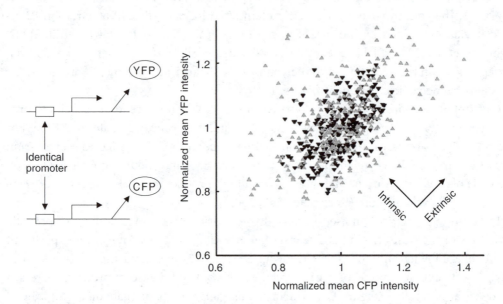

FIGURE D.2 An experimental measurement of extrinsic and intrinsic noise. Two almost identical genes encode proteins with different fluorescent colors, yellow and cyan fluorescent proteins (YFP and CFP). The two genes are expressed in the same cell from identical promoters. Extrinsic noise is the component of the noise shared by the genes due to upstream factors such as variations in regulators and the cells' metabolic capacity. Intrinsic noise is due to stochastic steps in transcription and translation of each gene. The measurements in the right panel are on *E. coli* cells, where each point is one cell. (From Elowitz et al., 2002.)

Regulatory circuits can affect the variability. For example, protein levels can be made to fluctuate less by means of negative feedback loops (see Chapter 3). Conversely, positive auto-regulation can increase cell-cell variability. Strong positive feedback can even lead to bistability (Figure D.3). Bistability often leads to a bi-modal

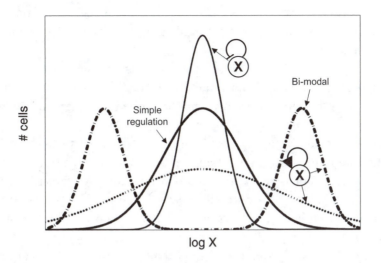

FIGURE D.3 Distribution of protein numbers per cell. Negative autoregulation can generally decrease the width of the distribution. Positive autoregulation widens the distribution and can lead to a bi-modal distribution.

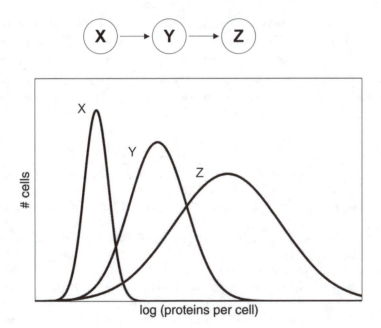

FIGURE D.4 Schematic distributions of protein concentrations in a regulatory cascade. Variability tends to increase with the steps in the cascade.

distribution, with two cell populations, with high and low expression (Novick and Weiner, 1957; Siegele and Hu, 1997; Ferrell and Machleder, 1998; Isaacs et al, 2002; Ozbudak et al., 2002). Noise can also be amplified by regulatory cascades: each step in the cascade receives variability from its upstream regulator (Figure D.4) (Blake et al., 2003; Hooshangi et al., 2005; and Pedraza et al., 2005). Rapidly degraded proteins can have narrower distributions than stable proteins, because stable proteins integrate the noise in production rates over longer times. As a rule of thumb, the faster the response time of a system, the smaller the fluctuations in the system.

One interesting observation is that the position of the noisiest step in a pathway can influence the overall noise (McAdams and Arkin, 1999; Ozbudak et al., 2002). This is because each step in the pathway usually amplifies noise in the previous steps. For example, consider two mechanisms that produce 100 proteins per hour: In mechanism A, one mRNA molecule is made on average per hour and is translated to 100 proteins on average. In mechanism B, 100 mRNAs are made per hour and are each translated to one protein on average. The fluctuations in protein production are much larger in mechanism A, because an average of one mRNA normally means that in some cells either 0 or 2 mRNAs will be made in a given hour, resulting in 0 or ~200 proteins. In mechanism B, there is little chance to make zero mRNAs during an hour, and fluctuations are smaller.

The chromosomal position of a gene can also affect noise, due to local differences in chromatin regulation (Blake et al., 2003; Becskei et al., 2005). Generally, noise level can be tuned over evolutionary timescales by changing the parameters of the noisy steps in the expression of each gene (Fraser et al., 2004). It appears that essential proteins and complex-forming proteins are less noisy than other proteins.

Noise in biological systems can be modeled using stochastic mathematical equations. Such theoretical treatment of stochastic effects is beyond the present scope. Excellent texts on stochastic processes, such as those by Gardiner and Van Kampen, can give access to the highly developed field of stochastic theory in physics, chemistry and engineering. Theory on biological noise has been reviewed (Paulsson, 2004; Kaern et al., 2005). Other theoretical studies are cited in the bibliography.

FURTHER READING

Acar, M., Becskei, A., and van Oudenaarden, A. (2005). Enhancement of cellular memory by reducing stochastic transitions. *Nature.* 435:228–232.

Blake, W.J., Kaern, M., Cantor, C.R., and Collins, J.J. (2003). Noise in eukaryotic gene expression. *Nature.* 422:633–637.

Elowitz, M.B., Levine, A.J., Siggia, E.D. and Swain, P.P. (2002). Stochastic gene expression in a single cell. *Science.* 297:1183–1186.

Gardiner, C.W. (2004). *Handbook of Stochastic Methods.* Springer.

Kaern, M., Elston, T.C., Blake, W.J., and Collins, J.J. (2005). Stochasticity in gene expression: from theories to phenotypes. *Nature. Rev. Gen.* 6:451–464.

McAdams, H.H. and Arkin, A. (1997). Stochastic mechanisms in gene expression. *Proc. Natl. Acad. Sci. U.S.A.* 94:814–819.

Novick, A. and Weiner, M. (1957). Enzyme induction as an all-or-none phenomenon. *Proc. Natl. Acad. Sci. U.S.A.* 43:553–566.

Ozbudak, E.M., Thattai, M., Kurtser, I., Grossman, A.D., and van Oudenaarden, A. (2002). Regulation of noise in the expression of a single gene. *Nat. Genet.* 31: 69–73.

Raser, J.M. and O'Shea, E.K. (2004). Control of stochasticity in eukaryotic gene expression. *Science.* 304:1811–1814.

Rosenfeld, N., Young, J.W., Alon, U., Swain, P.S., and Elowitz, M.B. (2005). Gene regulation at the single-cell level. *Science.* 307:1962–1965.

Glossary

Activator — A transcription factor that increases the rate of transcription of a gene when it binds a specific site in the gene's promoter.

Activation threshold — Concentration of activator in its active state needed for half-maximal activation of a gene.

Adaptation — Decreasing response to a stimulus that is applied continuously.

Adaptation time — Time for output to recover to 50% of prestimulus level following a step stimulus.

Allele — One of a set of alternative forms of a gene. In a diploid organism, such as most animal cells, each gene has two alleles, one on each of the two sister chromosomes.

Amino acid — A molecule that contains both an amino group (NH_2) and a carboxyl group (COOH). Amino acids are linked together by peptide bonds and serve as the constituents of proteins.

AND gate — A logic function of two inputs that outputs a one only if both inputs are equal to one.

Anti-motif — A pattern that occurs in a network less often than expected at random.

Antibody — A protein produced by a cell of the immune system that recognizes a protein present in or on invading microorganisms.

Antigen — A part of a protein or other molecule that is recognized by an antibody.

Arabinose — A sugar utilized by *E. coli* as an energy and carbon source, using the *ara* genes. These genes include metabolic genes *araBAD*, and the transporters *araE* and *araFGH*. Arabinose is not pumped into the cells if glucose, a better energy source, is present.

ATP (adenosine triphosphate) — A molecule that is the main currency in the cellular energy economy. The conversion of ATP to ADP (adenosine diphosphate) liberates energy.

B. subtilis (Bacillus subtilis) — A bacterium commonly found in the soil. It forms durable spores upon starvation.

Binomial distribution — A statistical distribution that describes, for example, the probability for k heads out of n throws of a coin that has probability p to give heads and 1-p to give tails.

Chemoreceptor — A receptor that responds to the presence of a particular chemical.

Chemotaxis — Movement up spatial gradients of specific chemicals (attractants), or down gradients of specific chemicals (repellents).

Chromosome — A strand of DNA, with associated proteins, found in the nucleus; carries genetic information.

Circadian rhythm — A daily rhythmical cycle of cellular activity. Generated by a biochemical oscillator in many different cells in animals, plants, and microorganisms. The oscillations can be entrained by periodic temperature and light signals. The oscillator runs also in the absence of entraining external signals (usually with a period somewhat different than 24 h).

Codon — Three consecutive letters on an mRNA. There are 64 codons (each made of three letters, A, C, G, and U). These code for the 20 amino acids (with most amino acids represented by more than one codon). Three of the codons signal translational stop (end of the protein).

Coherent feed-forward loop — A feed-forward loop in which the sign of the direct path from X to Z is the same as the sign of the indirect path from X through Y to Z.

Coherent pattern — A pattern with signs on the edges, in which, for each pair of nodes, the sign of all directed paths between the nodes is the same.

Cost–benefit analysis — A theory that seeks the optimal design such that the difference between the fitness advantage gained by a system (benefit) and fitness reduction due to the cost of its parts is maximal.

Cytoplasm — The viscous, semiliquid substance contained in the interior of a cell. The cytoplasm is densely packed with proteins.

Degree-preserving random networks — An ensemble of randomized networks that have the same degree sequence (the number of incoming and outgoing edges for each node in the network) as the real network. Despite the fact that the degree sequence is the same, the identity of which node connects to which other node is randomized. Such random networks can be generated on the computer by randomly switching pairs of edges, repeating the switching operations many times until the network is randomized. For a given real network, many thousands of different randomized degree-preserving networks can usually be readily generated.

Design — In this book, design means structure as related to function.

Developmental transcription networks — Networks of transcription interactions that guide changes in cell type. Important examples are networks that guide the selection of cell fate as cells in the embryo differentiate into tissues. Developmental transcription networks work on the timescale of cell generations and often make irreversible decisions. They stand in contrast to sensory transcription networks that govern responses to environmental signals.

Differentiation — The process in which a cell changes to a different type of cell.

DNA (deoxyribonucleic acid) — A long molecule composed of two interconnected helical strands. Contains the genetic information. Each strand in the DNA is made of four bases, A, C, T, and G. The two strands pair with each other so that A pairs with T, and C with G. Thus, DNA is made of a chain of base-pairs and can be represented by a string of four types of letters.

Dorsal — Side of an animal closer to its back.

Drosophila — Fruit fly, a model organism commonly used for biological research.

Edge — A link between two nodes in a network. Edges describe interactions between the components described by the nodes. Edges in most networks have a specific direction. Mutual edges are edges that link nodes in both directions. *See* transcription network for an example.

Endocytosis — Uptake of material into a cell.

Enzyme — A protein that facilitates a biochemical reaction. The enzyme catalyzes the reaction and does not itself become part of the end product.

ER (Erdos–Renyi) random networks — An ensemble of random networks with a given number of nodes, N, and edges, E. The edges are placed randomly between the nodes. This model can be used for comparison to real networks. A more stringent random model is the degree-preserving random network.

Error load — The reduction in the organism's fitness due to internal errors in a system.

E. coli (Escherichia coli) — A rod-shaped bacterium normally found in the colon of humans and other mammals. It is widely studied as a model organism.

Eukaryotic cells and organisms — Organisms made of cells with a nucleus. Includes all forms of life except for viruses and bacteria (prokaryotes). Yeast is a single-celled eukaryotic organism.

Exact adaptation (precise adaptation) — A property of an adapting sensory system in which the steady-state output is independent of the stimulus level.

Exponential phase — A phase of bacterial growth in which cells double with a constant cell generation time, resulting in exponentially increasing cell numbers. This occurs in a test tube when there are so few cells that nutrients are not depleted from the medium, and waste products do not accumulate to high levels. *See also* stationary phase.

Feedback — A process whereby some proportion or function of the output signal of a system is passed (fed back) to the input.

Feedback inhibition — A common control mechanism in metabolic networks, in which a product inhibits the first enzyme in the pathway that produces that product.

Feed-forward loop (FFL) — A pattern with three nodes, X, Y, and Z, in which X has a directed edge to Y and Z, and Y has a directed edge to Z. The FFL is a network motif in many biological networks, and can perform a variety of tasks (such as sign-sensitive delay, sign-sensitive acceleration, and pulse generation).

Fine-tuned property — A property of a biological circuit that depends sensitively on the biochemical parameters of the circuit (opposite to robust property).

First-order kinetics — Mathematical description of the rate of an enzymatic reaction in the limit where the substrate concentration is very low and is far from saturating the enzyme, such that the rate is equal to (v/K) E S, where v is the rate per enzyme, E is the enzyme concentration, K is the Michaelis constant, and S is the substrate concentration. *See also* Michaelis–Menten kinetics, zero-order kinetics.

Flagellum (plural flagella) — A long filament whose rotation drives bacteria through a fluid medium. Rotated by the flagellar motor.

Functionalism — The strategy of understanding an organism's structural or behavioral features by attempting to establish their usefulness with respect to survival or reproductive success.

Gene — The functional unit of a chromosome, which directs the synthesis of one protein (or several alternate forms of a protein). The gene is transcribed into mRNA, which is then translated into the protein. The gene is preceded by a regulatory DNA region called the promoter that includes binding sites for transcription factors that regulate the rate of transcription.

Gene circuit — A term used in this book to mean a set of biomolecules that interact to perform a dynamical function. An example is a feed-forward loop.

Gene product — The protein encoded by a gene. Sometimes, the RNA transcribed from the gene, when the RNA has specific functions.

Generation time — Mean time for an organism to produce offspring.

Genetic code — The mapping between the 64 codons and the 20 amino acids. The genetic code is identical in nearly all organisms (Figure 9.2).

Genetic drift — The statistical change over time of gene frequencies in a population due to random sampling effects in the formation of successive generations.

Genome — The total genetic information in a cell or organism.

Glucose — A simple sugar, a major source of energy in metabolism.

Hertz (Hz) — A cycle per second, a measure of frequency.

Homeostasis — The process by which the organism's substances and characteristics are maintained at their optimal level.

Homologous — Similar by virtue of a common evolutionary origin. Homologous genes generally show similarity in their sequence.

Hormone — A chemical substance liberated by an endocrine gland that has effects on target cells in other organs.

Immune system — The system by which the body protects itself from foreign proteins. In response to an infection, the white blood cells can produce antibodies that recognize and attack invading microorganisms.

Incoherent feed-forward loop — A feed-forward loop in which the sign of the direct path from X to Z is the opposite as the overall sign of the indirect path from X through Y to Z.

Incoherent pattern — A pattern with signs on the edges, in which there exists a pair of nodes with two different directed paths between these nodes, such that the overall sign of the paths is different.

Integral feedback — Feedback on a device in which the integral over time of the error (output minus the desired output) is negatively fed back into the input of the device. Integral feedback can lead to robust exact adaptation.

Lac operon — A group of three genes in *E. coli* that are adjacent on the chromosome and transcribed on the same mRNA. These genes are *lacZYA*, encoding for the metabolic enzyme LacZ, which cleaves lactose into glucose and galactose; the permease (pump) LacY, which pumps lactose into the cells; and LacA, whose function is unknown. Lactose is not pumped into the cells if glucose, a better energy source, is present, a phenomenon called "inducer exclusion". The *lac* operon is repressed by LacI and activated by CRP. LacI unbinds from the DNA and the system is induced in the presence of lactose (LacI binds a derivative of lactose called allo-lactose) or non-metaboliz-able analogs of lactose, such as IPTG.

Lactose — A sugar utilized by *E. coli* as an energy and carbon source, using the *lac* genes expressed from the *lac* operon.

Ligand — A molecule that specifically binds the binding site of a receptor.

Mathematically controlled comparison — A comparison between mechanisms that is carried out with equivalence of as many internal and external parameters as possible between the alter-native designs (Savageau, 1976). Internal parameters include biochemical parameters, such as the lifetime of the proteins that make up the circuit, and external parameters include desired output properties, such as steady-state levels.

Membrane — A structure consisting principally of lipid molecules that define the outer boundaries of a cell.

Membrane potential — The difference in electrical potential inside and outside of the cell expressed as voltage relative to the outside voltage. Membrane potential is maintained by protein pumps that transport ions across the membrane at the expense of energy supplied by ATP.

Michaelis–Menten kinetics — A mathematical description of the rate of an enzymatic reaction as a function of the concentration of the substrate. The rate is equal to $v\,E\,S/(K + S)$, where v is the rate per enzyme, E is the enzyme concentration, S is the substrate concentration, and K is the Michaelis constant. When $S \gg K$, one obtains zero-order kinetics (rate $= v\,E$), and when $S \ll K$, one obtains first-order kinetics (rate $= (v/K)\,E\,S$).

Micron — One millionth of a meter.

Modularity — A property of a system which can be separated into nearly independent sub-systems.

Morphogen — A molecule (protein) that determines spatial patterns. Morphogens bind specific recep-tors to trigger signal transduction pathways within the cells to be patterned. The signaling leads the cells to assume different cell fates according to the morphogen level.

Morphology — Physical shape and structure.

mRNA — A macromolecule made of a sequence of four types of bases, A, C, G, and U. Transcription is the process by which an RNA–polymerase enzyme produces an mRNA molecule that cor-responds to the base sequence on the DNA (where DNA T is mapped to RNA U). The mRNA is read by ribosomes, which produce a protein according to the mRNA sequence.

Mutation — A heritable change in the base-pair sequence of the chromosome.

Network motif — A pattern of interactions that recurs in a network in many contexts. Network motifs can be detected as patterns that occur much more often than in randomized networks.

Neuron (nerve cell) — Cell specialized to receive, transmit, and conduct signals in the nervous system.

Nucleus — A structure enclosed by a membrane found in eukaryotic cells (not in bacteria) that con-tains the chromosomes.

Operon — A group of genes transcribed on the same mRNA. Each gene is separately translated. Oper-ons are found only in prokaryotes.

Peptide — A chain of amino acids joined together by peptide bonds. Proteins are long peptides.

Point mutation — A change of a single letter (base-pair) in the DNA.

Poisson distribution — A distribution that characterizes a random process such as the number of heads in a coin-toss experiment, with many tosses, N, and a small probability for heads, $p \ll 1$. The mean number of heads is $m = p\,N$. The variance in a Poisson process is equal to the mean, $\sigma^2 = m$, and hence the standard deviation is the square root of the mean, $\sigma = \sqrt{m}$.

Promoter — A regulatory region of DNA that controls the transcription rate of a gene. The promoter contains a binding site for RNA polymerase (RNAp), the enzyme that transcribes the gene to produce mRNA. Each promoter also usually contains binding sites for transcription factor proteins; the transcription factors, when bound, affect the probability that RNAp will initiate transcription of an mRNA.

Protease — An enzyme that degrades proteins. Proteins are often targeted for degradation in biologically regulated ways. For example, many eukaryotic proteins are targeted for degradation in the proteosome by enzymes that attach a chain of ubiquitin molecules to the target protein. Different proteins can have different degradation rates.

Protein — A long chain of amino acids (on the order of tens to hundreds of amino acids) that can serve in a structural capacity or as an enzyme. Each protein is encoded by a gene. Proteins are produced in ribosomes, based on information encoded on an mRNA that is transcribed from the gene.

Protein kinase — An enzyme that attaches a phosphate (PO_4) group to a protein and thereby causes it to change its shape.

Receptor — A protein molecule, usually situated in the membrane of the cell, that is sensitive to a particular chemical. When the appropriate chemical (the ligand) binds to the binding site of the receptor, signal transduction cascades are triggered within the cell.

Repression threshold — Concentration of active repressor needed for half-maximal repression of a gene.

Repressor — A transcription factor that decreases the rate of transcription when it binds a specific site in the promoter of a gene.

Ribosome — A structure in the cytoplasm made of about 100 proteins and special RNA molecules that serves as the site of production of proteins translated from mRNA. In the ribosome, amino acids are assembled to form the protein chain according to an order specified by the codons on the mRNA. The amino acids are brought into the ribosome by tRNA molecules, which read the mRNA codons. Each tRNA is released when its amino acid is linked to the translated protein chain.

RNA Polymerase (RNAp) — A complex of several proteins that form an enzyme that transcribes DNA into RNA.

Robust Property — Property X is robust with respect to parameter Y, if X is insensitive to changes in parameter Y.

Sensitivity (Parameter sensitivity) — The parameter sensitivity coefficient of property X with respect to parameter Y upon a small relative change in Y is

$$S(X, Y) = \frac{d \log X}{d \log Y} = \frac{Y}{X} \frac{dX}{dY}$$

Sensory transcription networks — Transcription networks that respond to environmental and internal signals such as nutrients and stresses, and lead to changes in gene expression. These networks need to function rapidly, usually within less than a cell generation time, and usually make reversible decisions. They stand in contrast to developmental transcription networks.

Stationary phase — A state in which cells cease to divide and grow, that occurs when growth conditions are unfavorable, such as when the bacteria run out of an essential nutrient. *See also* exponential phase.

Teleology — The use of design or purpose as an explanation of natural phenomena.

Transcription factor — A protein that regulates the transcription rate of specific target genes. Transcription factors usually have two molecular states, active and inactive. They transit between these states on a rapid timescale (e.g., microseconds). When active, the transcription factor binds specific sites on the DNA to affect the rate of transcription initiation of target genes. Also called transcriptional regulator. *See* activator, repressor.

Transcription network — The set of transcription interactions in a cell. The network is made of nodes linked by directed edges. Each node represents a gene (or, in bacteria, an operon). Each edge is a transcriptional interaction. X → Y means that the protein encoded by gene X is a transcription factor that transcriptionally regulates gene Y.

XOR gate (exclusive OR) — A logic function of two inputs that outputs a one if either, but not both, inputs is equal to one.

Yeast — A single-celled eukaryote, a unicellular fungus. In this book, usually the budding yeast *Saccharomyces cerevisae*. Yeast is used for brewing and bread making and is a well-studied research model organism.

Zero-order kinetics — Mathematical description of the rate of an enzymatic reaction in the limit where the substrate concentration is saturating, such that the rate is equal to v E, where v is the rate per enzyme and E is the enzyme concentration. *See also* Michaelis–Menten kinetics, first-order kinetics.

Bibliography

References cited in the text, and additional references of interest.

Acar, M., A. Becskei, and A. van Oudenaarden. 2005. Enhancement of cellular memory by reducing stochastic transitions. *Nature* **435:** 228–232.

Ackers, G.K., A.D. Johnson, and M.A. Shea. 1982. Quantitative model for gene regulation by lambda phage repressor. *Proc Natl Acad Sci USA* **79:** 1129–1133.

Alberghina, L. and H.V. Westerhoff. 2005. *Systems Biology: Definitions and Perspectives*. Springer.

Albert, I. and R. Albert. 2004. Conserved network motifs allow protein-protein interaction prediction. *Bioinformatics* **20:** 3346–3352.

Albert, R. and H.G. Othmer. 2003. The topology of the regulatory interactions predicts the expression pattern of the segment polarity genes in Drosophila melanogaster. *J Theor Biol* **223:** 1–18.

Aldridge, P. and K.T. Hughes. 2002. Regulation of flagellar assembly. *Curr Opin Microbiol* **5:** 160–165.

Allen, R.J., P.B. Warren, and P.R. Ten Wolde. 2005. Sampling rare switching events in biochemical networks. *Phys Rev Lett* **94:** 018104.

Alon, U. 2003. Biological networks: the tinkerer as an engineer. *Science* **301:** 1866–1867.

Alon, U., M.G. Surette, N. Barkai, and S. Leibler. 1999. Robustness in bacterial chemotaxis. *Nature* **397:** 168–171.

Andrec, M., B.N. Kholodenko, R.M. Levy, and E. Sontag. 2005. Inference of signaling and gene regulatory networks by steady-state perturbation experiments: structure and accuracy. *J Theor Biol* **232:** 427–441.

Angeli, D., J.E. Ferrell, Jr., and E.D. Sontag. 2004. Detection of multistability, bifurcations, and hysteresis in a large class of biological positive-feedback systems. *Proc Natl Acad Sci USA* **101:** 1822–1827.

Arkin, A. 2000. Signal processing by biochemical reaction networks. *Self-Organized Biodynamics and Nonlinear Control*: 112–144.

Arkin, A. and J. Ross. 1994. Computational functions in biochemical reaction networks. *Biophysical Journal* **67:** 560–578.

Arkin, A. and J. Ross. 1995. Statistical construction of chemical reaction mechanisms from measured time-series. *J Phys Chem* **99:** 970–979.

Arkin, A., J. Ross, and H.H. McAdams. 1998. Stochastic kinetic analysis of developmental pathway bifurcation in phage-[lambda]-infected Escherichia coli cells. *Genetics* **149:** 1633–1648.

Arkin, A., P. Shen, and J. Ross. 1997. A test case of correlation metric construction of a reaction pathway from measurements. *Science* **277:** 1275–1279.

Arkin, A.P. 2001. Synthetic cell biology. *Curr Opin Biotech* **12:** 638–644.

Asakura, S. and H. Honda. 1984. Two-state model for bacterial chemoreceptor proteins. The role of multiple methylation. *J Mol Biol* **176:** 349–367.

Asthagiri, A.R. and D.A. Lauffenburger. 2000. Bioengineering models of cell signaling. *Annu Rev Biomed Eng* **2:** 31–53.

Atkinson, M.R., M.A. Savageau, J.T. Myers, and A.J. Ninfa. 2003. Development of genetic circuitry exhibiting toggle switch or oscillatory behavior in Escherichia coli. *Cell* **113:** 597–607.

Austin, D.W., M.S. Allen, J.M. McCollum, R.D. Dar, J.R. Wilgus, G.S. Sayler, N.F. Samatova, C.D. Cox, and M.L. Simpson. 2006. Gene network shaping of inherent noise spectra. *Nature* **439:** 608–611.

Azevedo, R.B., R. Lohaus, V. Braun, M. Gumbel, M. Umamaheshwar, P.M. Agapow, W. Houthoofd, U. Platzer, G. Borgonie, H.P. Meinzer, and A.M. Leroi. 2005. The simplicity of metazoan cell lineages. *Nature* **433:** 152–156.

Babu, M.M., N.M. Luscombe, L. Aravind, M. Gerstein, and S.A. Teichmann. 2004. Structure and evolution of transcriptional regulatory networks. *Curr Opin Struct Biol* **14:** 283–291.

Bagowski, C.P. and J.E. Ferrell, Jr. 2001. Bistability in the JNK cascade. *Curr Biol* **11:** 1176–1182.

Balaban, N.Q., J. Merrin, R. Chait, L. Kowalik, and S. Leibler. 2004. Bacterial persistence as a phenotypic switch. *Science* **305:** 1622–1625.

Balagadde, F.K., L. You, C.L. Hansen, F.H. Arnold, and S.R. Quake. 2005. Long-term monitoring of bacteria undergoing programmed population control in a microchemostat. *Science* **309:** 137–140.

Barabasi, A.L. and R. Albert. 1999. Emergence of scaling in random networks. *Science* **286:** 509–512.

Barabasi, A.L. and Z.N. Oltvai. 2004. Network biology: understanding the cell's functional organization. *Nat Rev Genet* **5:** 101–113.

Bargmann, C.I. 1998. Neurobiology of the Caenorhabditis elegans genome. *Science* **282:** 2028–2033.

Bar-Joseph, Z., G.K. Gerber, T.I. Lee, N.J. Rinaldi, J.Y. Yoo, F. Robert, D.B. Gordon, E. Fraenkel, T.S. Jaakkola, R.A. Young, and D.K. Gifford. 2003. Computational discovery of gene modules and regulatory networks. *Nat Biotechnol* **21:** 1337–1342.

Barkai, N. and S. Leibler. 1997. Robustness in simple biochemical networks. *Nature* **387:** 913–917.

Barkai, N. and S. Leibler. 2000. Circadian clocks limited by noise. *Nature* **403:** 267–268.

Barkai, N., M.D. Rose, and N.S. Wingreen. 1998. Protease helps yeast find mating partners. *Nature* **396:** 422–423.

Barkai, N. and B.Z. Shilo. 2002. Modeling pattern formation: counting to two in the Drosophila egg. *Curr Biol* **12:** R493–495.

Bar-Yam, Y. and I.R. Epstein. 2004. Response of complex networks to stimuli. *Proc Natl Acad Sci USA* **101:** 4341–4345.

Basu, S., R. Mehreja, S. Thiberge, M.T. Chen, and R. Weiss. 2004. Spatiotemporal control of gene expression with pulse-generating networks. *Proc Natl Acad Sci USA* **101:** 6355–6360.

Baumberg, S. 1999. *Prokaryotic Gene Expression*. Oxford University Press.

Becskei, A., M.G. Boselli, and A. van Oudenaarden. 2004. Amplitude control of cell-cycle waves by nuclear import. *Nat Cell Biol* **6:** 451–457.

Becskei, A., B. Seraphin, and L. Serrano. 2001. Positive feedback in eukaryotic gene networks: cell differentiation by graded to binary response conversion. *EMBO J* **20:** 2528–2535.

Becskei, A. and L. Serrano. 2000. Engineering stability in gene networks by autoregulation. *Nature* **405:** 590–593.

Beer, M.A. and S. Tavazoie. 2004. Predicting gene expression from sequence. *Cell* **117:** 185–198.

Berg, H.C. 1993. *Random Walks in Biology*. Princeton University Press.

Berg, H.C. 2003a. *E. coli in Motion*. Springer.

Berg, H.C. 2003b. The rotary motor of bacterial flagella. *Annu Rev Biochem* **72:** 19–54.

Berg, H.C. and D.A. Brown. 1972. Chemotaxis in Escherichia coli analysed by three-dimensional tracking. *Nature* **239:** 500–504.

Berg, H.C. and E.M. Purcell. 1977. Physics of chemoreception. *Biophys J* **20:** 193–219.

Berg, J. and M. Lassig. 2004. Local graph alignment and motif search in biological networks. *Proc Natl Acad Sci USA* **101:** 14689–14694.

Berg, J., M. Lassig, and A. Wagner. 2004. Structure and evolution of protein interaction networks: a statistical model for link dynamics and gene duplications. *BMC Evol Biol* **4:** 51.

Berg, O.G. 1978. A model for the statistical fluctuations of protein numbers in a microbial population. *J Theor Biol* **71:** 587–603.

Berg, O.G., R.B. Winter, and P.H. von Hippel. 1981. Diffusion-driven mechanisms of protein translocation on nucleic acids. 1. Models and theory. *Biochemistry* **20:** 6929–6948.

Bergmann, S., J. Ihmels, and N. Barkai. 2004. Similarities and differences in genome-wide expression data of six organisms. *PLoS Biol* **2:** E9.

Bernstein, J.A., A.B. Khodursky, P.H. Lin, S. Lin-Chao, and S.N. Cohen. 2002. Global analysis of mRNA decay and abundance in Escherichia coli at single-gene resolution using two-color fluorescent DNA microarrays. *Proc Natl Acad Sci USA* **99:** 9697–9702.

Bhalla, U.S. and R. Iyengar. 1999. Emergent properties of networks of biological signaling pathways. *Science* **283:** 381–387.

Bhalla, U.S. and R. Iyengar. 2001. Robustness of the bistable behavior of a biological signaling feedback loop. *Chaos* **11:** 221–226.

Bialek, W. and D. Botstein. 2004. Introductory science and mathematics education for 21st-Century biologists. *Science* **303:** 788–790.

Bialek, W. and S. Setayeshgar. 2005. Physical limits to biochemical signaling. *Proc Natl Acad Sci USA* **102:** 10040–10045.

Biggar, S.R. and G.R. Crabtree. 2001. Cell signaling can direct either binary or graded transcriptional responses. *EMBO J.* **20:** 3167–3176.

Bird, A.P. 1995. Gene number, noise reduction and biological complexity. *Trends Genet.* **11:** 94–100.

Blake, W.J., M. Kaern, C.R. Cantor, and J.J. Collins. 2003. Noise in eukaryotic gene expression. *Nature* **422:** 633–637.

Blewitt, M.E., S. Chong, and E. Whitelaw. 2004. How the mouse got its spots. *Trends Genet.* **20:** 550–554.

Bluthgen, N. and H. Herzel. 2003. How robust are switches in intracellular signaling cascades? *J Theor Biol* **225:** 293–300.

Bollobas, B. 1985. *Random Graphs*. Academic Press, New York.

Bolouri, H. and Davidson, E.H. (2002). Modeling transcriptional regulatory networks. *Bioessays*, 24: 1118–1129.

Booth, I.R. 2002. Stress and the single cell: intrapopulation diversity is a mechanism to ensure survival upon exposure to stress. *Int. J. Food Microbiol.* **78:** 19–30.

Borisov, N.M., N.I. Markevich, J.B. Hoek, and B.N. Kholodenko. 2005. Signaling through receptors and scaffolds: independent interactions reduce combinatorial complexity. *Biophys J* **89:** 951–966.

Borneman, A.R., J.A. Leigh-Bell, H. Yu, P. Bertone, M. Gerstein, and M. Snyder. 2006. Target hub proteins serve as master regulators of development in yeast. *Genes Dev.* **20:** 435–448.

Boyer, L.A., T.I. Lee, M.F. Cole, S.E. Johnstone, S.S. Levine, J.P. Zucker, M.G. Guenther, R.M. Kumar, H.L. Murray, R.G. Jenner, D.K. Gifford, D.A. Melton, R. Jaenisch, and R.A. Young. 2005. Core transcriptional regulatory circuitry in human embryonic stem cells. *Cell* **122:** 947–956.

Brandman, O., J.E. Ferrell, Jr., R. Li, and T. Meyer. 2005. Interlinked fast and slow positive feedback loops drive reliable cell decisions. *Science* **310:** 496–498.

Brauer, M.J., A.J. Saldanha, K. Dolinski, and D. Botstein. 2005. Homeostatic adjustment and metabolic remodeling in glucose-limited yeast cultures. *Mol Biol Cell* **16:** 2503–2517.

Bray, D. 1995. Protein molecules as computational elements in living cells. *Nature* **376:** 307–312.

Bray, D. 2002. Bacterial chemotaxis and the question of gain. *Proc Natl Acad Sci USA* **99:** 7–9.

Breitling, R. and D. Hoeller. 2005. Current challenges in quantitative modeling of epidermal growth factor signaling. *FEBS Lett* **579:** 6289–6294.

Brown, P.O. and D. Botstein. 1999. Exploring the new world of the genome with DNA microarrays. *Nat Genet* **21:** 33–37.

Bruggeman, F.J., H.V. Westerhoff, J.B. Hoek, and B.N. Kholodenko. 2002. Modular response analysis of cellular regulatory networks. *J Theor Biol* **218:** 507–520.

Buchler, N.E., U. Gerland, and T. Hwa. 2003. On schemes of combinatorial transcription logic. *Proc Natl Acad Sci USA* **100:** 5136–5141.

Bushman, F.D. and M. Ptashne. 1988. Turning lambda Cro into a transcriptional activator. *Cell* **54:** 191–197.

Bushman, F.D., C. Shang, and M. Ptashne. 1989. A single glutamic acid residue plays a key role in the transcriptional activation function of lambda repressor. *Cell* **58:** 1163–1171.

Carlson, J.M. and J. Doyle. 1999. Highly optimized tolerance: a mechanism for power laws in designed systems. *Phys Rev E* **60:** 1412–1427.

Carninci, P. T. et al., 2005. The transcriptional landscape of the mammalian genome. *Science* **309:** 1559–1563.

Carrier, T.A. and J.D. Keasling. 1997. Mechanistic modeling of prokaryotic mRNA decay. *J Theor Biol* **189:** 195–209.

Carrier, T.A. and J.D. Keasling. 1999. Investigating autocatalytic gene expression systems through mechanistic modeling. *J Theor Biol* **201:** 25–36.

Chalfie, M., J.E. Sulston, J.G. White, E. Southgate, J.N. Thomson, and S. Brenner. 1985. The neural circuit for touch sensitivity in Caenorhabditis elegans. *J Neurosci* **5:** 956–964.

Chow, S.S., C.O. Wilke, C. Ofria, R.E. Lenski, and C. Adami. 2004. Adaptive radiation from resource competition in digital organisms. *Science* **305:** 84–86.

Choy, H.E., S.W. Park, T. Aki, P. Parrack, N. Fujita, A. Ishihama, and S. Adhya. 1995. Repression and activation of transcription by Gal and Lac repressors: involvement of alpha subunit of RNA polymerase. *EMBO J* **14:** 4523–4529.

Christensen, B. and J. Nielsen. 2000. Metabolic network analysis. A powerful tool in metabolic engineering. *Adv Biochem Eng Biotechnol* **66:** 209–231.

Ciliberto, A., B. Novak, and J.J. Tyson. 2003. Mathematical model of the morphogenesis checkpoint in budding yeast. *J Cell Biol* **163:** 1243–1254.

Ciliberto, A., B. Novak, and J.J. Tyson. 2005. Steady states and oscillations in the p53/Mdm2 network. *Cell Cycle* **4:** 488–493.

Cimino, A. and J.-F. Hervagault. 1987. Experimental evidence for a zero-order ultrasensitivity in a simple substrate cycle. *Biochemical and Biophysical Research Communications* **149:** 615–620.

Cluzel, P., M. Surette, and S. Leibler. 2000. An ultrasensitive bacterial motor revealed by monitoring signaling proteins in single cells. *Science* **287:** 1652–1655.

Collado-Vides, J. and R. Hofestadt. 2004. *Gene Regulation and Metabolism : Post-Genomic Computational Approaches.* The MIT Press.

Collado-Vides, J., B. Magasanik, and J.D. Gralla. 1991. Control site location and transcriptional regulation in Escherichia coli. *Microbiol Rev* **55:** 371–394.

Conant, G.C. and A. Wagner. 2003. Convergent evolution of gene circuits. *Nat Genet* **34:** 264–266.

Cook, D.L., A.N. Gerber, and S.J. Tapscott. 1998. Modeling stochastic gene expression: implications for haploinsufficiency. *Proc Natl Acad Sci USA* **95:** 15641–15646.

Crow, J.F. and M. Kimura. 1970. *An Introduction to Population Genetics Theory.* Harper and Row, New York.

Csete, M. and J. Doyle. 2004. Bow ties, metabolism and disease. *Trends Biotechnol* **22:** 446–450.

Csete, M.E. and J.C. Doyle. 2002. Reverse engineering of biological complexity. *Science* **295:** 1664–1669.

Davidson, E.H. 2001. *Genomic Regulatory Systems: Development and Evolution.* Academic Press.

Davidson, E.H., J.P. Rast, P. Oliveri, A. Ransick, C. Calestani, C.H. Yuh, T. Minokawa, G. Amore, V. Hinman, C. Arenas-Mena, O. Otim, C.T. Brown, C.B. Livi, P.Y. Lee, R. Revilla, A.G. Rust, Z. Pan, M.J. Schilstra, P.J. Clarke, M.I. Arnone, L. Rowen, R.A. Cameron, D.R. McClay, L. Hood, and H. Bolouri. 2002. A genomic regulatory network for development. *Science* **295:** 1669–1678.

Dekel, E. and U. Alon. 2005. Optimality and evolutionary tuning of the expression level of a protein. *Nature* **436:** 588–592.

Dekel, E., S. Mangan, and U. Alon. 2005. Environmental selection of the feed-forward loop circuit in gene-regulation networks. *Phys Biol* **2:** 81–88.

Delbruck, M. 1945. Statistical fluctuations in autocatalytic reactions. *J Chem Phys* **8:** 120–124.

Demongeot, J., M. Kaufman, and R. Thomas. 2000. Positive feedback circuits and memory. *C R Acad Sci III* **323:** 69–79.

Dennis, P.P., M. Ehrenberg, and H. Bremer. 2004. Control of rRNA synthesis in Escherichia coli: a systems biology approach. *Microbiol Mol Biol Rev* **68:** 639–668.

Detwiler, P.B., S. Ramanathan, A. Sengupta, and B.I. Shraiman. 2000. Engineering aspects of enzymatic signal transduction: Photoreceptors in the retina. *Biophysical Journal* **79:** 2801–2817.

D'haeseleer, P., S. Liang, and R. Somogyi. 2000. Genetic network inference: From co-expression clustering to reverse engineering. *Bioinformatics* **16:** 707–726.

Dobrin, R., Q.K. Beg, A.L. Barabasi, and Z.N. Oltvai. 2004. Aggregation of topological motifs in the Escherichia coli transcriptional regulatory network. *BMC Bioinformatics* **5:** 10.

Dolmetsch, R.E., K. Xu, and R.S. Lewis. 1998. Calcium oscillations increase the efficiency and specificity of gene expression. *Nature* **392:** 933–936.

Doncic, A., E. Ben-Jacob, and N. Barkai. 2005. Evaluating putative mechanisms of the mitotic spindle checkpoint. *Proc Natl Acad Sci USA* **102:** 6332–6337.

Doyle, J. and M. Csete. 2005. Motifs, control, and stability. *PLoS Biol* **3:** e392.

Droge, P. and B. Muller-Hill. 2001. High local protein concentrations at promoters: strategies in prokaryotic and eukaryotic cells. *Bioessays* **23:** 179–183.

Dubrulle, J. and O. Pourquie. 2002. From head to tail: links between the segmentation clock and antero-posterior patterning of the embryo. *Curr Opin Genet Dev* **12:** 519–523.

Dueber, J.E., B.J. Yeh, R.P. Bhattacharyya, and W.A. Lim. 2004. Rewiring cell signaling: the logic and plasticity of eukaryotic protein circuitry. *Curr Opin Struct Biol* **14:** 690–699.

Duffield, G.E., J.D. Best, B.H. Meurers, A. Bittner, J.J. Loros, and J.C. Dunlap. 2002. Circadian programs of transcriptional activation, signaling, and protein turnover revealed by microarray analysis of mammalian cells. *Curr Biol* **12:** 551–557.

Duke, T.A., N. Le Novere, and D. Bray. 2001. Conformational spread in a ring of proteins: a stochastic approach to allostery. *J Mol Biol* **308:** 541–553.

Dyson, S. and J.B. Gurdon. 1998. The interpretation of position in a morphogen gradient as revealed by occupancy of activin receptors. *Cell* **93:** 557–568.

Durbin, R.M. (1987). Studies on the development and organization of the nervous system of *caenorhabitis elegans*. Ph.D. thesis, www.wormbase.org.

Eichenberger, P., M. Fujita, S.T. Jensen, E.M. Conlon, D.Z. Rudner, S.T. Wang, C. Ferguson, K. Haga, T. Sato, J.S. Liu, and R. Losick. 2004. The program of gene transcription for a single differentiating cell type during sporulation in Bacillus subtilis. *PLoS Biol* **2:** e328.

Eils, R. and C. Athale. 2003. Computational imaging in cell biology. *J Cell Biol* **161:** 477–481.

Eiswirth, M., A. Freund, and J. Ross. 1991. Operational procedure toward the classification of chemical oscillators. *J Phys Chem* **95:** 1294–1299.

El-Samad, H., Doyle, J.C., Gross, C.A., Khammash, M. 2005. Surviving heat shock: control strategies for robustness and performance. *Proc Natl Acad Sci USA* 102: 2736–41.

Eldar, A. and N. Barkai. 2005. Interpreting clone-mediated perturbations of morphogen profiles. *Dev Biol* **278**: 203–207.

Eldar, A., R. Dorfman, D. Weiss, H. Ashe, B.Z. Shilo, and N. Barkai. 2002. Robustness of the BMP morphogen gradient in Drosophila embryonic patterning. *Nature* **419**: 304–308.

Eldar, A., D. Rosin, B.Z. Shilo, and N. Barkai. 2003. Self-enhanced ligand degradation underlies robustness of morphogen gradients. *Dev Cell* **5**: 635–646.

Eldar, A., B.Z. Shilo, and N. Barkai. 2004. Elucidating mechanisms underlying robustness of morphogen gradients. *Curr Opin Genet Dev* **14**: 435–439.

Elena, S.F. and R.E. Lenski. 2003. Evolution experiments with microorganisms: the dynamics and genetic bases of adaptation. *Nat Rev Genet* **4**: 457–469.

Ellington, A.D. and J.W. Szostak. 1990. In vitro selection of RNA molecules that bind specific ligands. *Nature* **346**: 818–822.

Elowitz, M.B. and S. Leibler. 2000. A synthetic oscillatory network of transcriptional regulators. *Nature* **403**: 335–338.

Elowitz, M.B., A.J. Levine, E.D. Siggia, and P.S. Swain. 2002. Stochastic gene expression in a single cell. *Science* **297**: 1183–1186.

Elowitz, M.B., M.G. Surette, P.E. Wolf, J.B. Stock, and S. Leibler. 1999. Protein mobility in the cytoplasm of Escherichia coli. *J Bacteriol* **181**: 197–203.

Endy, D., L. You, J. Yin, and I.J. Molineux. 2000. Computation, prediction, and experimental tests of fitness for bacteriophage T7 mutants with permuted genomes. *Proc Natl Acad Sci USA* **97**: 5375–5380.

England, J.L. and J. Cardy. 2005. Morphogen gradient from a noisy source. *Phys Rev Lett* **94**: 078101.

Enver, T., C.M. Heyworth, and T.M. Dexter. 1998. Do stem cells play dice? *Blood* **92**: 352–351.

Epstein, I.R. 1995. The consequences of imperfect mixing in autocatalytic chemical and biological systems. *Nature* **374**: 321–327.

Erdos, P. and A. Renyi. 1959. On random graphs. *Publ. Math. (Debrecen)* **6**: 290.

Ettema, T., J. Van der Oost, and M. Huynen. 2001. Modularity in the gain and loss of genes: Applications for function prediction. *Trends in Genetics* **17**: 485–487.

Fall, C., E. Marland, J. Wagner, and J. Tyson. 2005. *Computational Cell Biology*. Springer.

Fell, D. 2003. *Understanding the control of metabolism*. Portland Press.

Ferrell, J.E. 1997. How responses get more switch-like as you move down a protein kinase cascade. *Trends in biochemical sciences* **22**: 288–289.

Ferrell, J.E. 2002. Self-perpetuating states in signal transduction: Positive feedback, double-negative feedback and bistability. *Current Opinion in Cell Biology* **14**: 140–148.

Ferrell, J.E., Jr. and E.M. Machleder. 1998. The biochemical basis of an all-or-none cell fate switch in Xenopus oocytes. *Science* **280**: 895–898.

Ferrell, J.E. and W. Xiong. 2001. Bistability in cell signaling: How to make continuous processes discontinuous, and reversible processes irreversible. *Chaos* **11**: 227–236.

Fiering, S., E. Whitelaw, and D.I. Martin. 2000. To be or not to be active: the stochastic nature of enhancer action. *Bioessays* **22**: 381–387.

Fischer, E. and U. Sauer. 2005. Large-scale in vivo flux analysis shows rigidity and suboptimal performance of Bacillus subtilis metabolism. *Nat Genet* **37**: 636–640.

Fisher, J., N. Piterman, E.J. Hubbard, M.J. Stern, and D. Harel. 2005. Computational insights into Caenorhabditis elegans vulval development. *Proc Natl Acad Sci USA* **102**: 1951–1956.

Forger, D.B. and C.S. Peskin. 2005. Stochastic simulation of the mammalian circadian clock. *Proc Natl Acad Sci USA* **102**: 321–324.

Fraser, H.B., A.E. Hirsh, G. Giaever, J. Kumm, and M.B. Eisen. 2004. Noise minimization in eukaryotic gene expression. *PLoS Biol.* **2:** e137.

Friedman, N., M. Linial, I. Nachman, and D. Pe'er. 2000. Using Bayesian networks to analyze expression data. *Journal of Computational Biology* **7:** 601–620.

Garcia-Ojalvo, J., M.B. Elowitz, and S.H. Strogatz. 2004. Modeling a synthetic multicellular clock: repressilators coupled by quorum sensing. *Proc Natl Acad Sci USA* **101:** 10955–10960.

Gardiner , C.W. 2004. *Handbook of Stochastic Methods*. Springer.

Gardner, T.S., C.R. Cantor, and J.J. Collins. 2000. Construction of a genetic toggle switch in Escherichia coli. *Nature* **403:** 339–342.

Gause, G.F. 1934. *The struggle for existence*. Dover Phoenix editions.

Gerhart, J. and M. Kirschner. 1997. *Cells, embryos and evolution*. Blackwell Science, Oxford.

Gerland, U., J.D. Moroz, and T. Hwa. 2002. Physical constraints and functional characteristics of transcription factor-DNA interaction. *Proc Natl Acad Sci USA* **99:** 12015–12020.

Gerlich, D. and J. Ellenberg. 2003. 4D imaging to assay complex dynamics in live specimens. *Nat Cell Biol* **Suppl:** S14–19.

Getz, G., E. Levine, and E. Domany. 2000. Coupled two-way clustering analysis of gene microarray data. *Proc Natl Acad Sci USA* **97:** 12079–12084.

Ghaemmaghami, S., W.K. Huh, K. Bower, R.W. Howson, A. Belle, N. Dephoure, E.K. O'Shea, and J.S. Weissman. 2003. Global analysis of protein expression in yeast. *Nature* **425:** 737–741.

Ghosh, B., R. Karmakar, and I. Bose. 2005. Noise characteristics of feed forward loops. *Phys Biol* **2:** 36–45.

Glass, L. and S.A. Kauffman. 1973. The logical analysis of continuous, non-linear biochemical control networks. *J Theor Biol* **39:** 103–129.

Goldbeter, A. 2002. Computational approaches to cellular rhythms. *Nature* **420:** 238–245.

Goldbeter, A., G. Dupont, and M.J. Berridge. 1990. Minimal model for signal-induced Ca2+ oscillations and for their frequency encoding through protein phosphorylation. *Proc Natl Acad Sci USA* **87:** 1461–1465.

Goldbeter, A. and D.E. Koshland Jr. 1981. An amplified sensitivity arising from covalent modification in biological systems. *Proc Natl Acad Sci USA* **78:** 6840–6844.

Goldbeter, A. and D.E. Koshland Jr. 1984. Ultrasensitivity in biochemical systems controlled by covalent modification. Interplay between zero-order and multistep effects. *J Biol Chem* **259:** 14441–14447.

Goldstein, B., J.R. Faeder, and W.S. Hlavacek. 2004. Mathematical and computational models of immune-receptor signalling. *Nat Rev Immunol* **4:** 445–456.

Gonze, D., J. Halloy, and A. Goldbeter. 2002a. Robustness of circadian rhythms with respect to molecular noise. *Proc Natl Acad Sci USA* **99:** 673–678.

Gonze, D., M.R. Roussel, and A. Goldbeter. 2002b. A model for the enhancement of fitness in cyanobacteria based on resonance of a circadian oscillator with the external light-dark cycle. *J Theor Biol* **214:** 577–597.

Goulian, M. 2004. Robust control in bacterial regulatory circuits. *Curr Opin Microbiol* **7:** 198–202.

Guelzim, N., S. Bottani, P. Bourgine, and F. Kepes. 2002. Topological and causal structure of the yeast transcriptional regulatory network. *Nat Genet* **31:** 60–63.

Guet, C.C., M.B. Elowitz, W. Hsing, and S. Leibler. 2002. Combinatorial synthesis of genetic networks. *Science* **296:** 1466–1470.

Guimera, R. and L.A. Nunes Amaral. 2005. Functional cartography of complex metabolic networks. *Nature* **433:** 895–900.

Gunsalus, K.C., H. Ge, A.J. Schetter, D.S. Goldberg, J.D. Han, T. Hao, G.F. Berriz, N. Bertin, J. Huang, L.S. Chuang, N. Li, R. Mani, A.A. Hyman, B. Sonnichsen, C.J. Echeverri, F.P. Roth,

M. Vidal, and F. Piano. 2005. Predictive models of molecular machines involved in Cae-norhabditis elegans early embryogenesis. *Nature* **436:** 861–865.

Han, J.D., N. Bertin, T. Hao, D.S. Goldberg, G.F. Berriz, L.V. Zhang, D. Dupuy, A.J. Walhout, M.E. Cusick, F.P. Roth, and M. Vidal. 2004. Evidence for dynamically organized modularity in the yeast protein-protein interaction network. *Nature* **430:** 88–93.

Hannenhalli, S. and S. Levy. 2002. Predicting transcription factor synergism. *Nucleic Acids Research* **30:** 4278–4284.

Harbison, C.T., D.B. Gordon, T.I. Lee, N.J. Rinaldi, K.D. Macisaac, T.W. Danford, N.M. Hannett, J.B. Tagne, D.B. Reynolds, J. Yoo, E.G. Jennings, J. Zeitlinger, D.K. Pokholok, M. Kellis, P.A. Rolfe, K.T. Takusagawa, E.S. Lander, D.K. Gifford, E. Fraenkel, and R.A. Young. 2004. Transcriptional regulatory code of a eukaryotic genome. *Nature* **431:** 99–104.

Hartl, D.L. and A.G. Clark. 1997. *Principles of Population Genetics.* Sinauer Associates Inc., Sunderland, MA.

Hartl, D.L. and Dykhuizen, D.E. 1984. The population genetics of *Escherichia coli. Ann Rev Genet.* 18: 31–68.

Hartl, D.L., E.N. Moriyama, and S.A. Sawyer. 1994. Selection intensity for codon bias. *Genetics* **138:** 227–234.

Hartwell, L.H., J.J. Hopfield, S. Leibler, and A.W. Murray. 1999. From molecular to modular cell biology. *Nature* **402:** C47–52.

Hasty, J., D. McMillen, and J.J. Collins. 2002. Engineered gene circuits. *Nature* **420:** 224–230.

Hasty, J., J. Pradines, M. Dolnik, and J.J. Collins. 2000. Noise-based switches and amplifiers for gene expression. *Proc Natl Acad Sci USA* **97:** 2075–2080.

Hayot, F. and C. Jayaprakash. 2005. A feedforward loop motif in transcriptional regulation: induction and repression. *J Theor Biol* **234:** 133–143.

Heinrich, R. and H.G. Holzhutter. 1985. Efficiency and design of simple metabolic systems. *Biomed Biochim Acta* **44:** 959–969.

Heinrich, R. and E. Klipp. 1996. Control analysis of unbranched enzymatic chains in states of maximal activity. *J Theor Biol* **182:** 243–252.

Heinrich, R., B.G. Neel, and T.A. Rapoport. 2002. Mathematical models of protein kinase signal transduction. *Mol Cell* **9:** 957–970.

Heinrich, R. and S. Schuster. 1996. *The Regulation of Cellular Systems.* Kluwer Academic Publishers.

Hershko, A. and A. Ciechanover. 1998. The ubiquitin system. *Annu Rev Biochem* **67:** 425–479.

Hertz, J., A. Krogh, and R.G. Palmer. 1991. *Introduction to the Theory of Neural Computation.* Perseus Books.

Hlavacek, W.S., A. Redondo, C. Wofsy, and B. Goldstein. 2002. Kinetic proofreading in receptor-mediated transduction of cellular signals: receptor aggregation, partially activated receptors, and cytosolic messengers. *Bull Math Biol* **64:** 887–911.

Hoffmann, A., A. Levchenko, M.L. Scott, and D. Baltimore. 2002. The IkappaB-NF-kappaB signaling module: temporal control and selective gene activation. *Science* **298:** 1241–1245.

Holland, P.W. and S. Leinhardt. 1975. Local structure in social networks. *Sociological Methodology.* Heise, D.R. (Ed) p. 1–45. Josey-Bass.

Hood, L. and D. Galas. 2003. The digital code of DNA. *Nature* **421:** 444–448.

Hooshangi, S., S. Thiberge, and R. Weiss. 2005. Ultrasensitivity and noise propagation in a synthetic transcriptional cascade. *Proc Natl Acad Sci USA* **102:** 3581–3586.

Hope, A.I., Ed. (1999). *C. elegans: A Practical Approach,* 1st ed. Oxford University Press.

Hopfield, J.J. 1974. Kinetic proofreading: a new mechanism for reducing errors in biosynthetic processes requiring high specificity. *Proc Natl Acad Sci USA* **71:** 4135–4139.

Hopfield, J.J., T. Yamane, V. Yue, and S.M. Coutts. 1976. Direct experimental evidence for kinetic proofreading in amino acylation of tRNAIle. *Proc Natl Acad Sci USA* **73:** 1164–1168.

Hornberg, J.J., F.J. Bruggeman, B. Binder, C.R. Geest, A.J. de Vaate, J. Lankelma, R. Heinrich, and H.V. Westerhoff. 2005. Principles behind the multifarious control of signal transduction. ERK phosphorylation and kinase/phosphatase control. *FEBS J* **272**: 244–258.

Houchmandzadeh, B., E. Wieschaus, and S. Leibler. 2002. Establishment of developmental precision and proportions in the early Drosophila embryo. *Nature* **415**: 798–802.

Huang, C.-Y.F. and J.E. Ferrell Jr. 1996. Ultrasensitivity in the mitogen-activated protein kinase cascade. *Proc Natl Acad Sci USA* **93**: 10078–10083.

Hume, D.A. 2000. Probability in transcriptional regulation and its implications for leukocyte differentiation and inducible gene expression. *Blood* **96**: 2323–2328.

Huynen, M.A. and E. van Nimwegen. 1998. The frequency distribution of gene family sizes in complete genomes. *Mol Biol Evol* **15**: 583–589.

Ibarra, R.U., J.S. Edwards, and B.O. Palsson. 2002. Escherichia coli K-12 undergoes adaptive evolution to achieve in silico predicted optimal growth. *Nature* **420**: 186–189.

Ideker, T. and D. Lauffenburger. 2003. Building with a scaffold: emerging strategies for high- to low-level cellular modeling. *Trends Biotechnol* **21**: 255–262.

Ideker, T., O. Ozier, B. Schwikowski, and A.F. Siegel. 2002. Discovering regulatory and signalling circuits in molecular interaction networks. *Bioinformatics* **18**.

Ideker, T., V. Thorsson, J.A. Ranish, R. Christmas, J. Buhler, J.K. Eng, R. Bumgarner, D.R. Goodlett, R. Aebersold, and L. Hood. 2001. Integrated genomic and proteomic analyses of a systematically perturbed metabolic network. *Science* **292**: 929–934.

Ideker, T.E., V. Thorsson, and R.M. Karp. 2000. Discovery of regulatory interactions through perturbation: inference and experimental design. *Pac Symp Biocomput.* 305–316.

Ihmels, J., S. Bergmann, and N. Barkai. 2004a. Defining transcription modules using large-scale gene expression data. *Bioinformatics* **20**: 1993–2003.

Ihmels, J., S. Bergmann, M. Gerami-Nejad, I. Yanai, M. McClellan, J. Berman, and N. Barkai. 2005. Rewiring of the yeast transcriptional network through the evolution of motif usage. *Science* **309**: 938–940.

Ihmels, J., G. Friedlander, S. Bergmann, O. Sarig, Y. Ziv, and N. Barkai. 2002. Revealing modular organization in the yeast transcriptional network. *Nature Genetics* **31**: 370–377.

Ihmels, J., R. Levy, and N. Barkai. 2004b. Principles of transcriptional control in the metabolic network of Saccharomyces cerevisiae. *Nat Biotechnol* **22**: 86–92.

Isaacs, F.J., J. Hasty, C.R. Cantor, and J.J. Collins. 2003. Prediction and measurement of an autoregulatory genetic module. *Proc Natl Acad Sci USA* **100**: 7714–7719.

Ishihara, A., J.E. Segall, S.M. Block, and H.C. Berg. 1983. Coordination of flagella on filamentous cells of Escherichia coli. *J Bacteriol* **155**: 228–237.

Ishihara, S., K. Fujimoto, and T. Shibata. 2005. Cross talking of network motifs in gene regulation that generates temporal pulses and spatial stripes. *Genes Cells* **10**: 1025–1038.

Itzkovitz, S. and U. Alon. 2005. Subgraphs and network motifs in geometric networks. *Phys Rev E* **71**: 026117.

Itzkovitz, S., R. Levitt, N. Kashtan, R. Milo, M. Itzkovitz, and U. Alon. 2005. Coarse-graining and self-dissimilarity of complex networks. *Phys Rev E* **71**: 016127.

Itzkovitz, S., R. Milo, N. Kashtan, G. Ziv, and U. Alon. 2003. Subgraphs in random networks. *Phys Rev E* **68**: 026127.

Jablanka, E. and A. Regev. 1995. Gene number, methylation and biological complexity. *Trends Genet.* **11**: 383–384.

Janes, K.A., J.G. Albeck, S. Gaudet, P.K. Sorger, D.A. Lauffenburger, and M.B. Yaffe. 2005. A systems model of signaling identifies a molecular basis set for cytokine-induced apoptosis. *Science* **310**: 1646–1653.

Jasuja, R., J. Keyoung, G.P. Reid, D.R. Trentham, and S. Khan. 1999. Chemotactic responses of Escherichia coli to small jumps of photoreleased L-aspartate. *Biophys J* **76**: 1706–1719.

Jeong, H., R. Albert, A.-L. Barabasi, B. Tombor, and Z.N. Oltval. 2000. The large-scale organization of metabolic networks. *Nature* **407**: 651–654.

Johnston, R.J., Jr., S. Chang, J.F. Etchberger, C.O. Ortiz, and O. Hobert. 2005. MicroRNAs acting in a double-negative feedback loop to control a neuronal cell fate decision. *Proc Natl Acad Sci USA* **102**: 12449–12454.

Kacser, H. and Burns, J.A. 1973. The control of flux. *Symp Soc Exp Biol.* **27**: 65–104.

Kaern, M., T.C. Elston, W.J. Blake, and J.J. Collins. 2005. Stochasticity in gene expression: from theories to phenotypes. *Nat Rev Genet* **6**: 451–464.

Kafri, R., A. Bar-Even, and Y. Pilpel. 2005. Transcription control reprogramming in genetic backup circuits. *Nat Genet* **37**: 295–299.

Kalir, S. and U. Alon. 2004. Using a quantitative blueprint to reprogram the dynamics of the flagella gene network. *Cell* **117**: 713–720.

Kalir, S., S. Mangan, and U. Alon, 2005. The coherent feed-forward loop with a SUM input function protects flagella production in Escherichia coli. *Molecular Systems Biology* **doi: 10.1038:** msb4100010.

Kalir, S., J. McClure, K. Pabbaraju, C. Southward, M. Ronen, S. Leibler, M.G. Surette, and U. Alon. 2001. Ordering genes in a flagella pathway by analysis of expression kinetics from living bacteria. *Science* **292**: 2080–2083.

Kao, K.C., Y.L. Yang, R. Boscolo, C. Sabatti, V. Roychowdhury, and J.C. Liao. 2004. Transcriptome-based determination of multiple transcription regulator activities in Escherichia coli by using network component analysis. *Proc Natl Acad Sci USA* **101**: 641–646.

Kashtan, N. and U. Alon. 2005. Spontaneous evolution of modularity and network motifs. *Proc Natl Acad Sci USA* **102**: 13773–13778.

Kashtan, N., S. Itzkovitz, R. Milo, and U. Alon. 2004a. Efficient sampling algorithm for estimating subgraph concentrations and detecting network motifs. *Bioinformatics* **20**: 1746–1758.

Kashtan, N., S. Itzkovitz, R. Milo, and U. Alon. 2004b. Topological generalizations of network motifs. *Phys Rev E* **70**: 031909.

Kepler, T.B. and T.C. Elston. 2001. Stochasticity in transcriptional regulation: origins, consequences, and mathematical representations. *Biophys J* **81**: 3116–3136.

Kerr, B., M.A. Riley, M.W. Feldman, and B.J.M. Bohannan. 2002. Local dispersal promotes biodiversity in a real-life game of rock-paper-scissors. *Nature* **418**: 171–174.

Kerszberg, M. 2004. Noise, delays, robustness, canalization and all that. *Curr Opin Genet. Dev* **14**: 440–445.

Keymer, J.E., R.G. Endres, M. Skoge, Y. Meir, and N.S. Wingreen. 2006. Chemosensing in Escherichia coli: Two regimes of two-state receptors. *Proc Natl Acad Sci USA* **103**: 1786–1791.

Kholodenko, B.N., S. Schuster, J. Garcia, H.V. Westerhoff, and M. Cascante. 1998. Control analysis of metabolic systems involving quasi-equilibrium reactions. *Biochim Biophys Acta* **1379**: 337–352.

Kierzek, A.M., J. Zaim, and P. Zielenkiewicz. 2001. The effect of transcription and translation initiation frequencies on the stochastic fluctuations in prokaryotic gene expression. *J. Biol. Chem.* **276**: 8165–8172.

Kirschner, M. and J. Gerhart. 1998. Evolvability. *Proc Natl Acad Sci USA* **95**: 8420–8427.

Kirschner, M.W. and J.C. Gerhart. 2005. *The Plausibility of Life: Resolving Darwin's Dilemma.* Yale University Press.

Kishony, R. and S. Leibler. 2003. Environmental stresses can alleviate the average deleterious effect of mutations. *J Biol* **2**: 14.

Kitano, H. 2001. *Foundations of Systems Biology.* The MIT Press.

Kitano, H. 2004. Biological robustness. *Nat Rev Genet* **5**: 826–837.

Kitano, H. 2006. Robustness from top to bottom. *Nat Genet* **38**: 133.

Kitano, H., K. Oda, T. Kimura, Y. Matsuoka, M. Csete, J. Doyle, and M. Muramatsu. 2004. Metabolic syndrome and robustness tradeoffs. *Diabetes* **53 Suppl 3:** S6–S15.

Klipp, E., R. Heinrich, and H.G. Holzhutter. 2002. Prediction of temporal gene expression. Metabolic opimization by re-distribution of enzyme activities. *Eur J Biochem* **269:** 5406–5413.

Klipp, E., R. Herwig, A. Kowald, C. Wierling, and H. Lehrach. 2005. *Systems Biology in Practice : Concepts, Implementation and Application.* John Wiley & Sons.

Kloster, M., C. Tang, and N.S. Wingreen. 2005. Finding regulatory modules through large-scale gene-expression data analysis. *Bioinformatics* **21:** 1172–1179.

Kmita, M. and D. Duboule. 2003. Organizing axes in time and space; 25 years of colinear tinkering. *Science* **301:** 331–333.

Knox, B.E., P.N. Devreotes, A. Goldbeter, and L.A. Segel. 1986. A molecular mechanism for sensory adaptation based on ligand-induced receptor modification. *Proc Natl Acad Sci USA* **83:** 2345–2349.

Ko, M.S. 1991. A stochastic model for gene induction. *J Theor Biol* **153:** 181–194.

Ko, M.S. 1992. Induction mechanism of a single gene molecule: stochastic or deterministic? *Bioessays* **14:** 341–346.

Kobayashi, H. 2004. Programmable cells: interfacing natural and engineered gene networks. *Proc Natl Acad Sci USA* **101:** 8414–8419.

Kolch, W., M. Calder, and D. Gilbert. 2005. When kinases meet mathematics: the systems biology of MAPK signalling. *FEBS Lett* **579:** 1891–1895.

Kollmann, M., L. Lovdok, K. Bartholome, J. Timmer, and V. Sourjik. 2005. Design principles of a bacterial signalling network. *Nature* **438:** 504–507.

Korobkova, E., T. Emonet, J.M. Vilar, T.S. Shimizu, and P. Cluzel. 2004. From molecular noise to behavioural variability in a single bacterium. *Nature* **428:** 574–578.

Kramer, B.P. 2004. An engineered epigenetic transgene switch in mammalian cells. *Nature Biotechnol* **22:** 867–870.

Kramer, B.P. and M. Fussenegger. 2005. Hysteresis in a synthetic mammalian gene network. *Proc Natl Acad Sci USA* **102:** 9517–9522.

Kriete, A. and R. Eils. 2005. *Computational Systems Biology.* Academic Press.

Kuang, Y., I. Biran, and D.R. Walt. 2004. Simultaneously monitoring gene expression kinetics and genetic noise in single cells by optical well arrays. *Anal Chem* **76:** 6282–6286.

Kupiec, J.J. 1997. A Darwinian theory for the origin of cellular differentiation. *Mol Gen Genet* **255:** 201–208.

Kurakin, A. 2005. Self-organization vs Watchmaker: stochastic gene expression and cell differentiation. *Dev Genes Evol* **215:** 46–52.

Kussell, E., R. Kishony, N.Q. Balaban, and S. Leibler. 2005. Bacterial persistence: a model of survival in changing environments. *Genetics* **169:** 1807–1814.

Kussell, E. and S. Leibler. 2005. Phenotypic diversity, population growth, and information in fluctuating environments. *Science* **309:** 2075–2078.

Lahav, G., N. Rosenfeld, A. Sigal, N. Geva-Zatorsky, A.J. Levine, M.B. Elowitz, and U. Alon. 2004. Dynamics of the p53-Mdm2 feedback loop in individual cells. *Nat Genet* **36:** 147–150.

Lamphier, M.S. and M. Ptashne. 1992. Multiple mechanisms mediate glucose repression of the yeast GAL1 gene. *Proc Natl Acad Sci USA* **89:** 5922–5926.

Laub, M.T., H.H. McAdams, T. Feldblyum, C.M. Fraser, and L. Shapiro. 2000. Global analysis of the genetic network controlling a bacterial cell cycle. *Science* **290:** 2144–2148.

Lauffenburger, D.A. 2000. Cell signaling pathways as control modules: Complexity for simplicity? *Proc Natl Acad Sci USA* **97:** 5031–5033.

Lawrence, P.A. (1995). *The Making of a Fly: The Genetics of Animal Design.* Blackwell Science, Ltd., The Alden Press.

Lazebnik, Y. 2002. Can a biologist fix a radio? -- Or, what I learned while studying apoptosis. *Cancer Cell*. 3:179–82.

Lee, T.I., N.J. Rinaldi, F. Robert, D.T. Odom, Z. Bar-Joseph, G.K. Gerber, N.M. Hannett, C.T. Harbison, C.M. Thompson, I. Simon, J. Zeitlinger, E.G. Jennings, H.L. Murray, D.B. Gordon, B. Ren, J.J. Wyrick, J.B. Tagne, T.L. Volkert, E. Fraenkel, D.K. Gifford, and R.A. Young. 2002. Transcriptional regulatory networks in Saccharomyces cerevisiae. *Science* **298:** 799–804.

Legewie, S., N. Bluthgen, R. Schafer, and H. Herzel. 2005. Ultrasensitization: switch-like regulation of cellular signaling by transcriptional induction. *PLoS Comput Biol* **1:** e54.

Lenski, R.E. 1998. Bacterial evolution and the cost of antibiotic resistance. *Int Microbiol* **1:** 265–270.

Lenski, R.E., J.A. Mongold, P.D. Sniegowski, M. Travisano, F. Vasi, P.J. Gerrish, and T.M. Schmidt. 1998. Evolution of competitive fitness in experimental populations of E. coli: what makes one genotype a better competitor than another? *Antonie Van Leeuwenhoek* **73:** 35–47.

Lenski, R.E., C. Ofria, T.C. Collier, and C. Adami. 1999. Genome complexity, robustness and genetic interactions in digital organisms. *Nature* **400:** 661–664.

Lenski, R.E., C. Ofria, R.T. Pennock, and C. Adami. 2003. The evolutionary origin of complex features. *Nature* **423:** 139–144.

Lev Bar-Or, R. 2000. Generation of oscillations by the p53-Mdm2 feedback loop: a theoretical and experimental study. *Proc Natl Acad Sci USA* **97:** 11250–11255.

Levchenko, A., J. Bruck, and P.W. Sternberg. 2000. Scaffold proteins may biphasically affect the levels of mitogen-activated protein kinase signaling and reduce its threshold properties. *Proc Natl Acad Sci USA* **97:** 5818–5823.

Levin, B.R. 2004. Microbiology. Noninherited resistance to antibiotics. *Science* **305:** 1578–1579.

Levin, M.D. 2003. Noise in gene expression as the source of non-genetic individuality in the chemotactic response of Escherichia coli. *FEBS Lett.* **550:** 135–138.

Levine, M. and E.H. Davidson. 2005. Gene regulatory networks for development. *Proc Natl Acad Sci USA* **102:** 4936–4942.

Levskaya, A., A.A. Chevalier, J.J. Tabor, Z.B. Simpson, L.A. Lavery, M. Levy, E.A. Davidson, A. Scouras, A.D. Ellington, E.M. Marcotte, and C.A. Voigt. 2005. Synthetic biology: engineering Escherichia coli to see light. *Nature* **438:** 441–442.

Lewis, J. 2003. Autoinhibition with transcriptional delay: a simple mechanism for the zebrafish somitogenesis oscillator. *Curr Biol* **13:** 1398–1408.

Li, F., T. Long, Y. Lu, Q. Ouyang, and C. Tang. 2004. The yeast cell-cycle network is robustly designed. *Proc Natl Acad Sci USA* **101:** 4781–4786.

Lipson, H., J.B. Pollack, and N.P. Suh. 2002. On the origin of modular variation. *Evolution Int J Org Evolution* **56:** 1549–1556.

Longabaugh, W.J., E.H. Davidson, and H. Bolouri. 2005. Computational representation of developmental genetic regulatory networks. *Dev Biol* **283:** 1–16.

Louis, M. and A. Becskei. 2002. Binary and graded responses in gene networks. *Sci. STKE* **2002:** PE33.

Luscombe, N.M., M.M. Babu, H. Yu, M. Snyder, S.A. Teichmann, and M. Gerstein. 2004. Genomic analysis of regulatory network dynamics reveals large topological changes. *Nature* **431:** 308–312.

Ma, H.W., J. Buer, and A.P. Zeng. 2004a. Hierarchical structure and modules in the Escherichia coli transcriptional regulatory network revealed by a new top-down approach. *BMC Bioinformatics* **5:** 199.

Ma, H.W., B. Kumar, U. Ditges, F. Gunzer, J. Buer, and A.P. Zeng. 2004b. An extended transcriptional regulatory network of Escherichia coli and analysis of its hierarchical structure and network motifs. *Nucleic Acids Res* **32:** 6643–6649.

Ma, L., J. Wagner, J.J. Rice, W. Hu, A.J. Levine, and G.A. Stolovitzky. 2005. A plausible model for the digital response of p53 to DNA damage. *Proc Natl Acad Sci USA* **102:** 14266–14271.

Ma'ayan, A., R.D. Blitzer, and R. Iyengar. 2005a. Toward predictive models of Mammalian cells. *Annu Rev Biophys Biomol Struct* **34:** 319–349.

Ma'ayan, A., S.L. Jenkins, S. Neves, A. Hasseldine, E. Grace, B. Dubin-Thaler, N.J. Eungdamrong, G. Weng, P.T. Ram, J.J. Rice, A. Kershenbaum, G.A. Stolovitzky, R.D. Blitzer, and R. Iyengar. 2005b. Formation of regulatory patterns during signal propagation in a Mammalian cellular network. *Science* **309:** 1078–1083.

Macnab, R.M. 2003. How bacteria assemble flagella. *Annu Rev Microbiol* **57:** 77–100.

Maloney, P.C. and B. Rotman. 1973. Distribution of suboptimally induces-D-galactosidase in Escherichia coli. The enzyme content of individual cells. *J Mol Biol* **73:** 77–91.

Mangan, S. and U. Alon. 2003. Structure and function of the feed-forward loop network motif. *Proc Natl Acad Sci USA* **100:** 11980–11985.

Mangan, S., S. Itzkovitz, A. Zaslaver, and U. Alon. 2006. The Incoherent Feed-forward Loop Accelerates the Response-time of the gal System of Escherichia coli. *J Mol Biol* **356:** 1073–1081.

Mangan, S., A. Zaslaver, and U. Alon. 2003. The coherent feedforward loop serves as a sign-sensitive delay element in transcription networks. *J Mol Biol* **334:** 197–204.

Maslov, S. and K. Sneppen. 2002. Specificity and stability in topology of protein networks. *Science* **296:** 910–913.

Maslov, S. and K. Sneppen. 2004. Detection of topological patterns in protein networks. *Genet Eng* **26:** 33–47.

Maslov, S. and K. Sneppen. 2005. Computational architecture of the yeast regulatory network. *Phys Biol* **2:** S94–S100.

Maslov, S., K. Sneppen, K.A. Eriksen, and K.K. Yan. 2004. Upstream plasticity and downstream robustness in evolution of molecular networks. *BMC Evol Biol* **4:** 9.

Maughan, H. and W.L. Nicholson. 2004. Stochastic processes influence stationary-phase decisions in Bacillus subtilis. *J Bacteriol* **186:** 2212–2214.

Maynard Smith, J. 1982. *Evolution and the Theory of Games.* Cambridge University Press.

Mayo, E., Y. Setty, S. Chalamish, A. Zaslaver, and U. Alon. 2006. Plasticity of the cis-Regulatory Input Function of a Gene. *PLoS Biol.*

Mazurie, A., S. Bottani, and M. Vergassola. 2005. An evolutionary and functional assessment of regulatory network motifs. *Genome Biol* **6:** R35.

McAdams, H.H. and A. Arkin. 1997. Stochastic mechanisms in gene expression. *Proc Natl Acad Sci USA* **94:** 814–819.

McAdams, H.H. and A. Arkin. 1999. It's a noisy business! Genetic regulation at the nanomolar scale. *Trends Genet* **15:** 65–69.

McAdams, H.H. and L. Shapiro. 2003. A bacterial cell-cycle regulatory network operating in time and space. *Science* **301:** 1874–1877.

McAdams, H.H., B. Srinivasan, and A.P. Arkin. 2004. The evolution of genetic regulatory systems in bacteria. *Nat Rev Genet* **5:** 169–178.

McKeithan, T.W. 1995. Kinetic proofreading in T-cell receptor signal transduction. *Proc Natl Acad Sci USA* **92:** 5042–5046.

McMillen, D., N. Kopell, J. Hasty, and J.J. Collins. 2002. Synchronizing genetic relaxation oscillators by intercell signaling. *Proc Natl Acad Sci USA* **99:** 679–684.

Mello, B.A., L. Shaw, and Y. Tu. 2004. Effects of receptor interaction in bacterial chemotaxis. *Biophys J* **87:** 1578–1595.

Mendes, P., D. Camacho, and A. de la Fuente. 2005. Modelling and simulation for metabolomics data analysis. *Biochem Soc Trans* **33:** 1427–1429.

Metzler, R. 2001. The future is noisy: the role of spatial fluctuations in genetic switching. *Phys Rev Lett* **8706:** 068103.

Meyer, T. and M.N. Teruel. 2003. Fluorescence imaging of signaling networks. *Trends Cell Biol* **13:** 101–106.

Mihalcescu, I., W. Hsing, and S. Leibler. 2004. Resilient circadian oscillator revealed in individual cyanobacteria. *Nature* **430:** 81–85.

Milo, R., S. Itzkovitz, N. Kashtan, R. Levitt, S. Shen-Orr, I. Ayzenshtat, M. Sheffer, and U. Alon. 2004. Superfamilies of evolved and designed networks. *Science* **303:** 1538–1542.

Milo, R., S. Shen-Orr, S. Itzkovitz, N. Kashtan, D. Chklovskii, and U. Alon. 2002. Network motifs: Simple building blocks of complex networks. *Science* **298:** 824–827.

Mittler, J.E. 1996. Evolution of the genetic switch in temperate bacteriophage. I. Basic theory. *J Theor Biol* **179:** 161–172.

Monk, N.A. 2003. Oscillatory expression of Hes1, p53, and NF-[kappa]B driven by transcriptional time delays. *Curr Biol* **13:** 1409–1413.

Monod, J., A.M. Pappenheimer, Jr., and G. Cohen-Bazire. 1952. The kinetics of the biosynthesis of beta-galactosidase in Escherichia coli as a function of growth. *Biochim Biophys Acta* **9:** 648–660.

Monod, J., J. Wyman, and J.P. Changeux. 1965. On the Nature of Allosteric Transitions: A Plausible Model. *J Mol Biol* **12:** 88–118.

Monsalve, M., M. Mencia, F. Rojo, and M. Salas. 1996. Activation and repression of transcription at two different phage phi29 promoters are mediated by interaction of the same residues of regulatory protein p4 with RNA polymerase. *Embo J* **15:** 383–391.

Moran, N.A. 2002. Microbial minimalism: genome reduction in bacterial pathogens. *Cell* **108:** 583–586.

Moran, N.A. 2003. Tracing the evolution of gene loss in obligate bacterial symbionts. *Curr Opin Microbiol* **6:** 512–518.

Morishita, Y. and K. Aihara. 2004. Noise-reduction through interaction in gene expression and biochemical reaction processes. *J Theor Biol* **228:** 315–325.

Morohashi, M., A.E. Winn, M.T. Borisuk, H. Bolouri, J. Doyle, and H. Kitano. 2002. Robustness as a measure of plausibility in models of biochemical networks. *J Theor Biol* **216:** 19–30.

Murray, J.D. 2004. *Mathematical Biology.* Springer.

Nelson, D.E., A.E. Ihekwaba, M. Elliott, J.R. Johnson, C.A. Gibney, B.E. Foreman, G. Nelson, V. See, C.A. Horton, D.G. Spiller, S.W. Edwards, H.P. McDowell, J.F. Unitt, E. Sullivan, R. Grimley, N. Benson, D. Broomhead, D.B. Kell, and M.R. White. 2004. Oscillations in NF-kappaB signaling control the dynamics of gene expression. *Science* **306:** 704–708.

Newman, M.E. 2004. Fast algorithm for detecting community structure in networks. *Phys Rev E* **69:** 066133.

Newman, M.E. and M. Girvan. 2004. Finding and evaluating community structure in networks. *Phys Rev E* **69:** 026113.

Newman, M.E., S.H. Strogatz, and D.J. Watts. 2001. Random graphs with arbitrary degree distributions and their applications. *Phys Rev E* **64:** 026118.

Nguyen, T.N., Q.G. Phan, L.P. Duong, K.P. Bertrand, and R.E. Lenski. 1989. Effects of carriage and expression of the Tn10 tetracycline-resistance operon on the fitness of Escherichia coli K12. *Mol Biol Evol* **6:** 213–225.

Noble, D. 2002. Modeling the heart--from genes to cells to the whole organ. *Science* **295:** 1678–1682.

Noble, D. 2004. Modeling the heart. *Physiology (Bethesda)* **19:** 191–197.

Novak, B., A. Csikasz-Nagy, B. Gyorffy, K. Chen, and J.J. Tyson. 1998. Mathematical model of the fission yeast cell cycle with checkpoint controls at the G1/S, G2/M and metaphase/anaphase transitions. *Biophysical Chemistry* **72:** 185–200.

Novick, A. and M. Weiner. 1957. Enzyme induction as an all-or-none phenomenon. *Proc Natl Acad Sci USA* **43:** 553–566.

Nowak, M.A. and K. Sigmund. 2002. Biodiversity: Bacterial game dynamics. *Nature* **418:** 138–139.

Nowak, M.A. and K. Sigmund. 2004. Evolutionary dynamics of biological games. *Science* **303:** 793–799.

Ochman, H., J.G. Lawrence, and E.A. Groisman. 2000. Lateral gene transfer and the nature of bacterial innovation. *Nature* **405:** 299–304.

Odom, D.T., N. Zizlsperger, D.B. Gordon, G.W. Bell, N.J. Rinaldi, H.L. Murray, T.L. Volkert, J. Schreiber, P.A. Rolfe, D.K. Gifford, E. Fraenkel, G.I. Bell, and R.A. Young. 2004. Control of pancreas and liver gene expression by HNF transcription factors. *Science* **303:** 1378–1381.

Orphanides, G. and D. Reinberg. 2002. A unified theory of gene expression. *Cell* **108:** 439–451.

Orrell, D. and H. Bolouri. 2004. Control of internal and external noise in genetic regulatory networks. *J Theor Biol* **230:** 301–312.

Ott, S., A. Hansen, S.Y. Kim, and S. Miyano. 2005. Superiority of network motifs over optimal networks and an application to the revelation of gene network evolution. *Bioinformatics* **21:** 227–238.

Ozbudak, E.M., M. Thattai, I. Kurtser, A.D. Grossman, and A. van Oudenaarden. 2002. Regulation of noise in the expression of a single gene. *Nature Genet* **31:** 69–73.

Ozbudak, E.M., M. Thattai, H.N. Lim, B.I. Shraiman, and A. Van Oudenaarden. 2004. Multistability in the lactose utilization network of Escherichia coli. *Nature* **427:** 737–740.

Paldi, A. 2003. Stochastic gene expression during cell differentiation: order from disorder? *Cell Mol Life Sci* **60:** 1775–1778.

Palsson, B.O. 2006. *Systems Biology : Properties of Reconstructed Networks*. Cambridge University Press.

Park, S.H., A. Zarrinpar, and W.A. Lim. 2003. Rewiring MAP kinase pathways using alternative scaffold assembly mechanisms. *Science* **299:** 1061–1064.

Paulsson, J. 2004. Summing up the noise in gene networks. *Nature* **427:** 415–418.

Paulsson, J., O.G. Berg, and M. Ehrenberg. 2000. Stochastic focusing: fluctuation-enhanced sensitivity of intracellular regulation. *Proc Natl Acad Sci USA* **97:** 7148–7153.

Paulsson, J. and M. Ehrenberg. 2001. Noise in a minimal regulatory network: plasmid copy number control. *Q Rev Biophys* **34:** 1–59.

Pawson, T. and J.D. Scott. 1997. Signaling through scaffold, anchoring, and adaptor proteins. *Science* **278:** 2075–2080.

Peccoud, J. and B. Ycart. 1995. Markovian modeling of gene-product synthesis. *Theor Popul Biol* **48:** 222–234.

Pedraza, J.M. and A. van Oudenaarden. 2005. Noise propagation in genetic networks. *Science* **307:** 1965–1969.

Penn, B.H., D.A. Bergstrom, F.J. Dilworth, E. Bengal, and S.J. Tapscott. 2004. A MyoD-generated feed-forward circuit temporally patterns gene expression during skeletal muscle differentiation. *Genes Dev* **18:** 2348–2353.

Petty, H.R. 2006. Spatiotemporal chemical dynamics in living cells: From information trafficking to cell physiology. *Biosystems* **83:** 217–224.

Pilpel, Y., P. Sudarsanam, and G.M. Church. 2001. Identifying regulatory networks by combinatorial analysis of promoter elements. *Nature Genet* **29:** 153–159.

Pirone, J.R. and T.C. Elston. 2004. Fluctuations in transcription factor binding can explain the graded and binary responses observed in inducible gene expression. *J Theor Biol* **226:** 111–121.

Pomerening, J.R., S.Y. Kim, and J.E. Ferrell, Jr. 2005. Systems-level dissection of the cell-cycle oscillator: bypassing positive feedback produces damped oscillations. *Cell* **122:** 565–578.

Pomerening, J.R., E.D. Sontag, and J.E. Ferrell, Jr. 2003. Building a cell cycle oscillator: hysteresis and bistability in the activation of Cdc2. *Nat Cell Biol* **5:** 346–351.

Postma, P.W., J.W. Lengeler, and G.R. Jacobson. 1993. Phosphoenolpyruvate:carbohydrate phosphotransferase systems of bacteria. *Microbiol Rev* **57:** 543–594.

Pourquie, O. 2003. The segmentation clock: converting embryonic time into spatial pattern. *Science* **301:** 328–330.

Pourquie, O. and A. Goldbeter. 2003. Segmentation clock: insights from computational models. *Curr Biol* **13:** R632–634.

Powell, E.O. 1958. An outline of the pattern of bacterial generation times. *J. Gen. Microbiol.* **18:** 382–417.

Prehoda, K.E. and W.A. Lim. 2002. How signaling proteins integrate multiple inputs: a comparison of N-WASP and Cdk2. *Curr Opin Cell Biol* **14:** 149–154.

Price, N.D., J.L. Reed, and B.O. Palsson. 2004. Genome-scale models of microbial cells: evaluating the consequences of constraints. *Nat Rev Microbiol* **2:** 886–897.

Prill, R.J., P.A. Iglesias, and A. Levchenko. 2005. Dynamic Properties of Network Motifs Contribute to Biological Network Organization. *PLoS Biol* **3:** e343.

Ptacek, J., G. Devgan, G. Michaud, H. Zhu, X. Zhu, J. Fasolo, H. Guo, G. Jona, A. Breitkreutz, R. Sopko, R.R. McCartney, M.C. Schmidt, N. Rachidi, S.J. Lee, A.S. Mah, L. Meng, M.J. Stark, D.F. Stern, C. De Virgilio, M. Tyers, B. Andrews, M. Gerstein, B. Schweitzer, P.F. Predki, and M. Snyder. 2005. Global analysis of protein phosphorylation in yeast. *Nature* **438:** 679–684.

Ptashne, M. 1992. *A genetic switch*. Cell Press & Blackwell Science.

Ptashne, M. and A. Gann. 2002. *Genes & Signals*. Cold Spring Harbor Library Press.

Ptashne, M. 2004. *Genetic Switch: Phage Lambda Revisited*, 3rd ed. Cold Spring Harbor Laboratory Press.

Rao, C.V. and A.P. Arkin. 2001. Control motifs for intracellular regulatory networks. *Annu Rev Biomed Eng* **3:** 391–419.

Rao, C.V., D.M. Wolf, and A.P. Arkin. 2002. Control, exploitation and tolerance of intracellular noise. *Nature* **420:** 231–237.

Rape, M., S.K. Reddy, and M.W. Kirschner. 2006. The processivity of multiubiquitination by the APC determines the order of substrate degradation. *Cell* **124:** 89–103.

Rappaport, N., S. Winter, and N. Barkai. 2005. The ups and downs of biological timers. *Theor Biol Med Model* **2:** 22.

Raser, J.M. and E.K. O'Shea. 2004. Control of stochasticity in eukaryotic gene expression. *Science* **304:** 1811–1814.

Ravasz, E., A.L. Somera, D.A. Mongru, Z.N. Oltvai, and A.-L. Barabasi. 2002. Hierarchical organization of modularity in metabolic networks. *Science* **297:** 1551–1555.

Reardon, J.T. and A. Sancar. 2004. Thermodynamic cooperativity and kinetic proofreading in DNA damage recognition and repair. *Cell Cycle* **3:** 141–144.

Rice, J.J., A. Kershenbaum, and G. Stolovitzky. 2005. Lasting impressions: motifs in protein-protein maps may provide footprints of evolutionary events. *Proc Natl Acad Sci USA* **102:** 3173–3174.

Rigney, D.R. and W.C. Schieve. 1977. Stochastic model of linear, continuous protein synthesis in bacterial populations. *J Theor Biol* **69:** 761–766.

Rodionov, D.A., I.L. Dubchak, A.P. Arkin, E.J. Alm, and M.S. Gelfand. 2005. Dissimilatory metabolism of nitrogen oxides in bacteria: comparative reconstruction of transcriptional networks. *PLoS Comput Biol* **1:** e55.

Ronen, M. and D. Botstein. 2006. Transcriptional response of steady-state yeast cultures to transient perturbations in carbon source. *Proc Natl Acad Sci USA* **103:** 389–394.

Ronen, M., R. Rosenberg, B.I. Shraiman, and U. Alon. 2002. Assigning numbers to the arrows: parameterizing a gene regulation network by using accurate expression kinetics. *Proc Natl Acad Sci USA* **99:** 10555–10560.

Rosenfeld, N. and U. Alon. 2003. Response delays and the structure of transcription networks. *J Mol Biol* **329:** 645–654.

Rosenfeld, N., M.B. Elowitz, and U. Alon. 2002. Negative autoregulation speeds the response times of transcription networks. *J Mol Biol* **323:** 785–793.

Rosenfeld, N., J.W. Young, U. Alon, P.S. Swain, and M.B. Elowitz. 2005. Gene regulation at the single-cell level. *Science* **307:** 1962–1965.

Ross, I.L., C.M. Browne, and D.A. Hume. 1994. Transcription of individual genes in eukaryotic cells occurs randomly and infrequently. *Immunol Cell Biol* **72:** 177–185.

Rossell, S., C.C. van der Weijden, A. Lindenbergh, A. van Tuijl, C. Francke, B.M. Bakker, and H.V. Westerhoff. 2006. Unraveling the complexity of flux regulation: A new method demonstrated for nutrient starvation in Saccharomyces cerevisiae. *Proc Natl Acad Sci USA.*

Rung, J., T. Schlitt, A. Brazma, J. Vilo, and K. Freivalds. 2002. Building and analysing genome-wide gene disruption networks. *Bioinformatics* **18.**

Russ, W.P., D.M. Lowery, P. Mishra, M.B. Yaffe, and R. Ranganathan. 2005. Natural-like function in artificial WW domains. *Nature* **437:** 579–583.

Russo, E., R. Martienssen, and A.D. Riggs. 1996. Epigenetic Mechanisms of Gene Regulation.

Sakata, S., Y. Komatsu, and T. Yamamori. 2005. Local design principles of mammalian cortical networks. *Neurosci Res* **51:** 309–315.

Samoilov, M., A. Arkin, and J. Ross. 2002. Signal processing by simple chemical systems. *Journal of Physical Chemistry A* **106:** 10205–10221.

Sanchez, L. and D. Thieffry. 2003. Segmenting the fly embryo: a logical analysis of the pair-rule cross-regulatory module. *J Theor Biol* **224:** 517–537.

Santillan, M. and M.C. Mackey. 2004. Influence of catabolite repression and inducer exclusion on the bistable behavior of the lac operon. *Biophys J* **86:** 1282–1292.

Sasai, M. and P.G. Wolynes. 2003. Stochastic gene expression as a many-body problem. *Proc Natl Acad Sci USA* **100:** 2374–2379.

Sato, K., Y. Ito, T. Yomo, and K. Kaneko. 2003. On the relation between fluctuation and response in biological systems. *Proc Natl Acad Sci USA* **100:** 14086–14090.

Sato, N., M. Nakayama, and K. Arai. 2004. Fluctuation of chromatin unfolding associated with variation in the level of gene expression. *Genes Cells* **9:** 619–630.

Sato, T.K., R.G. Yamada, H. Ukai, J.E. Baggs, L.J. Miraglia, T.J. Kobayashi, D.K. Welsh, S.A. Kay, H.R. Ueda, and J.B. Hogenesch. 2006. Feedback repression is required for mammalian circadian clock function. *Nat Genet* **38:**312–319.

Savageau, M.A. 1971. Parameter sensitivity as a criterion for evaluating and comparing the performance of biochemical systems. *Nature* **229:** 542–544.

Savageau, M.A. 1974a. Comparison of classical and autogenous systems of regulation in inducible operons. *Nature* **252:** 546–549.

Savageau, M.A. 1974b. Optimal design of feedback control by inhibition. Steady state considerations. *J Mol Evol* **4:** 139–156.

Savageau, M.A. 1976. *Biochemical systems analysis: a study of function and design in molecular biology.* Addison-Wesley, Reading, MA.

Savageau, M.A. 1977. Design of molecular control mechanisms and the demand for gene expression. *Proc Natl Acad Sci USA* **74:** 5647–5651.

Savageau, M.A. 1983. Regulation of differentiated cell-specific functions. *Proc Natl Acad Sci USA* **80:** 1411–1415.

Savageau, M.A. 1989. Are there rules governing patterns of gene regulation? In *Theoretical Biology* (eds. B. Goodwin and P. Saunders), pp. 42–66. Edinburgh University Press.

Savageau, M.A. 1998a. Demand theory of gene regulation. I. Quantitative development of the theory. *Genetics* **149:** 1665–1676.

Savageau, M.A. 1998b. Demand theory of gene regulation. II. Quantitative application to the lactose and maltose operons of Escherichia coli. *Genetics* **149**: 1677–1691.

Schaeffer, H.J. and M.J. Weber. 1999. Mitogen-activated protein kinases: specific messages from ubiquitous messengers. *Mol Cell Biol* **19**: 2435–2444.

Schimke, R. 1969. On the roles of synthesis and degradation in regulation of enzyme levels in mammalian tissues. *Curr Top Cell Regul* **1**: 77–124.

Schlosser, G. and G.P. Wagner. 2004. *Modularity in Development and Evolution*. University of Chicago Press.

Schneider, D. and R.E. Lenski. 2004. Dynamics of insertion sequence elements during experimental evolution of bacteria. *Res Microbiol* **155**: 319–327.

Schuster, S., B.N. Kholodenko, and H.V. Westerhoff. 2000. Cellular information transfer regarded from a stoichiometry and control analysis perspective. *Biosystems* **55**: 73–81.

Segal, E., N. Friedman, N. Kaminski, A. Regev, and D. Koller. 2005. From signatures to models: understanding cancer using microarrays. *Nat Genet* **37 Suppl:** S38–45.

Segal, E., M. Shapira, A. Regev, D. Pe'er, D. Botstein, D. Koller, and N. Friedman. 2003. Module networks: identifying regulatory modules and their condition-specific regulators from gene expression data. *Nat Genet* **34**: 166–176.

Segall, J.E., S.M. Block, and H.C. Berg. 1986. Temporal comparisons in bacterial chemotaxis. *Proc Natl Acad Sci USA* **83**: 8987–8991.

Segall, J.E., M.D. Manson, and H.C. Berg. 1982. Signal processing times in bacterial chemotaxis. *Nature* **296**: 855–857.

Sengupta, A.M., M. Djordjevic, and B.I. Shraiman. 2002. Specificity and robustness in transcription control networks. *Proc Natl Acad Sci USA* **99**: 2072–2077.

Setty, Y., A.E. Mayo, M.G. Surette, and U. Alon. 2003. Detailed map of a cis-regulatory input function. *Proc Natl Acad Sci USA* **100**: 7702–7707.

Shapiro, J.A. 1999. Transposable elements as the key to a 21st century view of evolution. *Genetica* **107**: 171–179.

Sharom, J.R., D.S. Bellows, and M. Tyers. 2004. From large networks to small molecules. *Curr Opin Chem Biol* **8**: 81–90.

Shen-Orr, S.S., R. Milo, S. Mangan, and U. Alon. 2002. Network motifs in the transcriptional regulation network of Escherichia coli. *Nature Genet* **31**: 64–68.

Shibata, T. 2003. Fluctuating reaction rates and their application to problems of gene expression. *Phys. Rev. E* **67**: 061906.

Shibata, T. and K. Fujimoto. 2005. Noisy signal amplification in ultrasensitive signal transduction. *Proc Natl Acad Sci USA* **102**: 331–336.

Shimizu, T.S., S.V. Aksenov, and D. Bray. 2003. A spatially extended stochastic model of the bacterial chemotaxis signalling pathway. *J Mol Biol* **329**: 291–309.

Shinar, G., E. Dekel, T. Tlusty, and U. Alon. 2006. Rules for gene regulation based on error minimization. *Proc Natl Acad Sci* **103**: 3999–4004.

Siegal, M.L. and A. Bergman. 2002. Waddington's canalization revisited: Developmental stability and evolution. *Proc Natl Acad Sci USA* **99**: 10528–10532.

Siegele, D.A. and J.C. Hu. 1997. Gene expression from plasmids containing the araBAD promoter at subsaturating inducer concentrations represents mixed populations. *Proc Natl Acad Sci USA* **94**: 8168–8172.

Simon, H.A. 1996. *The Sciences of the Artificial*. MIT press.

Simpson, M.L., C.D. Cox, and G.S. Sayler. 2003. Frequency domain analysis of noise in autoregulated gene circuits. *Proc Natl Acad Sci USA* **100**: 4551–4556.

Simpson, M.L., C.D. Cox, and G.S. Sayler. 2004. Frequency domain chemical Langevin analysis of stochasticity in gene transcriptional regulation. *J Theor Biol* **229**: 383–394.

Simpson, P. 1997. Notch signalling in development: on equivalence groups and asymmetric developmental potential. *Curr Opin Genet Dev* **7**: 537–542.

Singh, U.N. 1969. Polyribosomes and unstable messenger RNA: a stochastic model of protein synthesis. *J. Theor. Biol.* **25**: 444–460.

Smolen, P., D.A. Baxter, and J.H. Byrne. 1998. Frequency selectivity, multistability, and oscillations emerge from models of genetic regulatory systems. *American Journal of Physiology - Cell Physiology* **274**.

Smolen, P., D.A. Baxter, and J.H. Byrne. 2000. Modeling transcriptional control in gene networks--methods, recent results, and future directions. *Bull Math Biol* **62**: 247–292.

Smolen, P., D.A. Baxter, and J.H. Byrne. 2002. A reduced model clarifies the role of feedback loops and time delays in the Drosophila circadian oscillator. *Biophysical Journal* **83**: 2349–2359.

Snel, B., P. Bork, and M.A. Huynen. 2002. The identification of functional modules from the genomic association of genes. *Proc Natl Acad Sci USA* **99**: 5890–5895.

Socolich, M., S.W. Lockless, W.P. Russ, H. Lee, K.H. Gardner, and R. Ranganathan. 2005. Evolutionary information for specifying a protein fold. *Nature* **437**: 512–518.

Song, S., P.J. Sjostrom, M. Reigl, S. Nelson, and D.B. Chklovskii. 2005. Highly nonrandom features of synaptic connectivity in local cortical circuits. *PLoS Biol* **3**: e68.

Sontag, E., A. Kiyatkin, and B.N. Kholodenko. 2004. Inferring dynamic architecture of cellular networks using time series of gene expression, protein and metabolite data. *Bioinformatics* **20**: 1877–1886.

Sourjik, V. and H.C. Berg. 2002. Receptor sensitivity in bacterial chemotaxis. *Proc Natl Acad Sci USA* **99**: 123–127.

Sourjik, V. and H.C. Berg. 2004. Functional interactions between receptors in bacterial chemotaxis. *Nature* **428**: 437–441.

Spellman, P.T., G. Sherlock, M.Q. Zhang, V.R. Iyer, K. Anders, M.B. Eisen, P.O. Brown, D. Botstein, and B. Futcher. 1998. Comprehensive identification of cell cycle-regulated genes of the yeast Saccharomyces cerevisiae by microarray hybridization. *Mol Biol Cell* **9**: 3273–3297.

Spirin, V. and L.A. Mirny. 2003. Protein complexes and functional modules in molecular networks. *Proc Natl Acad Sci USA* **100**: 12123–12128.

Sporns, O. and R. Kotter. 2004. Motifs in brain networks. *PLoS Biol* **2**: e369.

Sprinzak, D. and M.B. Elowitz. 2005. Reconstruction of genetic circuits. *Nature* **438**: 443–448.

Spudich, J.L. and D.E. Koshland. 1976. Non-genetic individuality: chance in the single cell. *Nature* **262**: 467–471.

Stadler, B.M.R., P.F. Stadler, G.P. Wagner, and W. Fontana. 2001. The topology of the possible: Formal spaces underlying patterns of evolutionary change. *J Theor Biol* **213**: 241–274.

Stathopoulos, A. and M. Levine. 2005. Genomic regulatory networks and animal development. *Dev Cell* **9**: 449–462.

Stelling, J. 2004. Mathematical models in microbial systems biology. *Curr Opin Microbiol* **7**: 513–518.

Stelling, J. and E.D. Gilles. 2004. Mathematical modeling of complex regulatory networks. *IEEE Trans Nanobioscience* **3**: 172–179.

Stelling, J., U. Sauer, Z. Szallasi, F.J. Doyle, and J. Doyle. 2004. Robustness of cellular functions. *Cell* **118**: 675–685.

Sternberg, P.W. and M.A. Felix. 1997. Evolution of cell lineage. *Curr Opin Genet Dev* **7**: 543–550.

Steuer, R., C. Zhou, and J. Kurths. 2003. Constructive effects of fluctuations in genetic and biochemical regulatory systems. *Biosystems* **72**: 241–251.

Strogatz, S.H. 2001a. Exploring complex networks. *Nature* **410**: 268–276.

Strogatz, S.H. 2001b. *Nonlinear Dynamics and Chaos: With Applications to Physics, Biology, Chemistry and Engineering.* Perseus Books Group.

Stumpf, M.P.H., Z. Laidlaw, and V.A.A. Jansen. 2002. Herpes viruses hedge their bets. *Proc Natl Acad Sci USA* **99**: 15234–15237.

Surrey, T., F. Nedelec, S. Leibler, and E. Karsenti. 2001. Physical properties determining self-organization of motors and microtubules. *Science* **292**: 1167–1171.

Sveiczer, A., A. Csikasz-Nagy, B. Gyorffy, B. Novak, and J.J. Tyson. 2000. Modeling the fission yeast cell cycle: Quantized cycle times in wee1- cdc25? mutant cells. *Proc Natl Acad Sci USA* **97**: 7865–7870.

Swain, P.S. 2004. Efficient attenuation of stochasticity in gene expression through post-transcriptional control. *J Mol Biol* **344**: 965–976.

Swain, P.S., M.B. Elowitz, and E.D. Siggia. 2002. Intrinsic and extrinsic contributions to stochasticity in gene expression. *Proc Natl Acad Sci USA* **99**: 12795–12800.

Tanay, A., A. Regev, and R. Shamir. 2005. Conservation and evolvability in regulatory networks: the evolution of ribosomal regulation in yeast. *Proc Natl Acad Sci USA* **102**: 7203–7208.

Tao, Y. 2004. Intrinsic and external noise in an auto-regulatory genetic network. *J Theor Biol* **229**: 147–156.

Tavazoie, S., J.D. Hughes, G.M. Church, M.J. Campbell, and R.J. Cho. 1999. Systematic determination of genetic network architecture. *Nature Genet* **22**: 281–285.

Thattai, M. and B.I. Shraiman. 2003. Metabolic switching in the sugar phosphotransferase system of Escherichia coli. *Biophys J* **85**: 744–754.

Thattai, M. and A. van Oudenaarden. 2001. Intrinsic noise in gene regulatory networks. *Proc Natl Acad Sci USA* **98**: 8614–8619.

Thattai, M. and A. van Oudenaarden. 2002. Attenuation of noise in ultrasensitive signaling cascades. *Biophys J* **82**: 2943–2950.

Thattai, M. and A. van Oudenaarden. 2004. Stochastic gene expression in fluctuating environments. *Genetics* **167**: 523–530.

Thieffry, D., A.M. Huerta, E. Perez-Rueda, and J. Collado-Vides. 1998. From specific gene regulation to genomic networks: a global analysis of transcriptional regulation in Escherichia coli. *Bioessays* **20**: 433–440.

Thieffry, D. and D. Romero. 1999. The modularity of biological regulatory networks. *BioSystems* **50**: 49–59.

Thieffry, D. and R. Thomas. 1998. Qualitative analysis of gene networks. *Pac Symp Biocomput* **3**: 77–88.

Thomas, M.R. and E.K. O'Shea. 2005. An intracellular phosphate buffer filters transient fluctuations in extracellular phosphate levels. *Proc Natl Acad Sci USA* **102**: 9565–9570.

Thomas, R. and R. D'Ari. 1990. *Biological feedback*. CRC Press.

Thompson, A. 1998. *Hardware Evolution: Automatic design of electronic circuits in reconfigurable hardware by artificial evolution*. Springer.

Thorsen, T., S.J. Maerkl, and S.R. Quake. 2002. Microfluidic large-scale integration. *Science* **298**: 580–584.

Tlusty, T., R. Bar-Ziv, and A. Libchaber. 2004. High-fidelity DNA sensing by protein binding fluctuations. *Phys Rev Lett* **93**: 258103.

Tomioka, R., H. Kimura, J.K. T, and K. Aihara. 2004. Multivariate analysis of noise in genetic regulatory networks. *J Theor Biol* **229**: 501–521.

Tomlin, C.J. and J.D. Axelrod. 2005. Understanding biology by reverse engineering the control. *Proc Natl Acad Sci USA* **102**: 4219–4220.

Trusina, A., S. Maslov, P. Minnhagen, and K. Sneppen. 2004. Hierarchy measures in complex networks. *Phys Rev Lett* **92**: 178702.

Turner, P.E. and L. Chao. 1999. Prisoner's dilemma in an RNA virus. *Nature* **398**: 441–443.

Tyson, J.J., K. Chen, and B. Novak. 2001. Network dynamics and cell physiology. *Nat Rev Mol Cell Biol* **2**: 908–916.

Tyson, J.J., K.C. Chen, and B. Novak. 2003. Sniffers, buzzers, toggles and blinkers: dynamics of regulatory and signaling pathways in the cell. *Curr Opin Cell Biol* **15:** 221–231.

Tyson, J.J., A. Csikasz-Nagy, and B. Novak. 2002. The dynamics of cell cycle regulation. *Bioessays* **24:** 1095–1109.

Ueda, H.R., M. Hagiwara, and H. Kitano. 2001. Robust oscillations within the interlocked feedback model of Drosophila circadian rhythm. *J Theor Biol* **210:** 401–406.

Ueda, H.R., S. Hayashi, S. Matsuyama, T. Yomo, S. Hashimoto, S.A. Kay, J.B. Hogenesch, and M. Iino. 2004. Universality and flexibility in gene expression from bacteria to human. *Proc Natl Acad Sci USA* **101:** 3765–3769.

Van Kampen, N.G. 2001. *Stochastic Processes in Physics and Chemistry.* North Holland.

van Nimwegen, E. 2003. Scaling laws in the functional content of genomes. *Trends Genet* **19:** 479–484.

Variano, E.A., J.H. McCoy, and H. Lipson. 2004. Networks, dynamics, and modularity. *Phys Rev Lett* **92:** 188701.

Vilar, J.M., C.C. Guet, and S. Leibler. 2003. Modeling network dynamics: the lac operon, a case study. *J Cell Biol* **161:** 471–476.

Vilar, J.M., H.Y. Kueh, N. Barkai, and S. Leibler. 2002. Mechanisms of noise-resistance in genetic oscillators. *Proc Natl Acad Sci USA* **99:** 5988–5992.

Vilar, J.M. and S. Leibler. 2003. DNA looping and physical constraints on transcription regulation. *J Mol Biol* **331:** 981–989.

Voigt, C.A. and J.D. Keasling. 2005. Programming cellular function. *Nat Chem Biol* **1:** 304–307.

Voit, E.O. 2000. *Computational Analysis of Biochemical Systems : A Practical Guide for Biochemists and Molecular Biologists.* Cambridge University Press.

Volfson, D., J. Marciniak, W.J. Blake, N. Ostroff, L.S. Tsimring, and J. Hasty. 2005. Origins of extrinsic variability in eukaryotic gene expression. *Nature* 439: 861–4.

Von Dassow, G., E. Meir, E.M. Munro, and G.M. Odell. 2000. The segment polarity network is a robust developmental module. *Nature* **406:** 188–192.

Von Dassow, G. and G.M. Odell. 2002. Design and constraints of the Drosophila segment polarity module: Robust spatial patterning emerges from intertwined cell state switches. *J Experimental Zool* **294:** 179–215.

Vulic, M. and R. Kolter. 2001. Evolutionary cheating in Escherichia coli stationary phase cultures. *Genetics* **158:** 519–526.

Waddington, C.H. 1959. Canalization of development and genetic assimilation of acquired characters. *Nature* **183:** 1654–1655.

Wagner, A. 2005a. Energy constraints on the evolution of gene expression. *Mol Biol Evol* **22:** 1365–1374.

Wagner, A. 2005b. *Robustness and Evolvability in Living Systems.* Princeton University Press.

Wagner, G.P. and L. Altenberg. 1996. Complex adaptations and the evolution of evolvability. *Evolution* **50:** 967–976.

Wall, M.E., M.J. Dunlop, and W.S. Hlavacek. 2005. Multiple functions of a feed-forward-loop gene circuit. *J Mol Biol* **349:** 501–514.

Wang, E. and E. Purisima. 2005. Network motifs are enriched with transcription factors whose transcripts have short half-lives. *Trends Genet* **21:** 492–495.

Wang, W., J.M. Cherry, Y. Nochomovitz, E. Jolly, D. Botstein, and H. Li. 2005a. Inference of combinatorial regulation in yeast transcriptional networks: a case study of sporulation. *Proc Natl Acad Sci USA* **102:** 1998–2003.

Wang, Z.W., Z.H. Hou, and H.W. Xin. 2005b. Internal noise stochastic resonance of synthetic gene network. *Chem Phys Lett* **401:** 307–311.

Ward, J.P., J.R. King, A.J. Koerber, P. Williams, J.M. Croft, and R.E. Sockett. 2001. Mathematical modelling of quorum sensing in bacteria. *IMA Journal of Mathemathics Applied in Medicine and Biology* **18:** 263–292.

Wardle, F.C. and J.C. Smith. 2004. Refinement of gene expression patterns in the early Xenopus embryo. *Development* **131:** 4687–4696.

Wasserman, S. and K. Faust. 1994. *Social Network Analysis : Methods and Applications.* Cambridge University Press.

Waters, C.M. and B.L. Bassler. 2005. Quorum sensing: cell-to-cell communication in bacteria. *Annu Rev Cell Dev Biol* **21:** 319–346.

Watts, D.J. and S.H. Strogatz. 1998. Collective dynamics of 'small-world' networks. *Nature* **393:** 440–442.

Wernegreen, J.J. 2002. Genome evolution in bacterial endosymbionts of insects. *Nat Rev Genet* **3:** 850–861.

Wernicke, S. and F. Rasche. 2006. FANMOD: a tool for fast network motif detection. *Bioinformatics.*

West, G.B. and J.H. Brown. 2005. The origin of allometric scaling laws in biology from genomes to ecosystems: towards a quantitative unifying theory of biological structure and organization. *J Exp Biol* **208:** 1575–1592.

West, G.B., J.H. Brown, and B.J. Enquist. 1997. A general model for the origin of allometric scaling laws in biology. *Science* **276:** 122–126.

Weston, A.D. and L. Hood. 2004. Systems biology, proteomics, and the future of health care: toward predictive, preventative, and personalized medicine. *J Proteome Res* **3:** 179–196.

White, E.L. 1989. *Cortical Circuits.* Birkhauser.

White, J., E. Southgate, J. Thomson, and S. Brenner. 1986. The Structure of the Nervous System of the Nematode Caenorhabditis elegans. *Philos Trans R Soc London Ser B* **314:** 1.

Wilcox, J.L., H.E. Dunbar, R.D. Wolfinger, and N.A. Moran. 2003. Consequences of reductive evolution for gene expression in an obligate endosymbiont. *Mol Microbiol* **48:** 1491–1500.

Wiley, H.S., S.Y. Shvartsman, and D.A. Lauffenburger. 2003. Computational modeling of the EGF-receptor system: a paradigm for systems biology. *Trends Cell Biol* **13:** 43–50.

Wilke, C.O., J.L. Wang, C. Ofria, R.E. Lenski, and C. Adami. 2001. Evolution of digital organisms at high mutation rates leads to survival of the flattest. *Nature* **412:** 331–333.

Wilkins, A.S. 2001. *The Evolution of Developmental Pathways.* Sinauer Associates.

Winfree, A.T. 2001. *The Geometry of Biological Time.* Springer.

Woese, C.R. 2002. On the evolution of cells. *Proc Natl Acad Sci USA* **99:** 8742–8747.

Wolf, D.M. and A.P. Arkin. 2002. Fifteen minutes of fim: control of type 1 pili expression in E. coli. *OMICS* **6:** 91–114.

Wolf, D.M. and A.P. Arkin. 2003. Motifs, modules and games in bacteria. *Curr Opin Microbiol* **6:** 125–134.

Wolf, D.M. and F.H. Eeckman. 1998. On the relationship between genomic regulatory element organization and gene regulatory dynamics. *J Theor Biol* **195:** 167–186.

Wolpert, L. 1969. Positional information and the spatial pattern of cellular differentiation. *J Theor Biol* **25:** 1–47.

Wolpert, L., R. Beddington, R. Jessell, P. Lawrence, E. Meyerowitz, and S. J. 2002. *Principles of Development.* Oxford University Press.

Xiong, W. and J.E. Ferrell, Jr. 2003. A positive-feedback-based bistable 'memory module' that governs a cell fate decision. *Nature* **426:** 460–465.

Yagil, G. and E. Yagil. 1971. On the relation between effector concentration and the rate of induced enzyme synthesis. *Biophys J* **11:** 11–27.

Yang, A.S. 2001. Modularity, evolvability, and adaptive radiations: A comparison of the hemi- and holometabolous insects. *Evolution and Development* **3:** 59–72.

Yeger-Lotem, E., S. Sattath, N. Kashtan, S. Itzkovitz, R. Milo, R.Y. Pinter, U. Alon, and H. Margalit. 2004. Network motifs in integrated cellular networks of transcription-regulation and protein-protein interaction. *Proc Natl Acad Sci USA* **101:** 5934–5939.

Yi, T.M., Y. Huang, M.I. Simon, and J. Doyle. 2000. Robust perfect adaptation in bacterial chemotaxis through integral feedback control. *Proc Natl Acad Sci USA* **97:** 4649–4653.

Yokobayashi, Y., R. Weiss, and F.H. Arnold. 2002. Directed evolution of a genetic circuit. *Proc Natl Acad Sci USA* **99:** 16587–16591.

You, L., R.S. Cox, R. Weiss, and F.H. Arnold. 2004. Programmed population control by cell-cell communication and regulated killing. *Nature* **428:** 868–871.

Young, M.W. 2000. The tick-tock of the biological clock. *Sci Am* **282:** 64–71.

Yu, H., N.M. Luscombe, J. Qian, and M. Gerstein. 2003. Genomic analysis of gene expression relationships in transcriptional regulatory networks. *Trends Genet* **19:** 422–427.

Yuh, C.H., H. Bolouri, and E.H. Davidson. 1998. Genomic cis-regulatory logic: experimental and computational analysis of a sea urchin gene. *Science* **279:** 1896–1902.

Zarrinpar, A., S.H. Park, and W.A. Lim. 2003. Optimization of specificity in a cellular protein interaction network by negative selection. *Nature* 426:676–680.

Zaslaver, A., A.E. Mayo, R. Rosenberg, P. Bashkin, H. Sberro, M. Tsalyuk, M.G. Surette, and U. Alon. 2004. Just-in-time transcription program in metabolic pathways. *Nature Genet* **36:** 486–491.

Zhang, L.V., O.D. King, S.L. Wong, D.S. Goldberg, A.H. Tong, G. Lesage, B. Andrews, H. Bussey, C. Boone, and F.P. Roth. 2005. Motifs, themes and thematic maps of an integrated Saccharomyces cerevisiae interaction network. *J Biol* **4:** 6.

Zykov, V., E. Mytilinaios, B. Adams, and H. Lipson. 2005. Robotics: self-reproducing machines. *Nature* **435:** 163–164.

Index